The Primo Vascular System

Kwang-Sup Soh • Kyung A. Kang
David K. Harrison
Editors

The Primo Vascular System

Its Role in Cancer and Regeneration

Springer

Editors
Kwang-Sup Soh, Ph.D
Nano Primo Research Center
Advanced Institute of Convergence Technology
Seoul National University
Seoul, Korea
kssoh1@gmail.com

Kyung A. Kang
Department of Chemical Engineering
University of Louisville
Louisville, KY, USA
Kyung.Kang@louisville.edu

David K. Harrison, Ph.D
Institute of Cellular Medicine
Newcastle University
Newcastle upon Tyne, UK
d.k.harrison@ncl.ac.uk

ISBN 978-1-4899-9605-3 ISBN 978-1-4614-0601-3 (eBook)
DOI 10.1007/978-1-4614-0601-3
Springer New York Dordrecht Heidelberg London

Printed on acid-free paper

Springer is part of Springer Science+Business Media (www.springer.com)

Group Photo of ISPS 2010
September 17-18, 2010
Cheongpung Resort-Lake Hotel, Jecheon, Korea

Preface

The first International Symposium on Primo Vascular System 2010 (ISPS 2010) with special topics on cancer and regeneration was held in Jecheon, Korea during September 17–18, 2010.

The ISPS 2010 was devoted to topics related to biomedical findings on Primo Vascular System (PVS) that may be the anatomical structure corresponding to acupuncture meridians. Bong-Han Kim, in the early 1960's, published his discovery on a new, web-like vascular system. Unfortunately, his research results were not reproducible by others because he did not reveal the dye used in his study and thus neglected for a long time. In 2002, the Biomedical Physics Laboratory in the Department of Physics and Astronomy, Seoul National University launched a new PVS research project utilizing modern biomedical imaging techniques. Kim's claims were confirmed by the new study results, and functional aspects of PVS including its roles in the areas of regenerative medicine and cancer have been uncovered. The research results have also suggested the extensive roles of PVS in human, changing paradigm of medicine. With an expectation of Kim's claim, "PVS is acupuncture meridian" to be proven soon, the future PVS research may reveal the mechanisms of traditional oriental medicine practiced for several thousands years.

The symposium dealt with the past findings, current status, and future prospect of the PVS research in the context of cutting-edge investigation in oriental and occidental medicine, molecular biology, and biophysics. The symposium provided a FIRST international opportunity to exchange the research results on PVS among multidisciplinary experts. We are also happy to announce that, in August 2010, the International Society for Primo Vascular System (ISPVS) was formed.

The ISPS 2010 had 43 oral and 40 poster presentations and approximately 200 participants. The organizing committee would like to express our appreciation to the invited speakers, presenters, participants, and those who helped the symposium in various ways to make it very successful. We are also grateful to the authors who submitted manuscripts to this historic volume on PVS. We sincerely hope that we do not have to wait too long to have the second ISPS symposium.

Seoul, Korea

Kwang-Sup Soh

Organization of The International Symposium on Primo Vascular System 2010

Special Topics on Cancer, Regeneration, and Acupuncture

Organizing Committee:

President: Kwang-Sup Soh (Seoul National University)

Program Chairs: Chong-Kwan Cho (Dunsan Oriental Hospital, Daejeon University), Gil-Ja Jhon (Ewha Womans University), Chae Hun Leem (University of Ulsan), Pan Dong Ryu (Seoul National University)

Members: Eun-Yong Lee (Oriental Medicine Hospital in Chungju, Semyung University), Hesson Chung (Korea Institute of Science and Technology), Hi-Joon Park (Kyung Hee University), Minah Suh (Sungkyunkwan University), Min Su Kim (Chonbuk National University), Mison Chun (Ajou University), Sang-Hyun Park (KAIST Institute), Sang Joon Lee (Pohang University of Science and Technology), Seong Hun Ahn (Wonkwang University), Sunmi Choi (Korea Institute of Oriental Medicine), Yeonhee Ryu (Korea Institute of Oriental Medicine), Sang-Suk Lee (Sangji University)

Secretary: Jung Sun Yoo (Seoul National University)

Local Advisory Board:
Soo Sung Lee (Former Prime Minister of Korea), Si Jong Lee (Governor, Chungcheongbuk-do Province, Korea), Myeong Hyeon Choi (Mayor, Jecheon City, Korea), Jae Gap Kim (Secretary-General, 2010 World Oriental Medicine-Bio EXPO)

International Advisory Board:

Chair: Kyung A. Kang (University of Louisville, USA)

Members: Vitaly Vodyanoy (Auburn University, USA), Eduard Van Wijk (Leiden University, The Netherlands), Chris Zaslawski (University of Technology Sydney, Australia), Weibo Zhang (China Academy of Chinese Medical Science, China)

Hosts:
2010 World Oriental Medicine Bio-EXPO in Jecheon, Korea
College of Natural Sciences, Seoul National University

Sponsorship

2010 World Oriental Medicine Bio-EXPO in Jecheon, Korea
College of Natural Sciences, Seoul National University
MOBASE Co., Ltd.
Doraji Association Seoul
Korea Institute of Oriental Medicine
The Association of Korean Oriental Medicine
Korean Pharmacopuncture Institute
Wonkwang University, College of Oriental Medicine

ISPS 2010 Editors Thank the Reviewers

Seong Hun Ahn, Wonkwang University, Korea
Ping An, Renmin Hospital of Wuhan University, China
Ku Youn Baik, Kwangwoon University, Korea
Hesson Chung, Korea Institute of Science and Technology, Korea
David K. Harrison, Newcastle University, UK
Kyung A. Kang, University of Louisville, USA
Min Su Kim, Chonbuk National University, Korea
Larry Kwak, University of Texas, USA
Byoung Kwon, National Cancer Center, Korea
Hee Min Kwon, Seoul National University, Korea
Byung-Cheon Lee, Korea Advanced Institute of Science and Technology, Korea
Hyeran Lee, Washington University, USA
Donald Miller, University of Louisville, USA
Yoshiharu Motoo, Kanazawa Medical University, Japan
Vyacheslav Ogay, National Center for Biotechnology of the Republic of Kazakhstan, Kazakhstan
Pan Dong Ryu, Seoul National University, Korea
Kwang-Sup Soh, Seoul National University, Korea
Minah Suh, Sungkyunkwan University, Korea
Edward Van Wijk, Leiden University, The Netherlands
Peter Vaupel, University Medical Centre, Germany
Vitaly Vodyanoy, Auburn University, USA
Jung Sun Yoo, Seoul National University, Korea
Kurt Zaenker, University of Witten/Herdecke, Germany
Chris Zaslawski, University of Technology, Australia
Weibo Zhang, China Academy of Chinese Medical Science, China

Contents

Part II Primo Vascular System in Various Organs

Part I
Past, Present and Future of Primo Vascular System Research

Chapter 1
A Brief History of the Bong-Han Theory and the Primo Vascular System

Kwang-Sup Soh

Abstract A short history of the Bong-Han theory is presented. The original work by Bong-Han Kim in the Kyung-Rak Research Institute of North Korea in the early 1960s is described. The follow-up research by Fujiwara in Japan is briefly mentioned. Modern development since the rediscovery of the primo vascular system by the Seoul National University team is given in chronological order.

Bong-Han Kim was born in 1916, and graduated from the College of Medicine Seoul National University in 1941. He was an Associate Professor at Pyung Yang Medical School, Physiology Laboratory, when he announced his discovery of anatomical structures corresponding to acupuncture points and meridians at the Symposium at the Pyung Yang Medical School on August 18, 1961, which he published in 1962. There was no description of the method how the structures were found or identified.

The epoch-making discovery was made sometime between 1962 and 30 November 1963, when he published his second report as the director of the Kyung-Rak Research Institute (KRI). This Institute was known to be a National Institute of North Korea and was probably established in this period. The publications in the name of Bong-Han Kim were reports on the collective works of researchers in this Institute. Unlike journal papers, these reports did not have "Method" or "Materials" sections. In that paper, which was the second among a series of five articles with the name of Bong-Han Kim, he mentioned that he found the most important material, a blue tracing dye which revealed not only acupuncture meridians but their extensions into the body. Thereafter, his team established the existence of a new circulatory system running throughout an animal's body. This second paper was translated into

K.-S. Soh (✉)
Nano Primo Research Center, Advanced Institute of Convergence Technology, Seoul National University, Suwon 443-270, Korea
e-mail: kssoh1@gmail.com

K.-S. Soh et al. (eds.), *The Primo Vascular System: Its Role in Cancer and Regeneration*, DOI 10.1007/978-1-4614-0601-3_1, © Springer Science+Business Media, LLC 2012

English and other languages and distributed in most world-class libraries. The KRI published a third report which was a systematic investigation of the new circulatory system.

The fourth and the fifth reports in 1965 were about the "Sanal", which was a kind of microcell whose function was similar to embryonic-like stem cells in modern terminology. They described regeneration of damaged liver tissues and hematopoiesis via Sanals.

For some unknown reasons, the KRI was closed around 1966, and no official or any reliable sources gave any hint on the fate of Bong-Han Kim and other researchers of the Institute. Until to the present time, no traces have been found of the KRI or of Kim and his team.

The historical discovery of Bong-Han Kim in 1963 was widely publicized in daily newspapers of China, Japan, and Russia, and many teams tried to reproduce his results. Strangely enough, there was no record that any of them requested the mysterious staining blue dye of Bong-Han Kim, which was essential to confirming his results. It is not surprising at all that no one was able to either prove or disprove his claims.

Only one Japanese researcher, Satoru Fujiwara, who was an Assistant Professor of Anatomy at Osaka City University, was stubborn enough to observe the primo vascular system (PVS) in blood vessels and on the surface of organs of rabbits. He produced one journal article in 1967 and published a book, "Discovery of Acupuncture Meridian", both in Japanese. He recalled that it took nearly 6 months to find the (PVS) in his own way without knowing Kim's secret blue dye. His results were limited only to a few subsystems of the whole PVS, and he could not continue his research as people in Japan were sceptical of his pursuit after Kim's fall in North Korea. He opened a dental hospital on a small island near Osaka, but he kept his work records.

In the year 2002, Kwang-Sup Soh invited a Chinese veterinary student, Jiaowen Jiang, to begin a trial experiment for the test of Kim's claim on the existence of threadlike structures (primo vessels) in the large blood vessels of rabbits. In that summer (July–August), the task force team of Jiang was able to find the intravascular primo vessel, which was the start of PVS research. Obtaining long enough pieces of primo vessel from blood vessels for physiological analysis was not so easy; searching the PVS on organ surfaces seemed to be a better target for the analysis, and the SNU team spent nearly half a year observing them. After repeated failure, Soh looked for Professor Fujiwara and finally visited him in Osaka. Fujiwara kindly gave Soh the records of his research and a film showing experimental procedures.

Viewing the film gave momentum to efforts to observe the PVS on organ surfaces in the abdominal cavity, and further observations in other organs, such as in lymph vessels were successful. However, finding the PVS in other organs was not possible, so efforts to find the PVS in the whole body seemed impossible without Kim's secret blue dye.

A breakthrough occurred in November 2008 with the discovery of the Trypan blue technique by BC Lee, with which the weblike network of the PVS was observed in the omentum of a rat and in adipose tissue. The most striking progress was discovery of the PVS around cancer tissue with the Trypan blue technique, which attracted much attention from cancer researchers. The PVS around cancer tissue was not mentioned in Kim's or Fujiwara's works. Another great leap was the finding of the PVS floating in cerebrospinal fluid in the brain ventricle and the central canal of the spinal cord. This PVS can be visualized by injecting fluorescent nanoparticles into the lateral ventricle of the brain of a rat. The current goal is to find the PVS route from the skin to the brain via the peripheral nerve and spinal cord, which can be used for the diagnosis and the treatment, at the acupuncture point, of brain diseases, such as Alzheimer's disease or Parkinson's disease.

In September 17–18, an International Symposium of Primo Vascular System (ISPS 2010) with special topics on cancer, regeneration, and acupuncture was held in Jecheon, Korea. The topics suggested that the PVS was deeply related to cancer, regeneration/stem cells, and imaging of the acupuncture meridian. Based upon the detection techniques developed by the SNU group, applied research to gain a basic understanding of the PVS can be started, which is the second phase of the PVS research.

Chapter 2
Summary of Bong-Han Kim's Publications

Jungdae Kim, Jonghyun Jung, and Michael Potroz

Abstract We present a summary of the Bong-Han Kim's publications. His five articles were published in Korean in the *Journal of Jo Sun Medicine* from 1962 to 1965. The subjects of articles are about the studies on the reality of acupuncture meridian, the Kyungrak system, and the Sanal theory. Only the concluding parts of the articles were translated in English.

There were five articles published in the name of Bong-Han Kim. They were not research articles in a proper form but a kind of report of the Institute of Acupuncture Meridians which was a national research institute of the North Korean government, and of which Kim was the Director. Therefore, they had no "Materials and Method" sections and only described results. No "Analysis" or "Discussion" sections were given. These reports were all written in Korean but the second article was translated into English, Chinese, Russian, and Japanese, and was distributed to major libraries throughout the world. We have translated the "Conclusion" section of each article and present them here.

1 Study on the Reality of Acupuncture Meridian: *J Jo Sun Med* 1962;9:5–13

1.1 Conclusion

We found the physical substrate of acupuncture points (AP) and meridians (AM) by applying novel methods.

J. Kim (✉)
Biomedical Physics Laboratory, Department of Physics and Astronomy,
Seoul National University, Seoul 151-747, Korea
e-mail: tojdkim@gmail.com

K.-S. Soh et al. (eds.), *The Primo Vascular System: Its Role in Cancer and Regeneration*,
DOI 10.1007/978-1-4614-0601-3_2,

1. In general our findings of APs were in agreement with the position of classical APs of traditional oriental medicine. There were some new points which were different from the classical APs.
2. The AM is a bundle of vessels. They are distinctively different from the nervous, blood, or lymphatic systems with respect to histological and physiological properties.
3. The physical substrate of AMs is a novel anatomical system which has not been known untill the present time.

2 On the Kyungrak System (Primo Vascular System): *J Acad Med Sci DPR Korea* 1963 Dec 10;90:1–41

2.1 General Conclusion

All the results of a series of above-mentioned experiments on the Kyungrak system (primo vascular system (PVS)) show that the PVS is another, independent functional-morphological system.

1. The PVS consists of the primo nodes (Bonghan corpuscles) and the primo vessels (Bonghan ducts) linking them.
   The primo nodes exist not only in the skin but are widely distributed in the profunda of the organism as well.
   This coincides also with the experiences gained in the clinical acupuncture.
   In structure, however, the primo node in the skin (superficial primo node) is different from the profound primo node deep in the body.
   The superficial primo node consists of the outer layer of smooth muscles and the inner substance made of various cells.
   It is considered that this muscle layer is important in sending secretion to the primo vessel.
   It is also considered that there are various kinds of cells in the inner substance and they perform the secretory function.
   The results of histochemical and biochemical study prove that the inner substance has an abundance of nucleic acids, particularly DNA.
   In the profund primo node, specific cells are arranged in a definite order and materials, which are basophilic like the nucleus and varied in form, some rod-shaped and others thread-shaped, are irregularly located.
   These materials are arranged in the same direction as the path of the primo vessel and their DNA reaction proves positive histochemically. This is related to the fact that high concentration of DNA is contained in the primo vessel. The above-mentioned profund primo node has no outer muscle layer.
   The structure of the primo node is completely different from the other structures hitherto known.

2. The primo vessel has two forms of existence.
   One of the forms of its existence is that it runs inside the blood vessel or the lymphatic vessel and the other is that it runs outside the vessel.
   The intravascular primo vessel and the extravascular primo vessel take different directions from each other, but there is no difference between them in structure.
   The primo vessel comprises bundles of the primo lumens.
   The primo lumen is very soft and has a thin wall, which consists of endothelial cells of a single layer. It is difficult to discern clearly the internal structure of the nucleus of the endothelial cell by applying the usual staining method. It is of a peculiar rod shape.
   The contents of the primo lumen often appear in the shape of granule when they are stained by a routine method. Moreover, it has been established by cytochemical reaction that it contains DNA.
   The contents of the primo vessel are entirely different from those of the blood and lymphatic vessels. When stained with acridine orange, the primo vessel brightly fluoresces in yellowish green.
   This also clearly distinguishes the primo vessel from other tissues.
   Examination under a phase-contrast microscope of the primo vessel in the fresh specimen reveals that it has nuclei of a peculiar form and arrangement.
   The superficial primo vessels among the extravascular primo vessels are connected with the superficial primo nodes, while the profund primo vessels link together the intravascular primo vessels, the profund primo nodes, and organs.
3. Primo fluid circulates in the PVS.
   This has been substantiated by the method of dye injection into the primo node and the primo vessel and by the use of radioactive tracers.
   The speed of its circulation is slower than that of the blood, and is much slower in the extravascular primo vessel.
   Circulation in the intravascular primo vessel is considered to be maintained by the heart beat as is the case with the blood and lymph circulation. In other words, the circulation of the primo fluid is, it is considered, caused by the differences of pressure created around it, since the primo vessel lies in the blood current.
   It is therefore established that the primo fluid inside the primo vessel flows in one direction, in the same direction as the blood circulation.
   The contractile action of the smooth muscles of the outer layer of the primo node is believed to play a definite role in the circulation of the primo fluid in the system of extravascular primo vessels.
   The Kim Se Wook phenomenon (Phenomenon Kim Se Wook) to be observed when a needle is applied to a primo node shows the peculiar movement of the primo node.
4. The primo node has unique bioelectrical activity.
   A series of similar changes of electric potential are observed in the primo node even when various electrodes and induction systems are applied to it. These changes of electric potential are connected with the action of the living body, particularly with the action of the PVS.
   The electrogram of PVS directly induced from the primo node is different from the various electric changes so far induced from the skin.

It is presumed that "ㄱ" and "ㄴ" waves on the electrogram of PVS are directly connected with the action of the muscle layer of the primo node and "ㄷ" wave with the secretory action of the cells of the primo node.

The effect of a stimulus given to a primo node is conveyed to the next primo nodes through the same primo vessel.

It is confirmed through various functional experiments that the electrogram of PVS also reflects the general functioning of the organism.

5. A large quantity of nucleic acids, DNA in particular, is contained in the primo node and primo vessel.
   DNA in the primo vessel exists in a peculiar way, outside the nucleus in the homogeneous primo fluid.

   This is established not only by the results of biochemical experiments but also by Feulgen reaction and other histochemical methods and by the luminescent microscopic examination.

   In view of this, we consider that the action of the PVS is closely connected with nucleic acids.

   And the specific form of the existence of nucleic acids in the PVS also requires the study of the functions and metabolism of nucleic acids from a new viewpoint.

   Our new research achievements made public, we believe, have made a certain contribution to the comprehensive elucidation of the PVS, raised a series of important questions of principle in the field of modern biology and medicine, and opened up a new vista in the field.

   Publishing the results of our researches today, we extend our heartfelt gratitude with deep emotion to the Central Committee of the Workers' Party of Korea and to Comrade Kim Il Sung, our respected and beloved leader, who have always directed profound solicitude and concern to our scientific research work.

   We would also like to express our deep thanks to many scientists and friends at home and abroad who have actively supported and encouraged us in our research work.

   The Kyungrak Research Institute
   Pyongyang, Korea
   November 30, 1963

3 The Kyungrak System (Primo Vascular System): *J Jo Sun Med* 1965 June 5;108:1–38

3.1 Conclusion

1. The meridians have several structures.

   (a) All structures are commonly composed of primo vessels and primo nodes. All the primo nodes are connected by primo vessels. A primo vessel is a bundle of dozens of subducts.

(i) The Bonghan subducts (primo lumens) are made of thin endotheliocytes with rod-shape nuclei, smooth muscle cells, and adventitia with fine argyrophilic fibers. The space between the primo lumens is filled with a fibrous structure and amorphous substances. Groups of a dozen or so subducts are tightened by the surrounding membrane. The primo lumen contains basophilic corpuscles and small nuclei-shaped structures.

(ii) The primo nodes are constructed basically by extensions, divisions, and anastomosis of the primo vessels. The structure of the primo nodes is based on the adventitia of the primo lumen and the network substances between the primo lumens.

(b) The structures for the meridians are as follows.

(i) Structures for interior primo vessels.
These structures are composed of the vessels and nodes, and are systematically distributed inside the blood vessels, lymphatic vessels, and the heart. The interior primo vessels are very fragile with thin adventitia and an interstitial substance. The interior primo nodes have similar structures with those of hematosis organs. The networks contain cells affiliated with the bone marrow and the lymphatic system. They also gather similar cells with those of a series of real organs.

(ii) Structures for interior–exterior primo vessels.
These are composed of vessels and nodes separated from the surrounding organs. They are extended regardless of the morphological constitution of the blood vessels and the nervous system. The interstitial substance and adventitia of interior–exterior primo vessels are more developed than those of the interior primo vessels. Inside the interior–exterior primo vessels, there are basophilic structures as well as bright-cell traits.

(iii) Structures for exterior primo vessels.
These are composed of ducts and corpuscles along the blood vessels and nervous system. They are covered by a thick membrane of connective tissues. There are many chrome-affinitive granules.

(iv) Structures for neural primo vessels.
These are composed of ducts and corpuscles floating in the cerebrospinal fluid of the central nervous system. The branches are distributed in the peripheral nervous system as well as the central nervous system.

(v) Structures in the organs for the meridians.
There are also ducts and corpuscles inside the organs connected from the interior, exterior, and neural primo vessels. Every duct in an organ is combined into a terminal primo vessel with which all the cell nuclei in the organ are connected. Fine primo lumens are divided and connected with every cell in the tissue. All the structures for the meridians are interconnected: the interior primo vessels are connected through the blood vessel walls to the exterior primo nodes, which are again

connected by the exterior primo vessels. The interior–exterior primo vessels are connected with the exterior primo nodes to the exterior primo vessels and the neural primo vessels.

2. The meridians are a multicirculatory system for the primo fluid.

   (a) The biochemical components for the primo fluid are as follows.

      (i) A large quantity of nucleic acid and ribonucleic acid.
      (ii) Total nitrogen is 3.12–3.40%, and nonprotein nitrogen is 0.10–0.17%. Fat is 0.57–1.00%, and reducing sugar is 0.10–0.12%.
      (iii) Total hyaluronic acid is 170.4 mg%.
      (iv) There are more than 19 free amino acids including the essential amino acids.
      (v) There are more than 16 free mononucleotides.

      Unlike the pathways for blood circulation, the pathways for primo fluid are interconnected and are made with relatively independent multicirculation pathways. The staining dye and the radio-isotope injected into a Bonghan pathway circulate only in a specified region. But the primo fluid in a pathway can be transmitted to other pathways through the interconnections between the pathways.

   (b) Primo vessels with bioelectrical activity, excitatory conductivity, and mechanical motility.

      (i) The electrical changes occurring in the primo vessels are very slow and have the same wave characteristics as in the primo nodes. These electrical changes vary in relation to stimuli to the primo vessels.
      (ii) When the ducts are stimulated, bioelectrical changes are transmitted. Electrical changes with low amplitudes are transmitted faster (1–3 mm/s) while the changes with high amplitudes are transmitted more slowly.
      (iii) The ducts show spontaneously generated movements. These movements are transmitted and changed with stimuli to the ducts. The movements of the ducts are continuous and periodic with transverse and longitudinal wave modes. The results imply that the ducts have mechanisms to actively circulate the liquid.

   (c) All the cells of tissues are directly connected to the meridians.

      (i) All the nuclei of tissue cells are connected with fine terminal subducts. These subducts are connected to the primo vessels for the organs. The primo nodes for the organs are connected to the organ tissue cells within a specified range.
      (ii) All the primo nodes for the organs are connected to all the meridians. The structures of the meridians start from the primo nodes for the organs and end at the primo nodes for the organs.

(d) Results from the analysis for the circulatory pathway of primo fluid after injecting the radioactive material ^{32}P into various sites of the meridians.

   (i) Primo fluid from tissues circulates to superficial primo nodes.
   (ii) Primo fluid from superficial primo nodes goes to deep primo nodes.
   (iii) Primo fluid from deep primo nodes goes to tissue cells through primo nodes for organs. The same results were obtained from the staining dye injection experiments.

(e) The circulatory pathway for primo fluid is not simple and unitary.

3. Changes of the primo fluid circulation affect functions of organic tissues.

   (a) Stimuli to the primo vessels change the number of beats and power of the heart, and the intestinal canal movements. It also affects the fatigue curve for the skeletal muscles.
   (b) Cutting the primo vessels causes prominent changes to the cells of tissues.

      (i) If the primo vessels are disconnected, a kind of karyolysis occurs in the attached tissue cells and induces apoptosis.
      (ii) If the primo vessels in the peripheral nervous system are disconnected, excitability of nerves is prominently reduced.
      (iii) If the primo vessels in a motor nerve are disconnected and the motor nerve is stimulated repeatedly, there is no muscle movement.

4. Proliferation of the meridians takes place ahead of proliferation of any other organs, such as the blood vessels and the nervous system.
   Embryo development follows the following steps: the step for the formation of the primo vessel blast cell occurs 7–8 h after fertilization; the step for primordial primo vessel occurs 10 h after fertilization; the step for the formation of primitive primo lumens occurs 15 h after fertilization; and the final step for the completion of the primo lumens occurs 20–28 h after fertilization. The fact that the proliferation of the PVS precedes the formation of other structures suggests the PVS plays an important role during development of an organism.
5. The PVS seems to exist throughout the biological world.
   PVS may be found in invertebrates and vertebrates including the mammalians. They appear to exist in every multicellular living organism including plants. Based on the experimental results for the meridian system, the pathway for the primo fluid is as follows: Tissue cells → superficial primo nodes → deep primo nodes → primo nodes for organs → terminal primo nodes → tissue cells. The meridians are interconnected unified multicirculatory structures of these pathways, which facilitate the flow of primo fluid. Every tissue component for an organism is connected by the meridians, and they are arrayed in order following the meridians. Namely, all organisms are suggested to have meridians.

4 Sanal (Primo Microcell) Theory: *J Jo Sun Med* 1965 June 5;108:39–62

4.1 Conclusion

The above experimental results regarding sanals support Sanal theory.

1. Living organisms keep themselves alive via regeneration following the sanal–cell cycle.

   (a) Properties of sanals.

      (i) Sanals are spherical and their size is 0.8–2.4 μm.
      (ii) A sanal is composed of one sanalsome of various shapes and sanalplasm that surrounds the sanalsome.
      (iii) A sanalsome contains a large amount of DNA, and sanalplasm contains RNA.

      Their composition of bases and nucleotides is the same as those in a normal cell.

   (b) Sanals grow into cells and cells in turn become sanals.

      (i) Sanals grow into cells. Some sanalsomes were seen to come out of sanals, sanalplasm formed around them, and they grew into daughter sanals. In the same way, mother sanals gave rise to many daughter sanals, and they fused to make cell nuclei. After the formation of nuclei, cytoplasm formed around the nuclei.
      (ii) Each cell was sanalized to form many sanals. Sanals broke out of the nucleus and spread inside the cytoplasm as the nuclear membrane disintegrated. Sanals in the cytoplasm grew into mature sanals and were released from the cell with the rupture of the cytoplasm membrane.

   (c) Cells were renewed in the form of the Bonghan sanal–cell cycle.

      (i) Cells were generated not only by cell division but also by sanals.
      (ii) Cell generation by cell division was a special case of the Bonghan sanal–cell cycle. In other words, cell division could be regarded as a special form of sanalization following the process of sanalization. Cell division was a special aspect of the cycle, i.e., intracellular Bonghan sanal–cell cycle.
      (iii) Sanals moved around incessantly during the Bonghan sanal–cell cycle. In every stage of the cycle, the state of sanals changed. The states of sanals varied according to their situation: outside cells, proliferating to form nuclei, inside nuclei, or coming out of cells after nuclei were sanalized. When sanals were in nuclei, they did not have all of the substances that they carried when they were outside cells. These substances were supplied by the nuclei and cytoplasm when the sanals came out of the nuclei and cells.

(iv) The cell is a particular step among the sanal cyclic processes.
(v) Sanalsomes are a kind of chromosome that form when cells divide. The DNA amount in a sanal was seen to be similar to that of one chromosome. The number of sanalsomes during the sanalization of a nucleus was the same as the chromosome number.
The dyeing characteristics of chromosomes which emerged in the metaphase of cell division were the same as that of sanalsomes.

2. All structural elements in living organisms regenerated ceaselessly.
If physiological regeneration processes in organisms only depend on cell division, very limited scope in the regeneration processes could be recognized. However, given that the concept, Bonghan sanal–cell cycle, applies to the regeneration process, it was recognized that all tissue cells regenerated continually. This process was also clearly observed in culture conditions.
Continuous self-renewal of organisms generally occurred not only in the molecular and individual level but also in the cellular level. In other words, continuous self-renewal processes of structural elements took place with continuous metabolism in organisms.
(a) By comparing a specimen of normal tissue with an in vivo specimen, many processes of the sanal–cell cycle occurred more actively in normal tissues. Both the sanalization of cells (2–4% in the liver of a rabbit) and the sanals forming cells (1–3% in the liver of a rabbit) happened synchronously.
(b) In the recovery process of injured tissue, the renewal process was very active, and the sanal–cell cycle was shown clearly.

3. The self-renewal process of an organism was managed by the PVS.
(a) The sanal presented only in the PVS.
(i) The sanal and its growth stages could be observed in every primo vessel and corpuscle.
(ii) On the other hand, there were few sanals in blood, lymph, and tissue fluid.
(iii) All cells had sanal-like structures, and therefore they can be sanalized.
(b) Sanals from the sanalization of tissue cells always circulated and matured via the circulatory path of primo fluid.
(i) According to the analysis of the comparison of normal tissue samples and in vivo samples, sanals produced by the sanalization of cells went into the primo vessels, and their nucleus-like structures grew and matured there and migrated to local tissues.
(ii) When sanals labeled with the isotope ^{32}P were injected into primo vessels, it was observed that they flowed with the primo fluid and grew into cells of corresponding tissues.
(iii) Sanals extracted from different regions of primo vessels were under corresponding stages of cell formation.

(iv) Many types of cells were formed after sanals extracted from superficial primo nodes were cultured.

(c) Primo fluid contained various chemicals that were required to grow sanals.

(i) The liquid included plentiful free amino acids, free mononucleotides, hyaluronic acids, and various hormones, along with other proteins, sugars, and lipids.
(ii) Sanals became cells if they were cultured in primo fluid or a culture solution which has a similar composition to primo fluid.
(iii) A sanal does not grow into a cell when it is in blood, lymph, or tissue fluid.

(d) All of the tissue cells were linked to the PVS.

(i) The tissue cells were directly connected with the meridian system.
(ii) The tissue cells died after the dissolution of the nucleus when primo vessels connected to them were cut.

The PVS appeared to control and dominate the formation, maintenance, and death of tissue cells. All of the fundamental processes of life appear to be based on the sanals' movement. So studying the law of this movement should become a fundamental subject of biology.

The Kyungrak Research Institute
Pyongyang, Korea
April 15, 1965

5 Sanals and Hematopoiesis: *J Jo Sun Med* 1965 Oct 8; Volume Number Unknown:1–6

5.1 Conclusion

1. The self-renewal process of blood cells is conducted by the sanal–cell cycle.

(a) Erythrocytes, granulocytes, and lymphocytes were sanalized (sanalization: transformation of a cell into sanals) under artificial culture conditions. They grew to their previous parent cells again by culture.

(i) The sanalization of granulocytes and lymphocytes and the formation of cells by sanals (abbreviated as FCS) were similar to that of normal tissue cells. The sanalization of granulocytes occurs in this sequence: sanals, basophilic structures, round nuclear-like structures, and granulocytes. In the process of changing from the round nuclear-like structures to the granulocytes, the early stages of granulocytes maturation, which was described before, were rarely observed.

(ii) The sanalization and the FCS of nonnucleated erythrocytes were special. The sanals of nonnucleated erythrocytes lacked sanalsomes. In the course of sanalization, the erythrocyte sanals seemed to proliferate radially outward. But artificially constricted erythrocyte sanals showed the fusion of them. The nonnucleated erythrocyte sanals did not have DNA. They had relatively more RNA than erythrocytes, and the amount of it decreased as the sanals matured. However, the amount of hemoglobin increased, and the activity of oxidative enzymes decreased.

(b) Erythrocytes, granulocytes, and lymphocytes also underwent self-renewal processes in the process of the sanal–cell cycle in vivo. The sanalization and FCS of blood cells in vivo were similar to that in artificial culture conditions. The renewal of blood cells is conducted in two separate ways, i.e., cell division, or the sanal–cell cycle inside a cell, and the sanal–cell cycle of cells. However, the latter is the principal method and the former is very uncommon. The latter ensures the abundant renewal of blood cells. Mammals have two kinds of erythrocyte sanals. One is without a nucleus and the other is with a nucleus. These sanals have different and independent sanal–cell cycles. After birth the transition between the cycles is improbable. The sanal–cell cycle of nucleated erythrocyte sanals resembles that of normal tissue cells.

(c) By analyzing each stage of the formation of a cell by blood cell sanals, the blood cells, understood in relation to different stages of differentiation and maturation, have different and independent sanal–cell cycles.

2. The sanal–cell cycle of blood cells, or the self-renewal process of them, is conducted by the meridian system. Blood cell sanals usually flow inside intravascular primo vessels and mature to blood cells in intravascular primo nodes. Some of them were observed in other systems of primo nodes. Hematopoietic organs such as bone marrow, the spleen, and lymphatic nodes have well-developed primo vessels whose structure and function are the same as primo nodes. In the vessels, the formation of cells by blood cell sanals is abundant.

(a) The formation of cells by blood cell sanals is carried out in the meridian system.

(b) The sanalization of blood cells also occurs in the meridian system. That is, blood cells are sanalized in each primo node inside blood and lymphatic vessels.

The Kyungrak Research Institute
Pyongyang, Korea
October 8, 1965

Chapter 3
A Follow-up Study on the Morphological Characteristics in Bong-Han Theory: An Interim Report

Satoru Fujiwara and Sun-Bong Yu

Abstract Morphological characteristics in Bong-Han theory was confirmed in the interior-exterior primo vessel system of rabbits, dogs, and rats. Especially, a superficial primo node in the abdominal skin of a rabbit was observed.

For the first time, in the 1960s, a new meridian theory was suggested in which the meridian has physical substances. The so-called "Bong-Han theory" has been studied by Prof. Bong-Han Kim and his colleagues. According to the theory, the meridian system exists as a third circulatory system besides the blood, the lymphatic, and the nervous systems in the body. This is composed of "primo nodes" and "primo vessels," which are distributed over the blood and lymphatic vessels, the nervous tissue, organ surfaces, inside organs, and inside the skin. The meridian system is a kind of circulatory system with a large quantity of DNA in the "primo fluid." They reported that the meridian system plays an important and substantial role in the manifestations of vital phenomena. Moreover, they suggested that this meridian system was directly related to the processes of cell formation and destruction and that conventional cell proliferation theory was included in their point of view. We cannot help but think that this revolutionary theory requires overall reexaminations of the propositions of conventional biology and medicine. Their five reports are not experiment reports, but rather general remarks with many conclusive descriptions. Even though their reports contain difficult formats for follow-up studies, we carried out morphological studies on a model of the important suggestions [1].

In Japan, there are several reports, which are all in affirmative directions for the Bong-Han theory, such as the interior-exterior primo node by Mr. Hatai and the superficial primo node by Mr. Fujiwara. In this interim report, we present our data

S. Fujiwara (✉)
Department of Anatomy, Osaka City University, Osaka, Japan
e-mail: hibiyakoube@hanmail.net

K.-S. Soh et al. (eds.), *The Primo Vascular System: Its Role in Cancer and Regeneration*, DOI 10.1007/978-1-4614-0601-3_3, © Springer Science+Business Media, LLC 2012

on the organ surface primo vessel system and include an example of the superficial primo node. Our work is complementary to the previous work of Fujiwara.

As for our experimental animals, we used mostly rabbits, as well as dogs, rats, etc. A series of structures, which are described in this paper, were unknown before Kim's reports. However, it was absolutely difficult to make a direct follow-up study only with his descriptions. In light of our experiences with purposeful searches for such fine structures, we outline their observation methods in most of this paper.

Our study methods included gross anatomy with a loupe and stereo-microscope, as well as histology with smashed tissue samples. As for the smashed samples, we observed them with the Giemsa, Daria, Feulgen, and chromaffin reaction staining after mainly alcohol fixations. After the tissue samples had been fixed in 10% formalin and embedded in paraffin, they consecutively were cut into slice 3–15-μm thick. We used various staining methods for observations such as HE staining, Azan staining, PAS staining, PAM staining, and hematoxylin staining. We also used the staining-solution spreading method in order to recognize the existence of the superficial primo nodes on the skin membrane.

From our follow-up study on morphological characteristics, we obtained some definite results, as mentioned before. However, judging from the broad and diverse contents of the theory, our data contains only preliminary and partial verification. An evaluation of the entire theory, we think, requires multifaceted research including physiological experiments. Through this work, we only cross-checked our data with those of Prof. Kim.

Starting with the rabbit experiments, we observed a series of net-shaped structures which were distributed independently from the blood vessels and which were separated from the surfaces of various organs in all experimental animals. Determining the names of the structures, we kept the two following points in mind: The first point is whether the structure was a pathological product or not. If the structure was a pathological product, we would generally reach unrealistic and illogical conclusions. Since we found the existence in every animal, we decided that this structure always existed physiologically. The second point is whether the structure was explainable with current knowledge of biology and medicine or not. In our consideration, we were not able to completely explain the structure of definite systemicity and morphology with conventional knowledge. Based on the agreement with Kim's description of the morphological characteristics of the structure, we concluded that the confirmed structures in our study belonged to the interior-exterior primo vessel system. Also, the corpuscle and the threadlike structures corresponded to the primo node and the primo vessel, respectively.

We could not designate all the details because our data were partial and preliminary and because Kim gave a general description of the structure. For example, we confirmed histologically that the primo vessel is composed of a bundle of primo lumens, while we could not clarify the details of the compositions in the subduct's wall. Even though we think that portions of the empty space and of the porus type structures match primo vessel lumens, further results from future research need to be more precise. In this study, we found two kinds of fiber structures; one is the net structure that forms the baseline structure of the corpuscle and is distributed over the whole corpuscle tissue except portion of the ducts. The other is the microfibril structure that penetrates into the corpuscle through the outside membrane and produces fine nets. Bong-Han Kim also described the structure of the network for the formation of corpuscle substance. The relations between his network and our fiber structures should be appreciated and considered in future study.

As for the large cells with an epithelial shape, Bong-Han Kim described their existence in the interior primo node (the primo node inside the blood vessels and lymphatic vessels), not in the interior-exterior primo node. As for the superficial primo node, even though Fujiwara has already reported on it briefly, we also report a complementary revision corresponding to the supplement of Kim's opinion before the publication of his third-paper. Based on the size, the shape, the position, the existence of the exoplasm in the smooth muscle-like tissue, the well-developed vessel network, and the smooth muscle-like structure in the endoplasm, our observed structure is thought to correspond to the superficial primo node. However, chromaffin cells and epithelial structures were not observed in this study. Moreover, it was not sufficient to confirm the primo lumen or primo node sinus. In our opinion, the small ductule structure between the smooth muscle cells may be the primo lumen, and the crevice in the tissue in the smooth muscle structure may correspond to the primo node sinus. This might be confirmed by dye injections into the body, but we have not started this type of experiment yet. Related to this, it is important to find a method to identify the portions of the superficial primo node from the skin. We briefly discussed the staining-solution spreading method, which we used in this report.

The summary is as follows:

1. We obtained several affirmative results from a follow-up study on the morphological characteristics for the Bong-Han theory, as a new meridian theory.
2. We succeeded in observing a series of structures with every experimental animal and concluded that the structures belong to the interior-exterior primo vessel system based on their morphological characteristics.
3. In the skin near the linea alba of the abdomen, we found one example of a structure that might be called a superficial primo node based on the histology.
4. We discussed on our preliminary trials to recognize the area of the superficial primo node, using the staining-solution spreading method.

Reference

1. Fujiwara S, Yu SB (1967) 'Bonghan theory' morphological studies. Igaku no Ayumi (Other title: Journal of Clinical and Experimental Medicine, Medicine in Progress; ISSN: 0039-2359) 60:567–577 [In Japanese]

Chapter 4
Recollection of Early Research on Primo Vascular System: Ultimate Implication of Bong-Han Theory

Jong-Su Lee

Abstract The author recollected his early research on the primo vascular system in the 1970s. He speculated on possible impacts of the Bong-Han theory upon modern medicine. He also presented the hypothesis linking the cause of cancer to the primo vascular system.

Whenever I think about Bong-Han theory, a big stone compress on my chest. What happened to Prof. Kim Bong-Han and why is North Korea keeping silence about Bong-Han theory for last half-century? I got the book *Geiracuno-Haken, The Theory of Acupuncture System*, in 1971. I experimented on Bong-Han theory in rabbit to find if the theory of Prof. Kim Bong-Han was correct.

What Is Bong-Han Theory? First of all, all living cells of living beings, when they get old, they dissolve to chromosomes of cells, and then they circulate in Bong-Han system. After receiving some energy, they become cells, and they regenerate damaged tissues. All cells are connected to Bong-Han system, Bong-Han canal penetrate to nuclei of cells. If we have problems of hematological or endocrine systems, Bong-Han systems solve these problems. So what Prof. Kim Bong-Han found was a new circulatory system in our body besides blood vessel system.

The Bong-Han system solves many problems of modern medicine. One of the fundamental principles of modern western medicine is mitosis. Mitosis cannot explain those matters, but mitosis is included as part of Bong-Han theory. Plant cells live by photosynthesis. Animal cells, I would like to say, live by Bong-Han synthesis.

According to the Chinese acupuncture theory, we have five organs, six viscera. All these eleven organs have their own special Bong-Han system. For instance, the Bong-Han system of lung begins in large intestine, middle intestine, stomach, diaphragm, lung, both of upper extremities and it finally ends at finger.

J.-S. Lee (✉)
Lee Jong Su Clinic, TaeGu, Korea
e-mail: jhkim10@snu.ac.kr

K.-S. Soh et al. (eds.), *The Primo Vascular System: Its Role in Cancer and Regeneration*, DOI 10.1007/978-1-4614-0601-3_4,

When I synthesize Bong-Han theory to Chinese acupuncture system, there appear a hint about the cause of cancer. For an instance, virus infects the pulmonary Bong-Han system, then the extra, pathological cells of lung are produced and they result in the lung cancer.

Now, all medical students should be taught that cancer metastasises through Bong-Han canal in vessel, the hormone distributed through Bong-Han canal in vessel, and all brain cells are being renewed continuously. I like to insist that we doctors should carry acupuncture needle instead of stethoscope.

Many of the medical textbook must be rewritten, because many of the medical problems can be solved by the shift of paradigm due to Bong-Han theory. Even though, many doctors are still skeptical at the Bong-Han theory.

Finally, I would like to present my hypothesis about the cause of cancer. The cause of cancer is the pathological change of Bong-Han system. I repeat that the cause of cancer is the pathological change of Bong-Han system.

Thank you.

Chapter 5
Current State of Research on the Primo Vascular System

Kwang-Sup Soh

Abstract We provide reviews on the current state of primo vascular system (PVS) research from two different perspectives. The first is about the places where the PVS was observed: nerve system, cardio-vascular system, lymphatic system, fascia in the abdominal cavity, adipose tissue, generative system (testis), skin and abdominal wall, primo fluid and microcells, egg vitelline membrane, and cancer. The second sorts out which parts of Bong-Han Kim's claims have been confirmed, and which have not yet been confirmed. New findings and methods that were not in Kim's reports are listed as well. This review is intended to provide a bird's eye view on PVS research to those who plan to embark on this novel area.

1 Introduction

Since the extensive reinvestigation of the Bong-Han Theory began in the year 2002, until 2008, essentially only one laboratory at Seoul National University (SNU) was involved. Then, suddenly six or so teams in Korea, three teams in China, one in Kazachstan, and three teams in USA started some experiments on the primo vascular system (PVS) in 2009 and 2010. Many more researchers throughout the world showed keen interest in the PVS, and the International Symposium on the PVS, with special topics on cancer, regeneration, and acupuncture, was held in September 2010. At this symposium, the International Society on PVS (ISPVS) was founded.

The SNU group has emphasized developing methods to visualize the PVS in various organs of an animal, principally rabbits, rats, and mice. A major step in the methodology was the discovery in the year 2008 of the Trypan blue technique for

K.-S. Soh (✉)
Nano Primo Research Center, Advanced Institute of Convergence Technology,
Seoul National University, Suwon 443-270, Korea
e-mail: kssoh1@gmail.com

K.-S. Soh et al. (eds.), *The Primo Vascular System: Its Role in Cancer and Regeneration*,
DOI 10.1007/978-1-4614-0601-3_5,

the specific visualization of the PVS. Before this, the SNU group was groping its way searching for the PVS in various organs. With the Trypan blue technique, the PVS was found around cancer tissues, in adipose tissues, and in brain ventricles. Especially, the observation of a cancer-PVS and a fat-PVS was the first finding not mentioned in Kim's reports and suggests a far more important significance of the PVS in medical subjects, such as cancer, obesity and diabetes.

Considering the rapidly expanding group of researchers in biology, medicine, and other bio-related fields, it is timely to provide a bird's eye view of the scope of research achievement. This article summarizes various aspects of PVS research: the organs in which the PVS has been observed, the parts of Kim's reports that have been confirmed and those that are still unconfirmed, and new findings made by the SNU group.

2 Systems and Organs in Which the PVS Has Been Observed

2.1 Brain, Spinal Cord, and Sciatic Nerve

Primo vessels (PVs) and Primo nodes (PNs) in the third ventricle, the fourth ventricle, and the cerebral aqueduct of the brain of a rabbit were visualized by hematoxylin staining. This PV ran along the central canal of the spinal cord of a rabbit. The average diameters of the spinal cord, the central canal, and the threadlike structure were 5,000, 150, and 30 μm, respectively. The PVs were not attached, but were freely floating in cerebrospinal fluid (CSF) [1].

In the case of a rat, PVs and PNs were observed in the third and fourth ventricles and in the spinal cord by using the Trypan blue staining technique. They were also observed on the arachnoid mater, and especially weblike nets of PVs were found on the surface of the cerebellum of the rat [2]. A slightly different method of injecting Trypan blue into the lateral ventricle through a hole in the skull also revealed the PVs and the PNs in the lateral and third ventricle of a rat [3].

The visualization of PVs in the perineurium and the endoneurium of a rat sciatic nerve were achieved by using the Trypan blue staining technique [2]. PVs in the perineurium and epineurium were visualized by injecting fluorescent nanoparticles (FNP) into the hyperdermis at the acupoint Zusanli (St-36) of a rat [4].

2.2 Cardio-Vascular System

2.2.1 Bovine Heart

Networks of PVs (about 20 μm in thickness) and nodes (40–100 μm in diameter) were visualized with Trypan blue staining in the bovine heart atrium, and they were freely moving in the endocardium. A morphological study with confocal laser scanning microscopy and electron microscopy showed the characteristic rod-shaped nuclei and collagen fibers of extracellular matrices [5].

2.2.2 Blood Vessels

PVs in blood vessels of rabbits, rats, and mice were first observed in various large blood vessels, such as the caudal vena cava, the hepato vein, the hepato portal vein, the femoral vein, the abdominal artery, and the aorta [6]. Most of the samples were not long, and the longest one ever taken was about 4 cm in a rat, whose trunk length was about 10 cm [7]. In these early works, no staining dye was used. Instead, a perfusion method was used. Dextrose (10%) was infused into a femoral vein of a rat at a speed of 20 drops/min by gravity for about 50 min.

The most difficult part was to distinguish the PV from similar looking and more abundant fibrin strings of blood coagulation. Our contribution was the development of a method to discern the PV from the fibrin strings by using acridine-orange fluorescent dye to reveal the characteristic distribution of rod-shaped nuclei [8].

2.3 *Lymphatic Systems*

The only place where the PVS is visible in vivo, in situ without a staining dye is inside a lymph vessel. The large lymph vessel along the caudal vena cava of a rabbit or a rat was investigated to search for a PVS floating in the lymph flow. The PVS is transparent and hardly visible without some visualization technique. In the early stage, staining dyes were injected into the lymph flow, and the PVS absorbed the dyes, preferentially to the wall of lymph vessels, and became visible. The effective dyes were Janus Green B [9], Alcian blue [10], and fluorescent magnetic nanoparticles [11, 12]. Later, with a carefully aligned illumination with spectral adjustment, the PVS became visible without any external chemical dyes [13]. The primo vessel does not run only in the lymph vessel, but comes out of the vessel and is connected to the PVS on organ surfaces.

Until the present time, the required skill was too difficult to apply to a smaller animal, for example a mouse. Histological examinations of the lymph PVS were performed, but analysis at a cellular or molecular level has not been done mainly because the amount of sample is too small.

2.4 *Fascia in the Abdominal Cavity*

The PVS was detected as a freely movable threadlike structure floating over the surfaces of various internal organs, such as liver, stomach, small and large intestine, and bladder. They are thin semitransparent threads with associated primo nodes and are often visible under stereomicroscope without applying any staining dyes. The primo vessels are strong enough to resist tensions due to holding and lifting with forceps. They are elastic and snap away if cut. They are uniform in thickness and have branches and nodes. The Trypan blue technique turns out to be the most useful one for detecting the PVS even in mice. The PVS in the abdominal cavity was most extensively studied in rabbits [14–17] and rats [7, 18], and sometimes in mice [19]. Few case studies were reported for pigs [20] and dogs [21].

Frequently, the tracking of the PVS ended in the fascia of the abdominal cavity wall or in the peritoneum surrounding an internal organ, and it was not possible to trace it further because a method to differentiate the primo vessel from a membranous structure has not yet been developed. Only very rarely has a primo vessel been observed to enter internal organ tissues [22].

A weblike net of primo vessels with many primo nodes at the branching points of the vessels was observed on the great omentum of a rat [18]. Similar net structures were also observed in the great omentum of a dog [21], the bovine heart [5], the brain of a rat [2], and the superficial fascia in the hypodermal layer of a rat [23].

Histological studies [14], ultrastructural analyses with various types of electron microscopy [15], proteomics analysis of the PVS and primo fluid [24], and flow speed measurements of the liquid in the primo vessels [16] were performed with the PVS in the abdominal cavity. The primo microcells were also obtained here, and some of their basic properties were studied [25].

There are several difficulties in studying the PVS in the abdominal cavity. First of all, the PVS is not regularly distributed, and sometimes no PVS is detected in an animal. We do not understand the reason the PVS are easily detected in some animals, and not in other animals. Probably, the development of the PVS depends upon the physiological or health states of the subject animals. Second, bleeding must be controlled; otherwise, blood coagulation forms strings of fibrin that look deceptively similar to primo vessels. Third, a torn peritoneum also looks like primo vessels, which requires careful and skilful operation. In addition, drying would make detection of the PVS more difficult. Finally, the small volume of the primo vessel or nodes makes applying quantitative analyses, such as proteomics, genomics, or other molecular or chemical analysis, very difficult.

2.5 Fats (Adipose Tissues)

We were often frustrated in our efforts to trace the PVS by its escaping and hiding in adipose tissues because the semitransparent primo vessel was not visible. By using the Trypan blue staining technique, those in the adipose tissues could be detected if they were in the shallow region below the surface, less than about 100 μm [26]. In the fat layer just above the superficial fascia of the hypodermal skin of a rat, a primo vessel and primo node were detected by staining with Trypan blue. A more striking case is the fat band in the abdominal wall in the abdominal cavity side, where a set of primo nodes aligned along the central line reside. These nodes were not completely covered by fat and were easily identified [27].

2.6 Generative System: Testis

Bong-Han Kim emphasized that the PVS was an important generative system, and he studied the ovary extensively [28]. We conjectured that a testis might be similarly important. Using a testis has an advantage in that staining dyes can be injected into the testis without laparascopic surgery.

Injection of FNP into a testis of a rat from outside of the skin was done with a syringe with needle gauge 31, and laparascopic surgery was done after 24 h. The PVS over the surfaces of internal organs in the abdominal cavity were observed with a fluorescent microscope, and the samples were isolated to examine the flow of FNP [22]. This result strongly suggested that the observed PVS was a flow path starting from a testis, but there was no direct proof from the immediate surface of a testis.

In the ensuring experiment, we improved the injection method to find the PVS starting from the immediate surface of a testis. In this way we proved that a flow path starting from the testis, which is distinctively different from blood or lymph vessels, existed [29]. Further study to elucidate in detail the PVS entering the testis is necessary.

2.7 Skin and Abdominal Wall

In the midline of the abdominal wall of a rat, there is a band of adipose tissues which we named the conception vessel (CV) fat line. Along this CV fat line, we can see a large vein and artery running from the xiphoid through the navel to the bladder. According to the chart of human acupuncture meridians, there is a CV meridian, and the WHO nomenclature named the acupoints on this meridian as CV14 at the xiphoid and CV8 at the navel; other points between these two acupoints are located at equal distances. Primo nodes at CV12, 10, 8 were observed, and basic histological study with H&E and Mason's trichrome revealed that they were different from lymph nodes. By injecting FNP into the primo nodes, we traced the flow of nanoparticles along the CV line to the ligament wrapping the bladder in the primo vessels. Thus, we established the presence of extravascular primo vessels along the blood vessels just outside the connective tissues of the blood vessel [30]. In this experiment, the PVS ran along the CV fat line to the bladder.

The acupoint number 23 in the bladder line (BL-23) is supposed to control the function of the kidney. We injected Alcian blue into the putative BL-23 of a rat and hypothesized that the dye would flow to the kidney. However, we did not find any evidence showing such flow. Nevertheless, the Alcian blue appeared in the PVS suspended over the organs in the abdominal cavity [31]. This experiment suggested that the acupoints in the dorsal skin are connected to the PVS in the abdominal cavity and form a circulatory system.

Another acupoint we tried was Zusanli at the stomach line (St-36). The fluorescence nanoparticles injected into the hypodermal layer at the St-36 flowed along the sciatic nerve toward the spine. A PVS surrounded with connective tissue at the epineuria of the sciatic nerve was revealed by the fluorescence of the FNP and was examined to check for the characteristic nuclei distribution of a primo vessel [4]. The FNP also flowed toward the foot along the putative ST-line in the fine networks of the PVS lying in the hypodermal superficial fascia [23].

A fat layer exists in the hypodermal layer of the ventral rat skin. Primo vessels and nodes in the fat layer were detected by using a Trypan blue staining technique, but systematic observation of their distribution has not been performed. The result only suggests that an extensive fine network of the PVS exists in the superficial fascia and fats of the hypodermis.

2.8 *Fluid System*

2.8.1 Hormone

According to Bong-Han Kim in the liquid that circulated in the PVS were hormones, hyaluronic acid, and mononucleotides [28]. We confirmed the existence of adrenalin- and noradrenalin-producing and -storing cells in the primo nodes taken from the PVS in the abdominal cavity of a rabbit [32]. The ELISA method was used to assay the hormones [33, 34].

2.8.2 Primo Microcells

Wound healing and cell therapy for regeneration of damaged tissues were the claimed function of the PVS, and they were carried out by primo microcells [35]. Primo microcells were obtained from the primo nodes in the abdominal cavities of rabbits and rats. The random motion of the primo microcells was measured and analyzed to compute the viscosity of the liquid [36]. An increase in the average speed with illumination of UV-A (360 nm) was observed [25], which needs further study to establish and elucidate the mechanism. The primo microcells were spherical, 1–2 μm in diameter, and their DNA was fragmented [37]. Transmission electron microscopy and atomic force microscopy have been used to investigate their specific morphology [38]. The budding of the primo microcells as the first step of their proliferation was observed by using scanning electron microscopy and atomic force microscopy [39].

Very small embryonic-like (VSEL) stem cells, which were similar to primo microcells in shape, size, and their putative functions, were found by Ratajczak et al. [40]. Whether they are of the same kind or not is a most important question.

2.9 *Egg Vitelline Membrane*

Based upon the close relation between the PVS and the fascia [41], we conjectured that there is the PVS-like structure in the membrane between yellow yolk and white albumin of a chicken egg. Indeed, we observed nets of primo vessel-like structures stained by Trypan blue [42]. In the primo vessel-like threads, there were microparticles with DNA, which is consistent with Bong-Han Kim's claim that in the chicken egg, basophilic particles appear in early stage primo vessels [28].

2.10 *Cancer*

One of the most significant findings with the Trypan blue staining technique was the discovery of the PVS on the fascia of tumor tissues. Tumor tissues were grown in the skin of nude mice after subcutaneous inoculation with human lung cancer cells. Trypan blue staining revealed a visualization of the PVS connected to tumor tissues [43]. In the case of intraperitoneal inoculation, the PVS was also connected to tumor tissues grown in internal organs [19]. The PVS may be utilized as a drug delivery path for cancer or as a novel route of acupuncture treatment for cancer. There is an adverse side of the PVS as well: it may be a hitherto unknown path of cancer metastasis. Indeed, we observed such metastatic cases that are even stronger than the lymphatic route [44]. This suggests that tumor metastasis control requires knowledge of the PVS, which will open a new research area in cancer biology.

Similar to the PVS in the fascia of a tumor, other conduits have been found inside tumor tissues, namely, the vasculagenic mimicry (VM), first found by Hendrix [45]. It is an imminent question how the tumor PVS and VM are related.

3 Confirmed and Not Yet Confirmed Parts of Bong-Han Kim's Reports and New Findings by the SNU Group

3.1 *Subject Animals (Normal)*

The subject animal throughout Kim's work [28, 35, 46–48] was the rabbit, and no data on human subject was presented. His first work [49] was apparently on human acupuncture points even though he did not explicitly mention that. In his third work [28], he hinted that the PVS was observed in various mammalians, including humans, avians, amphibians, fishes, and invertebrata (hydra), without mentioning any specific animal species. He presented explicit data on the PVS in the sciatic nerve of frogs, and the development of a PVS in the chicken egg [28]. Furthermore, he claimed the PVS was present in plants and specifically mentioned Hellanthus annual (sunflower) with respect to primo microcells [35].

Confirmed cases were rabbits [14–18]. Not confirmed cases were humans, avians, reptiles, amphibians, fishes, and invertebrata. The plant has not been tested. Development of a chicken egg was only partially confirmed [42].

The SNU group worked mostly with rats and mice [18, 19]. Other groups have worked with bovine hearts [5], pigs [20], and dogs [21].

3.2 *Subject Animals (Disease Model)*

Kim induced anemia with phenyl hydrazine and inferred a hematopoietic function for intravascular primo nodes [28]. He studied regeneration of damaged liver tissues of rabbits through the stem cell-like function of primo microcells [35]. The damage was incurred by piercing the liver with a glass tube whose diameter was 2 mm.

The SNU group used phenyl hydrazine in rabbits or rats and observed an increase in the sizes of the primo nodes and the primo vessels on the organ surfaces. No systematic experiments or statistical analysis were made, and no reports were presented.

The most important new discovery in modern reinvestigations was the PVS in the surrounding capsule of cancer tissue [43] and its possible role as an additional metastasis route of cancer [44]. The PVS can possibly be used as a delivery path of anticancer drugs, instead of intravascular or digestive administration. The SNU group found, for the first time, the PVS in adipose tissues [26], which naturally raises the role of the PVS in obesity, diabetes and origin of stem cells in adipose tissues.

3.3 *Methodology*

3.3.1 Tracer

Kim found a mysterious blue dye that flowed in the primo vessel to reveal the entire network in the body. He mentioned its use in many places, but did not give any information on either its substance or the procedure for using it. This is the most essential key element to the discovery of the PVS without which it is extremely difficult to observe or identify the PVS. He also used radioisotopes mainly to demonstrate the circulatory function of the PVS [46, 47] and to trace the flow of primo microcells [35].

Without knowing the substance and the method of the blue dye, the SNU group tried various dyes and obtained partially fruitful results with Janus Green B [9], Alcian blue [10, 31], and Trypan blue [18]. Among these, Trypan blue was the most useful because it specifically visualizes the PVS among lymph vessels, blood vessels, fascia, nerves, muscles, and fat tissues. It opened the way to discover the PVS on cancer tissues [43] and to observe the PVS in the brain and the spine [2]. The most frequently used tracers were FNP [22, 23] and magnetic nanoparticles [11, 12]. For the cancer migration study, quantum dots [44] and GFP cancer cells were used.

3.3.2 Histology

Kim used various standard staining methods to characterize the PVS: H&E, Mason's trichrome, Verfroff, silver staining, Feulgen reaction, and Acridine orange, all of which were also used by the SNU group. The SNU team applied more modern techniques, which were not available at Kim's time. PI and DAPI were used for nuclei identification, DiI and DiO for lipid membrane, and Phalloidin for f-actin.

More importantly, the SNU group used an immunohistochemical method with various antibodies. In particular, LYVE-1 was useful to differentiate the PVS from lymph vessels.

3.3.3 Instruments

Kim used various optical microscopes, such as a stereomicroscope and phase-contrast, inverted, and fluorescent microscope. He even had an early model of a TEM. The SNU group used a modern optical microscope system and a confocal microscope. For electromicroscopy, it had SEM, including FIB-SEM, and high voltage TEM. In addition, it used AFM for the physical study of primo microcells. Furthermore, X-ray microscopy at the Pohang linear accelerator was applied to study the inner structure of the primo vessel, but this technique has not yet been fully utilized. MRI and CT were tried, but no important results have been obtained until now.

3.4 Physiology

In addition to the unknown blue dye flow, Kim's team used radioisotopes to prove the circulatory function of the PVS [46, 47]. They studied the electrical conductance, wave transmission, and mechanical motion of primo vessels. They stimulated the PN and examined the heart motion, the peristaltic motion of the intestines, and the skeletal muscle motion. After cutting PVs, they investigated the effect upon the nerve system [28]. They found a hematopoietic function for the intravascular PVS inside lymph vessels and blood vessels [49].

The SNU group only tested the flow of Alcian blue in the primo vessel on the surfaces of internal organs [16]. The electrical rest potential and some action potentials were measured using an intracellular method, but further study is needed because the statistics were not high enough [50].

3.5 Chemical Analysis of Primo Fluid

The PVS as a hormone path for adrenalin/noradrenalin, female hormones and others was Kim's claim, and the SNU group confirmed the presence of adrenalin/noradrenalin in a PN by using ELISA technique [32–34]. Until now, the SNU group has not determined the chemical components of the primo fluid, which Kim's team claimed were hyaluronic acid, free amino acids, free mononucleotides, sugar, N_2, and lipids.

3.6 Structure of PV and PN Cells

The SNU group confirmed the basic structure of the PV, thin surrounding membrane, bundle of multiple lumens, intercellular matter of reticular fibers, and rod-shaped nuclei of endothelial cells, that Kim described [28, 47]. The chromaffin cells, basophilic nuclear bodies, endothelial cells, and smooth muscle-like cells that Kim depicted have not yet been clearly identified. DNA-containing microparticles in egg albumin and in brain-PV were newly found, as were many immune cells in the PV and the PN taken from organ surfaces [15, 42].

3.7 Primo Microcells (Sanals)

Primo microcells were extracted from organ surface PNs, and the spherical shapes, the sizes (about 1–2 μm), and the DNA-containing of sanals were in agreement with Kim's report [35–39]. Cultivation, chemical components, circulation of sanals, regeneration of damaged function, formation of cells out of sanals, and formation of sanals from cells have not yet been tested. The effect of UV-A (360 nm) on the speed of sanal motion and investigation of sanal surfaces of membranes with an atomic force microscope are new results [25, 39].

4 Summary

Most of the SNU group's work was on the visualization and the detection of the PVS in various organs and tissues, and these are summarized in Table 5.1. This work provides a point from which further studies, such as physiological functions, mapping and imaging, and medical applications of PVS, can be started. Table 5.2 gives a summary of the confirmed and not yet confirmed claims of Kim's reports, as new findings by the SNU group. This table will give a bird eye's view of the current state of PVS research.

Table 5.1 Systems and organs where primo vascular system (PVS) have been detected

System	Organs	Animals	Dye or equipment
Neural system	Brain	Rabbit	Hematoxylin
		Rat	Trypan blue
	Spinal cord	Rabbit	Hematoxylin
		Rat	
	Sciatic nerve	Rat	Trypan blue
			Fluorescent nanoparticles (FNP)
Cardio-vascular system	Heart	Cow	Trypan blue
	Artery and vein	Rabbit	Acridine orange
		Rat	Alcian blue
		Mouse	
Lymph	Lymph vessel near the caudal vena cava	Rabbit	JGB
		Rat	FNP
			Alcian blue
			Optical method
Fascia	Abdominal wall	Rabbit	Trypan blue
		Rat	
		Mouse	
	Omentum	Dog	Trypan blue
		Rat	
	Peritonia (liver, stomach, intestine, bladder)	Rabbit	Alcian blue
		Rat	Trypan blue
		Mouse	
Adipose tissue	Abdominal cavity hypodermal layer	Rat	Trypan blue
Generation system	Testis	Rat	FNP
			Cancer cells
Skin and abdominal wall	CV Fat band	Rat	FNP
	BL-23	Rat	Alcian blue
	St-36	Rat	FNP
	Hypodermal	Rat	Trypan blue
Fluid	Hormone	Rat	ELISA
	Primo microcells (sanals)	Rabbit	TEM
		Rat	AFM
			SEM
Egg	Vitelline membrane	Chicken egg	Trypan blue
Cancer	Skin	Nude mouse	Trypan blue
	Intra-peritonia		

Table 5.2 Confirmed and not yet confirmed claims of Bong-Han Kim's reports

Category	BH Kim's reports: Confirmed	BH Kim's reports: Not yet confirmed	New findings and instruments
Subject animals (normal)	Rabbit Chicken egg development	Humans Mammals Avians Reptiles Amphibians (frog) Fish Invertebrata (hydra)	Cow Pig Dog Rat Mouse
Disease model	Anemia	Liver damage	Cancer Adipose tissue
Methods			
Tracer		Blue staining dyes Radio isotope	Janus Green B Alcian blue Trypan blue Fluorescent nanoparticles Magnetic nanoparticles Quantum dots GFP cancer cells
Histology	Acridine orange H&E Feulgen reaction Trichrom Verhoff Silver		DAPI, PI DiI, DiO Immunohistochemistry (LYVE-1, VWF, etc.)
Instruments	Optical microscopes (phase, inverted, fluorescent) TEM		Confocal microscope SEM HV TEM AFM X-ray microscopy
Technique		Injection technique into PN	Lateral ventricle injection Distinction of fibrin string and primo vessel
Physiology	Flow (local)	Circulation with radio isotopes Electric conductance of PV Stimulation of PVs and its effects upon heart motion, peristaltic motion and skeletal muscle The effects of PV cutting on nerve system Hematopoietic function of the PN inside a lymph vessel Hematopoietic function (RBC)	Extracellular electrical properties (action potential, rest potential)

(continued)

Table 5.2 (continued)

	BH Kim's reports		
Category	Confirmed	Not yet confirmed	New findings and instruments
Chemical analysis of primo fluid	Adrenalin/ noradrenalin	Other hormones Hyalurone Free amino acids Free mono nucleotide Sugar, N_2 and lipid	Proteomics ELISA (adrenalin/ noradrenalin)
Structure and cells	Reticular fibers	Chromaffin cells	DNA-containing microparticles in egg and brain PVS
	Rod-shaped nuclei Surrounding thin membrane	Basophilic nuclear-like bodies Endothelial cells	Immune cells in the organ surface PVS
	Bundle of lumens	Smooth muscle-like cells	
Primo microcell (sanal)	Random motion	Extraction	"360-nm" Effect (increase of speed)
	Size/shape Presence of DNA	Cultivation Chemical components Sanal circulation (P^{32} isotope method) Regeneration of damaged liver tissue Formation of cells out of sanals Formation of sanals from cells	AFM images of surface

New findings by the Seoul National University (SNU) group

Acknowledgments This research was supported by a "Systems Biology Infrastructure Establishment Grant" from the Gwangju Institute of Science and Technology and by the Association of Korean Oriental Medicine.

References

1. Lee BC, Kim SK, Soh KS (2008) Novel anatomic structures in the brain and spinal cord of rabbit that may belong to the Bonghan system of potential acupuncture meridians. J Acupunct Meridian Stud 1:29–35
2. Lee BC, Eom KH, Soh KS (2010) Primo vessels and primo nodes in rat brain, spine and sciatic nerve. J Acupunct Meridian Stud 3:111–115
3. Dai JX, Lee BC, An P et al (2011) In vivo in situ staining of the primo vascular system in ventricles and subarachnoid space of brain by injecting Trypan blue into the lateral ventricle. Neural Regeneration Research, accepted
4. Jia ZF, Lee BC, Eom KH et al (2010) Fluorescent nanoparticles for observing primo vascular system along sciatic nerve. J Acupunct Meridian Stud 3:150–155
5. Lee BC, Kim HB, Sung B et al (2011) Structure of the sinus in the primo vessel inside the bovine cardiac chambers. ISPS Proc (Please, check the current edition)

6. Jiang X, Kim HK, Shin HS et al (2002) Method for observing intravascular Bonghan duct. Korean J Orient Prev Med 6:162–166
7. Baik KY, Lee BC, Johng HM et al (2004) Long threadlike structure inside the blood vessels of rats. Newest Med 47:18–22
8. Lee BC, Baik KY, Johng HM et al (2004) Acridine orange staining method to reveal the characteristic features of an intravascular threadlike structure. Anat Rec B New Anat 278:27–30
9. Lee BC, Yoo JS, Baik KY et al (2005) Novel threadlike structures (Bonghan ducts) inside lymphatic vessels of rabbits visualized with a Janus Green B staining method. Anat Rec B New Anat 286:1–7
10. Lee C, Seol SK, Lee BC et al (2006) Alcian blue staining method to visualize Bonghan threads inside large caliber lymphatic vessels and X-ray microtomography to reveal their microchannels. Lymphat Res Biol 4:181–190
11. Johng HM, Yoo JS, Yoon TJ et al (2007) Use of magnetic nanoparticles to visualize threadlike structures inside lymphatic vessels of rats. Evid Based Complement Alternat Med 4:77–82
12. Yoo JS, Johng HM, Yoon T et al (2007) In vivo fluorescence imaging of threadlike tissues (Bonghan ducts) inside lymphatic vessels with nanoparticles. Curr Appl Phys 4:342–348
13. Lee BC, Soh KS (2008) Contrast-enhancing optical method to observe a Bonghan duct floating inside a lymph vessel of a rabbit. Lymphology 41:178–185
14. Shin HS, Johng H, Lee BC et al (2005) Feulgen reaction study of novel threadlike structures on the surface of rabbit livers. Anat Rec B New Anat 284:35–40
15. Lee BC, Yoo JS, Ogay V et al (2007) Electron microscopic study of novel threadlike structures on the surfaces of mammalian organs. Microsc Res Tech 70:34–43
16. Sung B, Kim MS, Lee BC et al (2008) Measurement of flow speed in the channels of novel threadlike structures on the surfaces of mammalian organs. Naturwissenschaften 95:117–124
17. Yoo JS, Kim MS, Sung B et al (2007) Cribriform structure with channels in the acupuncture meridian-like system on the organ surfaces of rabbits. Acupunct Electrother Res 32:130–132
18. Lee BC, Kim KW, Soh KS (2009) Visualizing the network of Bonghan ducts in the omentum and peritoneum by using Trypan blue. J Acupunct Meridian Stud 2:66–70
19. Yoo JS, Hossein Ayati M, Kim HB et al (2010) Characterization of the primo-vascular system in the abdominal cavity of lung cancer mouse model and its differences from the lymphatic system. PLoS One 5:e9940
20. Hossein Ayati MH, Yu-Ying T, Tao H et al (2011) Finding a novel threadlike structure on the intra-abdominal organ surface of small pigs by using in-vivo Trypan blue staining. ISPS Proc
21. Jia Z, Soh KS, Zhou Q et al (2011) Observation of the primo vascular system on the fascia of dogs. ISPS Proc
22. Han HJ, Ogay V, Park SJ et al (2010) Primo vessels as new flow paths for intratesticular injected dye in rats. J Acupunct Meridian Stud 3:81–88
23. Lee BC, Su Z, Sung B et al (2011) Network of the primo vascular system in the rat hypodermis. ISPS Proc
24. Lee SJ, Lee BC, Nam CH et al (2008) Proteomic analysis for tissues and liquid from Bonghan ducts on rabbit intestinal surfaces. J Acupunct Meridian Stud 1:97–109
25. Sung B, Ogay V, Yoo JS et al (2005) UV-A-induced activation of Bonghan granules in motion. J Int Soc Life Inf Sci 23:297–301
26. Lee BC, Bae KH, Jhon GJ et al (2009) Bonghan system as mesenchymal stem cell niches and pathways of macrophages in adipose tissues. J Acupunct Meridian Stud 2:79–82
27. Bae KH, Su Z, Soh KS et al (2011) Histological comparison of primo nodes and lymph nodes of rat. ISPS Proc
28. Kim BH (1965) The Kyungrak system [in Korean]. J Jo Sun Med 108:1–38
29. Lim JK, Lee SW, Soh KS et al (2011) Primo vessel and lymph vessel from testis (unpublished)
30. Eom KH (2010) Imaging of the primo vascular system in the conception vessel line by using fluorescent nanoparticles. Thesis of Master's degree, Seoul National University, Seoul
31. Han HJ, Sung B, Ogay V et al (2009) The flow path of Alcian blue from the acupoint BL23 to the surface of abdominal organs. J Acupunct Meridian Stud 2:182–189

32. Yoo JS, Choi K, Baik KY et al (2005) Liquid-phase microextraction method in capillary electrophoresis to detect adrenaline in Bonghan lipid. J Int Soc Life Inf Sci 23:292–295
33. Kim JD, Ogay V, Lee BC et al (2008) Catecholamine producing novel endocrine organ: Bonghan system. Med Acupunct 20:97–102
34. Ogay V, Kim MS, Seok HJ et al (2008) Catecholamine-storing cells at acupuncture points of rabbits. J Acupunct Meridian Stud 1:83–90
35. Kim BH (1965) Sanal theory [in Korean]. J Jo Sun Med 108:39–62
36. Sung B, Kim MS, Corrigan A et al (2009) In situ microextraction method to determine the viscosity of biofluid in threadlike structures on the surfaces of mammalian organs. Phys Rev E 79:022901, 1–3
37. Ogay V, Baik KY, Lee BC et al (2006) Characterization of DNA-containing granules flowing through the meridian-like system on the internal organs of rabbits. Acupunct Electrother Res 31:13–31
38. Kwon JH, Baik KY, Lee BC et al (2007) Scanning probe microscopy study of microcells from the organ surface Bonghan corpuscle. Appl Phys Lett 90:173903, 1–3
39. Baik KY, Ogay V, Jeoung SC et al (2009) Hypothesis on the origin of adult stem cells and Bonghan microcells observed with an electron microscope and an atomic force microscope. J Acupunct Meridian Stud 2:124–129
40. Ratajczak MZ, Kucia M, Ratajczak J et al (2009) A multi-instrumental approach to identify and purify very small embryonic like stem cells (VSELs) from adult tissues. Micron 40:386–393
41. Yu B, Lin Y, Soh KS et al (2010) Possible applications for fascial anatomy and fasciaology in traditional Chinese medicine. J Acupunct Meridian Stud 3:125–132
42. Lee SY, Lee BC, Soh KS et al (2011) Development of the putative primo vascular system before the formation of vitelline vessels in chick embryos. ISPS Proc
43. Yoo JS, Kim HB, Ogay V et al (2009) Bonghan ducts as possible metastasis-paths of cancer. J Acupunct Meridian Stud 2:118–123
44. Yoo JS, Kim HB, Won N et al (2010) Evidence for an additional metastatic route: in vivo imaging of cancer cells in the primo-vascular system around tumors and organs. Mol Imaging Biol. doi:10.1007/s11307-010-0366-1
45. Maniotis AJ, Folberg R, Hess A et al (1999) Vascular channel formation by human melanoma cells in vivo and in vitro: vasculogenic mimicry. Am J Pathol 155:739–752
46. Kim BH (1963) On the acupuncture meridian system [in Korean]. J Jo Sun Med 90:6–35
47. Kim BH (1963) On the Kyungrak system. J Acad Med Sci DPR Korea 90:1–41
48. Kim BH (1965) Sanal and hematopoiesis [in Korean]. J Jo Sun Med 108:1–6
49. Kim BH (1962) Study on the reality of acupuncture meridians [in Korean]. J Jo Sun Med 9:5–13
50. Park SH, Lee BC, Choi CJ et al (2009) Bioelectrical study of Bonghan corpuscles on organ surfaces in rat. J Korean Phys Soc 55:688–693

Chapter 6
Primo Vascular System: Basic and Applied Research Outline

Michael Potroz and Kwang-Sup Soh

Abstract Based on research surrounding the independent fluid-conducting system known as the primo vascular system, we have identified the main areas necessary for future research and some of the key development opportunities. Mapping, imaging, and monitoring: As the primo vascular system forms a network that extends throughout the body it is necessary to develop a comprehensive physical model and techniques for quantitatively measuring the functional characteristics thereof. This will allow for the identification of normal primo vascular system characteristics and eventually to the identification of abnormal characteristics that relate to specific health conditions and the study of various direct and indirect stimuli leading to many opportunities in the fields of diagnostics and treatment. Brain, spine, nervous system: The primo vascular system extends throughout the brain, spine, and nervous system and due to its role in regeneration is hypothesized to play a key role in the development of brain, spine, and nervous system conditions. Developing methods to see the brain, spine, and nervous system's primo vascular system in living animals and humans will create many opportunities for research and the development of new diagnostics and treatment regimes. Cancer: The primo vascular system has been identified to form new structures on the surface of tumors and to enter the inside of tumors. It is necessary to determine how the primo vascular system influences tumor formation, growth, and spread. There are opportunities for a new cancer drug delivery path, diagnostic protocols, and a wide range of possible new treatment modalities. Aging and regeneration: The primo vascular system carries hormones and immune cells and appears to play a key role in natural body regeneration. It is necessary to determine the primo vascular systems role in physiological functions, such as immune function and regeneration. This will lead to developments in tissue engineering and biomaterials, stem cells and cellular

M. Potroz (✉)
Biomedical Physics Laboratory, Department of Physics and Astronomy,
Seoul National University, Seoul 151-747, Korea
e-mail: mikepotroz@gmail.com

K.-S. Soh et al. (eds.), *The Primo Vascular System: Its Role in Cancer and Regeneration*,
DOI 10.1007/978-1-4614-0601-3_6,

therapy, and endocrine and immune system diagnostics and treatment. Due to the early nature of this research, it is difficult to consider all of the potential implications but our preliminary study suggests that the primo vascular system will likely have a significant impact in the future of medicine and industry.

1 Basic Research

1.1 Mapping, Imaging, and Monitoring

The focus of this research will be to develop methods for visual detection of the primo vascular system through one of various staining techniques, and imaging and monitoring by utilizing existing technologies.

With regards to visual detection, the most likely paths for development are the creation of a highly specific stain, such as that used in immunohistochemical staining, and/or the development of a transgenic animal line with a visible or fluorescent primo vascular system. The ability to visualize the primo vascular system and primo nodes will allow for the completion of primo vascular system identification, the study of the primo vascular system relationship to existing health conditions, and the development of entirely new treatment regimes.

With regards to imaging and monitoring, the most promising and well developed technologies are optical and ultrasound. A combination of these in the form of photoacoustic imaging is likely to be the best candidate. The ability to image and monitor the primo vascular system and specifically the activity of primo nodes, will allow for the study of the primo vascular system relationship to existing health conditions and body functions, as well as the development of new diagnostics equipment and treatment regimes.

1.2 Brain, Spine, and Nervous System

The focus of this research will, first, be to develop imaging methods and equipment for the noninvasive imaging of the primo vascular system in these regions, and second, to develop new diagnosis and treatment regimes for existing brain, spine, and nerve conditions.

With regards to the development of imaging methods and equipment, research will need to focus on determining the most appropriate existing imaging method, such as MRI, CT, or PET and once this is defined, then a nontoxic contrast agent will need to be developed along with a delivery system and other necessary hardware and software. The ability to image the primo vascular system in relation to the brain and entire nervous system will add an entirely new dimension to the study of brain and nerve development, healing, injury, and conditions.

With regards to the development of new diagnosis and treatment regimes for existing brain, spine, and nerve conditions, the ability to image these structures in detail, will allow researchers to determine new treatment modalities for existing conditions such as Parkinson's and Alzheimer's and to optimize new and existing forms of treatment to heal injuries and cure disease.

1.3 *Cancer*

The focus of this research will relate to, the role of the primo vascular system in cancer formation and spread, cancer diagnosis, the development of novel targeted drug delivery systems, and the development of new treatment protocols.

With regards to the role of the primo vascular system in cancer formation and spread, research will need to focus on determining the exact role of the primo vascular system in angiogenesis and metastasis, as well as, the potential relationships of the primo vascular system to vasculogenic mimicry and primo microcells to cancer stem cells. Research has already shown that the primo vascular system forms on tumors, potentially enters tumors, and acts as a pathway for metastasis.

With regards to new methods in cancer diagnosis, research will need to determine if there are any biomarkers in primo fluid, or that relate to the primo vascular system, that indicate the presence of cancer, the stage of cancer, the type of cancer, the most promising course of treatment, and/or the response of cancer to treatment.

With regards to the development of a novel targeted drug delivery system, research will need to focus on the mode of drug delivery, such as directly into the primo vascular system on or near the tumor, or, at specific points on the skin which are connected to a specific tumor via the primo vascular system. Research will also need to focus on new targets for drug delivery. This may allow for significant improvements in drug efficacy and significant reductions in undesirable side effects.

With regards to the development of new cancer treatment protocols, once researchers have developed an understanding of the relationship between the primo vascular system and cancer there will be a wide range of new possibilities for surgical interventions, drug development, and other forms of treatment and monitoring.

1.4 *Aging and Regeneration*

This is a very broad area of research, but, initially the primary areas to focus on relate to: the primo vascular system relationship to the endocrine system, biomaterials and tissue engineering, stem cells and cellular therapy, as well as, drug delivery and diagnostics.

With regards to the primo vascular system relationship to the endocrine system, research will need to determine the role of the primo vascular system in hormone transport and signaling. Primo fluid carries many hormones and if the primo vascular system is shown to play a significant role in the endocrine system this will have a huge impact on existing theory and treatment and in the development of new diagnostic and treatment protocols.

With regards to biomaterials and tissue engineering, research will need to determine how the primo vascular system influences and regulates tissue growth. As the primo vascular systems' key role is assumed to involve regenerative processes, this may be the missing factor explaining some of the unpredictability in reliably initiating and controlling the growth of replacement tissue and organs.

With regards to stem cells and cellular therapy, research will need to focus on analyzing primo microcells as well as identifying their role in the body and relationship to stem cells and VSEL stem cells. Primo microcells could prove to be a very important key in understanding and utilizing stem cells and therefore open a wide range of new treatment modalities.

With regards to drug delivery and diagnostics, research needs to determine how the primo vascular system can be effectively utilized to simplify drug delivery and improve drug efficacy. Also, it is necessary to identify what new biomarkers the primo vascular system offers and which ones are the most useful to complement existing diagnostics.

2 Applied Research

2.1 Real Time Health Monitoring

The focus of this research will be to develop noninvasive ways to monitor the overall health and the functional state of specific internal structures of patients, in real time. It will be necessary to determine the relationship between the autonomic nervous system and the primo vascular system. By analysis of primo node activity, it may prove possible to determine the functional state of related internal organs and therefore the overall health of the body.

With regards to the autonomic nervous system and the primo vascular system relationship, it is accepted that the autonomic nervous system connects all of the body's internal organs, and therefore, certain changes in bioelectrical characteristics should relate to changes in internal organ function. Research has shown that the primo vascular system forms on the nervous system and therefore the degree to which the two systems are interlinked needs to be determined.

With regards to primo node and internal organ relationships, traditional acupuncture teaches a clear functional relationship between specific internal organs and specific acupuncture points in a similar way to that of the accepted spine/autonomic nervous system and internal organ relationship.

The measurement of electrical properties or physical characteristics of primo nodes at specific acupuncture points may provide a clear indication of health status and of a wide variety of health conditions. If achievable, measuring primo node activity may become as widely utilized as measuring blood pressure and may allow for the development of a wide range of personalized consumer diagnostics and treatment devices.

2.2 *Drug Delivery*

The focus of this research will be to determine how the primo vascular system can be best utilized to improve drug delivery. How the primo vascular system can be best used with existing drug delivery methods, whether the primo vascular system offers a targeted pathway to internal structures from primo nodes at acupuncture points, and what new methods of targeting drugs are possible by utilizing the primo vascular system, are all key areas of study.

With regards to existing drug delivery methods, research needs to determine whether injectable, transdermal, implantable, or other common methods of drug delivery can be more effectively utilized in conjunction with the primo vascular system. If primo nodes are shown to be linked to specific internal structures then there will be many opportunities to utilize consumer friendly targeted transdermal drug delivery products.

With regards to the use of the primo vascular system as a targeted delivery pathway, research will need to determine if compounds introduced into the primo vascular system at one point in the body reliably travel to other specific points in the body. Acupuncture theory relates many specific points under the skin to specific points in the body and if a specific structure can be identified to explain these connections it will create the many opportunities for new reliable drug products. Also, this may provide a novel solution to delivering drugs to the brain and central nervous system which is normally hampered by the blood brain barrier.

2.3 *In Vitro Diagnostics*

The focus of this research will be to identify new predictive biomarkers, diagnostic biomarkers, prognostic biomarkers, and drug-related biomarkers for a wide range of existing health conditions and then determine which of these can be detected in other easily obtainable bodily fluids.

With regards to the detection of new biomarkers, new molecular biologic techniques such as proteomics and genomics offer the possibility to identify disease markers without the need for detailed insight into the disease mechanisms. Biomarker identification is playing a key role in all areas of health care, especially the rapidly developing fields of preventive healthcare, personalized healthcare, and companion diagnostics.

With regard to the detection of primo fluid biomarkers in other body fluids, the primo vascular system has been shown to exist throughout the body and intravascularly within blood and lymphatic vessels and therefore it seems likely that some of the biomarkers found within primo fluid will be detectable at varying concentrations in other more easily collectable bodily fluids such as blood and urine. This will enable the practical use of new primo vascular system-related biomarkers in normal clinical settings.

2.4 Pharmaceutical Drug Development

The focus of this research will be to look at ways in which the primo vascular system can assist in the identification and development of new drugs. In particular, the primo vascular system may provide a key to increase the rate of identifying the active components in botanicals such as those used in traditional Chinese medicine (TCM) or traditional oriental medicine (TOM), for example.

With regard to the identification and development of new drugs, as many drugs are designed to assist the body's natural healing mechanisms, by analyzing changes in primo fluid composition it may be possible to identify the mechanism of action (MOA) of many herbal and drug compounds.

With regard to identifying active components in botanicals, for botanicals that are traditionally used in conjunction with acupuncture meridians, such as those applied to acupuncture points or those believed to stimulate certain meridians and organs, it would seem likely that they act in someway on the primo vascular system. From these botanicals, it would seem logical that the active compounds would enter into the primo fluid. Therefore, by analysis of the botanical and analysis of primo fluid from a patient to whom the botanical has been administered, then cross referencing the results to determine which compounds have been transferred into the primo fluid, it would be possible to determine the active compounds and their respective concentrations to take into account the synergistic effects. A process such as this utilizing some form of spectrographic analysis or perhaps genomics or proteomics may significantly increase the rate at which new compounds can be identified.

3 Conclusion

The above basic and applied research outlines are the result of in-depth discussions with Prof. Soh about his team's work, past and present, and comes after much work preparing a detailed research plan for the future. This outline only provides a brief introduction to the potential future of this new field of research and only highlights a few of the possible industry applications that will likely emerge.

Chapter 7
The Primo Vascular System: Facts, Open Questions, and Future Perspectives*

David K. Harrison and Peter Vaupel

Abstract Based on research surrounding the independent fluid-conducting system known as the primo vascular system, we have identified the main areas necessary for future research and some of the key development opportunities. Mapping, imaging, and monitoring: As the primo vascular system forms a network that extends throughout the body, it is necessary to develop a comprehensive physical model and techniques for quantitatively measuring the functional characteristics thereof. Brain, spine, and nervous system: The primo vascular system extends throughout the brain, spine, and nervous system and due to its role in regeneration is hypothesized to play a key role in the development of brain, spine, and nervous system conditions. Cancer: The primo vascular system has been identified to form new structures on the surface of tumors and to enter the inside of tumors. It is necessary to determine how the primo vascular system influences tumor formation, growth, and spread. Aging and regeneration: The primo vascular system carries hormones and immune cells and appears to play a key role in natural body regeneration. This will lead to developments in tissue engineering and biomaterials, stems cells and cellular therapy, and endocrine and immune system diagnostics and treatment. Due to the early nature of this research it is difficult to consider all of the potential implications but our preliminary study suggests that the primo vascular system will likely have a significant impact in the future of medicine and industry.

*With contributions by Samuel Achilefu, Larry Kwak, Pan Dong Ryu, and Christopher Zaslawski. This chapter is a summary of the impressions of the panelists at the close of the ISPS2010-meeting.

D.K. Harrison (✉)
Institute of Cellular Medicine, Newcastle University, Newcastle upon Tyne NE1 7RU, UK

Heilig-Kreuz-Strasse 19, 39030 St Lorenzen, Italy
e-mail: d.k.harrison@ncl.ac.uk

K.-S. Soh et al. (eds.), *The Primo Vascular System: Its Role in Cancer and Regeneration*, DOI 10.1007/978-1-4614-0601-3_7, © Springer Science+Business Media, LLC 2012

1 Introduction

The first general observation concerns the transdisciplinary nature of the symposium. We have had a wide range of presentations and posters, from the history and development of the primo vascular system (PVS) in the early 1960s by Bong Han Kim to "high tech" imaging, acupuncture, cutting-edge molecular biology, and innovative cancer treatment all hoping to elucidate some aspect related to the PVS. Experts in some knowledge areas were naïve beginners in others yet each participant was willing to listen closely to each other and offer constructive and helpful criticism where necessary. The second observation is the internationalization of the research. While Korea can rightly claim to the original development of the PVS, it has been heartening to see the support of many individuals from many countries willing to be involved and support research in the area. It is only by internationalizing that the true value of the research can be determined and the scientific validity confirmed. The third observation was the willingness to cross cultures and enter into dialogue. Researchers from Korea, China, and Japan have had to present and discuss in the English language in order that true communication occurs across national boundaries. Those presenters should be applauded for their willingness to achieve what is often a difficult, complex, and technical task. The final general observation is of the enormity of the task ahead. The PVS has potential to advance understanding of the human body and medicine. Many tantalizing proposals have been put forward during the symposium and only concerted research effort will be able to confirm or refute some of these hypotheses. Time will tell whether some of these proposals will come to fruition.

2 The Anatomy and Function of the Primo Vascular System

The ISPS2010 was the first international meeting for the studies on the PVS, and numerous experimental results and stimulating ideas were newly introduced during the two full-day meeting. P.D. Ryu summarized those results and ideas related to the PVS itself, and possible short- and long-term directions in PVS research.

In the opening remark, K.S. Soh presented an overview on the current status of the PVS including the history of PVS research during the 1960s and after 2002, the known structure of PVS and its distribution in the animal body, and the recent progress in identification of PVS (see Soh KS (2009) Bonghan circulatory system as an extension of acupuncture meridians. J Acupunct Meridian Stud 2:93–106). Particularly, he envisaged the PVS as a novel regeneration system of the living body as well as an extension of acupuncture meridians.

In the two oral sessions on the PVS, the presence of PVS was demonstrated in the bovine heart (B.C. Lee) and abdominal cavity of the pig (W. Zhang). The subducts of organ surface primo vessels were identified by microscopic analysis of dye staining (Y. Ryu) and phase contrast X-image (M.S. Kim). The functional role of PVS was indicated in the recovery of hepatic tissue (S.Y. Ahn) and oxygenation of murine melanoma tissue (M. Suh). Caution was expressed regarding possible methodological artifacts that may appear to identify structures similar to the PSV (C.H. Leem and Y. Ryu). The recollections of two pioneering PVS researchers in the late 1960s to the early 1970s revealed their passion and devotion on the PVS research and the importance of the PVS research for the future (oriental) medicine (S. Fujiwara and J.S. Lee). Dr Fujiwara pointed out the importance of the primo nodes in skin in relation to the meridian points on the body surface.

In the Regeneration Session, V. Ogay showed that primo-microcells (2–3 μm) have unique structure, and express Oct4 and Nanog, markers for embryonic stem cells, and S.H. Ahn showed that the primo-microcells isolated from SD rats were growing and differentiating to a colony of cell-like structures in culture. The similarity between the primo-microcells and very small embryonic-like stem cells (M. Kucia) remains to be studied.

In the Poster Session, the majority of the posters were on the PVS structure. The selected posters are as follows: A novel model for acupuncture meridian – the PVS combined with the fascia (B.C. Lee and K.S. Soh), the unique fiber structures of the PVS (J.H. Jung et al.), the coiled blood vessels of primo node along the meridian (J. Kim and K.S. Soh), the identification of PVS in mesentery (P. An et al.), the expression of keratin 10 in the PVS (S.R. Kim et al.), development of a laparoscope for in vivo observation of the organ surface PVS (J. Lim et al.), identification of the organ surface PVS in dogs (W.H. Yu et al.), flow of nanoparticles from ST36 point to the PVS around the sciatic nerve (Z.F. Jia et al.), selective marking of the PVS with phalloidin (Z. Su et al.), the microcells with short process or bud (K.Y. Baik et al.), visualization of the PVS in chicken (S.Y. Lee et al.), the PVS network identified at the acupuncture meridians (B.C. Lee et al.), spontaneous action potentials recorded in the cells of primo vessels by intracellular recording (C.J. Choi et al.), TEA-sensitive currents in primo node cells by patch clamp recording (J.H. Choi et al.), role of PVS in tumor tissue (H.B. Kim et al.), and differences of the PVS from blood and lymphatic vascular tissues (J.S. Yoo et al.)

As P.D. Ryu has pointed out, new approaches introduced in this meeting will be useful in future studies on the PVS. They are based on the oxygenation status (oral presentations by M. Suh and by Y. Lee), the unique fiber structures of the PVS (J.H. Jung et al.), the coiled blood vessels of primo node along the meridians (J. Kim and K.S. Soh), selective marking of the PVS with phalloidin (Z. Su et al.), and the primo-microcells with short processes (K.Y. Baik et al.). In addition, the notion that the PVS is the anatomical structure of the acupuncture meridian is further supported by the following presentations: the identification of the primo nodes underneath the acupoints (CV8–CV12, H.M. Kwon), the coiled blood vessels of primo node along the meridian (J. Kim and K.S. Soh), and the PVS network identified at the acupuncture meridian (B.C. Lee et al.). In the electrophysiological studies, the most PVS cells were round and un-excitable (P.D. Ryu, C.J. Choi et al., J.H. Choi et al.), but some cells are excitable and spontaneously active (C.J. Choi et al.). These results will be of help for better understanding the structure and function of the whole network of the PVS.

3 Medical Imaging and the Primo Vascular System

Medical imaging will play an invaluable role in validating the existence and understanding the functions of the PVS. At the first International Symposium on PVS, diverse imaging methods were used to demonstrate the existence and potential functions of PVS. Of particular interest were the discussions on methods to image PVS and strategies to develop molecular probes for noninvasive imaging of PVS.

A variety of delivery routes that includes retro-orbital and intradermal approaches for imaging the uptake of trypan blue in mouse PVS demonstrated exciting potential to visualize the system with minimal trauma to the target organ. From the questions and presentations, it was clear that another important area of research is the imaging of the dynamics of the PVS and its fluid. This topic was addressed by developing an X-ray-based technique to monitor the biodynamics of PVS fluid in vivo. Extensive study along this line will elucidate the source and directionality of PVS fluid flow in future. To accurately localize PVS and molecular processes within this vessel, there is a need to develop imaging agents that are highly specific for PVS. Some researchers have developed highly luminescent nanomaterials for this purpose. Moreover, initial studies suggest that cells labeled with quantum dots migrate to distal organs from the source of injection through the PVS. Another impressive direction is the development of an elegant method to screen a library of fluorescent dyes to select high binders to a target tissue. Although the initial binding affinity may be low, the approach presents a new technique to rapidly identify potential PVS binders and optimize the product for in vivo use. Finally, the potential of using aptamers and multimodal imaging platforms to target PVS was presented. The vision of how different components of the study can be integrated into a comprehensive imaging illustrates the paradigm-shifting applications of imaging to PVS.

There are clearly many unmet needs in the field of medical imaging of PVS and the diverse nature of medical imaging requires immediate engagement of imaging scientists in PVS. An immediate need is to provide an image-based atlas of the PVS in vertebrates and invertebrates. For this goal, imaging systems with high temporal and spatial resolutions as well as high dynamic range are needed. In rodents, the PVS is in the order of 10–40 μm. This level of resolution is achievable by optical imaging and intravital microscopy methods for in vivo applications. Ex vivo imaging will essentially be possible with most available biological imaging techniques such as electron and X-ray microscopy. The close association of PVS with other vessels requires new image segmentation and deconvolution methods to delineate PVS in a mesh of structures. Expertise in physics, engineering, image reconstruction, segmentation, and algorithm development is imperative to achieve the earlier mentioned goal. For large animals and humans, existing clinical imaging systems may have adequate spatial resolution to image PVS because the vessels are larger and readily identifiable. In all cases, multimodal imaging platform may be necessary to report functions and anatomical landscape simultaneously.

Another imaging need is the development of highly-specific imaging agents and molecular probes for PVS. Current efforts have demonstrated that trypan blue is effective in staining PVS. However, the exact mechanism of uptake has to be elucidated. In addition, there is an urgent need to develop more specific molecular probes for noninvasive imaging of PVS. These efforts will range from the incorporation of aptamers, antibodies, and ligands to PVS proteins and other PVS binding materials. Chemists, biologists, and materials scientists will undoubtedly facilitate this area of research. These efforts are already underway and we expect researchers to report their finding in the near future.

S. Achilefu pointed out, that the way of the future of research in PVS research will be to understand the functions of PVS by asking the appropriate biological questions. Progress will depend on our ability to tap into and harness the expertise of physicians, biologists, and allied scientists and professionals. Extensive collaboration will allow researchers to develop imaging tools that will be sensitive to specific functional and molecular processes pertinent to PVS. The future will also require the development of new and exciting transgenic animal models and genetically encoded molecular probes. Image guidance will be needed to optimize PVS-mediated drug delivery and to monitor tumor trafficking through the PVS. Ultimately, the wealth of information should lead us to our goal of translating these findings from animals to humans. The aspiration and hope of this inaugural symposium is to integrate medical imaging into PVS research, which will accelerate the acceptance of PVS and spur the realization of its full potential in the management of health and better yet, the prevention of human disease.

4 Acupuncture and the Primo Vascular System

There are several points to be made regarding acupuncture and the interface with the PVS. First, while the initial motivation for the PVS research resulted from the desire to discover the anatomical basis of acupuncture channels (jingluo) to help explain some of the effects observed following acupuncture, we are not yet to that point where the two theories can be linked together. Traditional channel theory had its origins in a culture which had a proto-science perspective of the world embedded in a philosophy based on harmony and integration of humankind with the surrounding world. This is vastly different to the scientific worldview we have been contemplating during the symposium that utilizes high tech imaging techniques and molecular science that has enabled humankind to overcome many diseases and offer hope for a cancer cure in the future. Yet, both systems, that of acupuncture and scientifically oriented PVS, have similarities. Both systems lay claim to promoting self-healing, the PVS by promoting regeneration by small stem cells circulating in the vessels and nodes, and acupuncture by facilitating smooth circulation of the ki (qi) around its 12 main channels. Both also have networks that extend across all parts of the body from the superficial dermis to the wrapping around intestines and internal organs. Certainly a case was put forward by two of the presenters (E. van Wijk and Y. Lin) in the acupuncture session that fascia (connective tissue that occupies nearly all spaces of the internal body) may be crucial to linking these two disparate theories together. Other presenters in the session also attempted to make connections. S.H. Yi discussed his research on moxibustion (burning of the herb mugwort) on acupoints sites involving either insulation of the treatment site whereby moxa is burnt upon a slice of garlic/ginger or held directly over the site as a cigar. Careful monitoring and reporting of the heat gradients produced under such circumstances gave way to speculation on how these factors may affect PVS nodes (acupoints). Finally, M. Ohkuma also drew parallels between his research and his clinical experiences of lymphedema and the possible interaction between the PVS and the lymph system.

Yet, many questions remain. How do acupuncture needles physically stimulate the PVS vessels and nodes? What stimulation variables such as needling depth, the amount of needle manipulation, and the number of needling sites affect the mechanical aspect of stimulating the PVS? And do the PVS nodes correlate to the traditional location of acupoints? These are only a few questions that lie ahead for PVS researchers who want to correlate these two systems.

C. Zaslawski's perception is that the current state of PVS research is at the later stage of an early phase, and while some basic questions have been answered there are many more that remain. While this has been the first international symposium on the PVS, a second has been proposed in the not too distant future. Hopefully, progress will have been made and some of these questions can be answered.

5 Cancer and the Primo Vascular System

As pointed out by L. Kwak, there are at least three potential directions for future research to explore the relevance of PVS to cancer:

(a) PVS as a potential conduit for tumor cell metastasis
If confirmed, this previously unrecognized mechanism of tumor spread would be of enormous importance: K. Zaenker illustrated how tumor cells may gain access to the blood, lymphatics, and possibly to the PVS by active process of invasion. This may also be relevant to the propensity of pancreatic cells to invade and metastasize under stress. Concerning the ability of a herbal compound, MSB0052, to shrink murine melanoma cells by reducing angiogenesis, it was speculated that a similar effect may occur on PVS. Finally, one could speculate that the PVS might serve as a sanctuary site for low grade lymphoma cells. This possibility could explain the failure to effect cures in up to 50% of patients with stage I follicular lymphoma treated with local radiotherapy. These patients, without evidence for distant disease at diagnosis, even by careful staging, nevertheless are observed to relapse sometime many years after primary therapy.

(b) PVS as a novel route for drug delivery
One example discussed during the symposium was a scFv antibody inhibitor of the LMO-Protein complex. Such an "intrabody," when expressed as a protein drug could be small enough to be delivered through the PVS.

(c) PVS and the immune system
L. Kwak was intrigued about the possible role of PVS in trafficking of immune effector cells. The availability of another conduit for effector cells could be exploited by immunotherapy strategies. Alternatively, negative regulatory T cells (Treg) could also gain access to this conduit and dominate the immune phenotype, such as they do in the blood of cancer patients, representing an obstacle that needs to be overcome. Either way, the answer would be significant and would have implications for immunotherapy strategies.

6 Physiology of the PVS and Its Relation to Oxygen

There were five presentations in the session relating to oxygen and physiology in the primo vascular system, and a further presentation involving oxygen measurements during the second session on the primo vascular system. The first paper described the application of tissue oxygen saturation measurements to investigate the consequences of cancer treatment in patients. It was suggested that this could be a non-invasive tool that could be used to study the possible role of the primo vascular system in oxygen supply to tissue.

The second paper on oxygen measurements described the use of a polarographic electrode to measure oxygen pressures on the skin surface at acupuncture sites and control locations. It was found that there was a significantly higher (normalized) oxygen pressure at the acupuncture sites than at the control locations. A similar observation was made in the presentation of the results of measurements using a very fine needle electrode (15 nm tip diameter) in murine melanomas. Here, the oxygen partial pressures close to the PVS vessels were much higher than in the tumor mass, and higher even than in the vicinity of blood vessels.

In the three presentations involving electrophysiology and proteomics, the first involved the measurements of the effect of protein kinase A activation on the voltage dependence of the sodium–potassium and sodium–calcium exchange currents in cardiac myocytes. It was found that there was no effect on the Na–K pump but there was on the Na–Ca antiporter. An investigation into the electrophysiological properties of cells in the primo vascular system using the patch clamp technique showed that two different levels of resting membrane potentials could be found among the cells, together with multiple types of ion channels. The final paper described the first proteomic analysis of the Bonghan duct at the surface of the rabbit intestine. Categorization according to biological processes, molecular function, and cellular compartment showed enrichment of proteins involved in metabolism. The use of proteomic analysis could prove to be a powerful tool in elucidating the nature and function of the primo vascular system.

The main direction of the panel discussion, headed by D.K. Harrison, centered on the oxygen measurements at the acupuncture sites in normal skin and in murine melanomas. Unfortunately, neither author could be present for the discussion which involved the possible mechanism by which the measurements could possibly be explained. It was pointed out that the differences in normal skin were small and that further investigations of the heterogeneity of oxygenation in normal skin should be investigated. More controversy surrounded a possible mechanism for the needle electrode measurement of higher oxygen partial pressures in the vicinity of PVS vessels than adjacent to blood vessels in the melanomas. One explanation may be that proteins, with high affinities for oxygen, other than hemoglobin, may be responsible – although these have yet to be identified. It was pointed out that the high O_2 consumption rate of the tumor cells would cause very steep diffusion gradients in the tissue such that diffusion of oxygen would occur only over very limited distances even in the presence of a high affinity source. There was also the question of delivery in the primo vascular system that seems to be comprised mainly of endothelial cells. Clearly, more investigations are required, preferably using multiple techniques for the measurement of oxygen supply to tissue, in order to explain these phenomena. Finally, the question: "Does the primo vascular system contribute to the oxygen supply to tissue?" remains to be answered.

7 Conclusions

There are many important questions remaining untouched although a large amount of the novel results on the PVS were presented in this meeting: (a) The physiological roles of the PVS (during pre-birth, development and growth, and aging), (b) the roles of PVS in diseases including metabolic diseases, cardiovascular diseases, and diseases of central nervous system, (c) the formation of primo-microcells and their functions in regeneration, (d) relations between the effects of PVS stimulation and acupuncture therapy, and (e) classification and generalization of the PVS in vertebrates and invertebrates. It can be expected that the results on these questions will be appearing more and more in the future meetings on the PVS.

Part II
Primo Vascular System in Various Organs

Chapter 8
Structure of the Sinus in the Primo Vessel Inside the Bovine Cardiac Chambers

Byung-Cheon Lee, Hong Bae Kim, Baeckkyoung Sung, Ki Woo Kim, Jamin Sohn, Boram Son, Byung-Joon Chang, and Kwang-Sup Soh

Abstract We report the structure of sinuses in the primo vessels on the surfaces of endocardia of atriums and ventricles in the bovine heart. About 1–5 sinuses (0.5–7 μm in diameter) were observed in the cross sections of the primo vessels (20–50 μm in diameter). The boundary of the sinus was clear and was regularly surrounded mainly by collagenous fibers (~30 nm in diameter) and partly by 1–2 cells.

1 Introduction

Primo vessels (Bonghan ducts) have been observed in blood and lymph vessels [1–5], in brain ventricles and spinal cords [6], and on the organ surfaces [7–9] in small animals. All these primo vessels showed common structural characteristics in that they had longitudinally aligned rod-shaped nuclei and multiple (or rarely single) sinuses. Especially, based on some physiological evidence, the sinuses have been supposed to function as microcirculatory channels [10, 11]. Generally, three morphological types are known for the structures of the sinuses in primo vessels: (1) sinus (2–40 μm in diameter) with smooth or irregular inner surfaces composed of extracellular matrices (ECMs) (including collagenous fibers) and sometimes partly covered by cell membranes [8–10], (2) sinuses (5–10 μm in diameter) surrounded by a single layer of endothelial cells [12], and (3) numerous small pores (<1 μm in diameter) irregularly distributed in the overall region of the cross section of the primo vessel [9, 10]. In this article, we report the structure of sinuses in the primo vessels inside bovine cardiac chambers by using optical and electron microscopy.

B. Sung (✉)
Biomedical Physics Laboratory, Department of Physics and Astronomy,
Seoul National University, Seoul 151-747, Korea
e-mail: baeckkyoung@gmail.com

K.-S. Soh et al. (eds.), *The Primo Vascular System: Its Role in Cancer and Regeneration*,
DOI 10.1007/978-1-4614-0601-3_8, © Springer Science+Business Media, LLC 2012

2 Materials and Methods

Fresh bovine hearts (Korean bovine, male, 650–700 kg (24–28 months)) were taken from a public slaughterhouse established by the National Agricultural Cooperative Federation of Korea and were carried in a cold solution of phosphate buffered saline (PBS; 0.1 M, pH 7.4) to the laboratory in 30 min. As soon as they had arrived at the laboratory, the hearts were rapidly washed with a cold solution of PBS (0.1 M, pH 7.4) three times; then, fat tissues were removed from the surfaces of the hearts. The hearts were then longitudinally sectioned surgically, and their endocardia were exposed. A 0.2–0.4% Trypan blue (Sigma) solution was applied to the inner surfaces of the cardiac chambers. At first, the in situ staining procedure started with a 0.2% Trypan blue solution, and after, when the visualization of the vessels and nodes is not sufficient under a stereomicroscope, we additionally applied a 0.4% Trypan blue solution. A Trypan blue solution for cell culture was used, or its dilution with PBS was used. About 3–5 mL of Trypan blue solution was carefully spread on the endocardia by using a Pasteur pipette. After one and half minutes, the endocardia were directly and softly washed with PBS (or sometimes with saline) by using a Pasteur pipette. The washing was done more quickly when the concentration of Trypan blue was higher. All the staining processes were done at room temperature. The staining and the observation processes were conducted with a stereomicroscope (SZX12, Olympus, Japan).

For TEM examination, tissues were fixed with 2.5% glutaraldehyde in a 0.1-M sodium-cacodylate buffer at 4°C for 4 h, postfixed in 1% OsO_4 in a 0.2-M sodium-cacodylate buffer for 1 h, dehydrated with ethanol and propylene oxide, and embedded in epoxy resin (Epon 812). (Semi-thin sections were stained with Toluidine blue and were collected on slide glasses to be observed under a phase-contrast microscope (Axiophot, Carl Zeiss, Germany).) Ultrathin sections were collected on large-scale copper grids, contrasted using 2% uranyl acetate and Reynolds' lead citrate, and examined in a transmission electron microscope (JEM 1010, JEOL, Japan) at an accelerating voltage of 80 kV, with images being obtained by using a digital CCD camera (ES1000W, Gatan, USA) and its software processing (Gatan, USA).

3 Results

Threadlike structures were found, stained by Trypan blue, on the surfaces of endocardia in atriums and ventricles. In the stereo-microscopic image (Fig. 8.1), the threadlike structures were composed of nodal structures (primo nodes; 100–150 μm in diameter) and threads linking them to each other (primo vessels; 20–50 μm in diameter). The primo vessels and nodes were interconnected and were only partly adhered to the endocarduim.

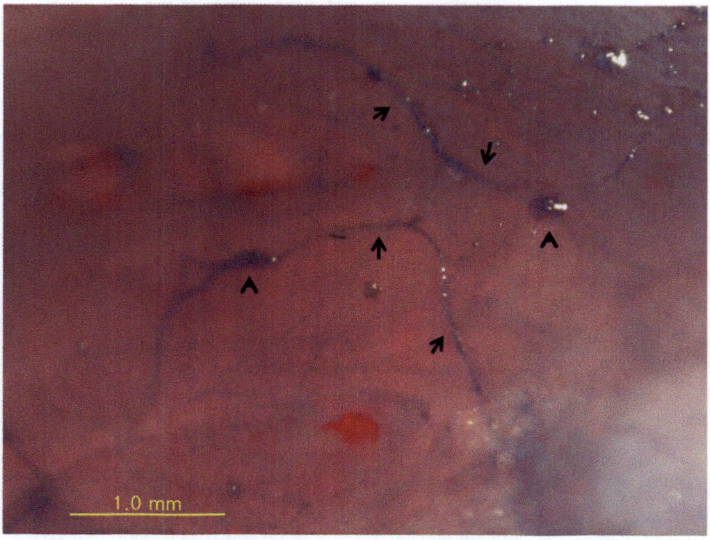

Fig. 8.1 Stereomicroscopic image of the primo vessels (arrows) and nodes (*arrow heads*) on the surface of the bovine endocardium. The primo vessels and nodes are interconnected and are only partly adhered to the endocarduim. They could be visualized by using an in situ Trypan blue staining method. Scale bar = 1 mm

A phase-contrast microscopic image of a cross section (semi-thin section; 1 μm in thickness) of the primo vessel is presented in Fig. 8.2a. The semi-thin section, stained with Toluidine blue, showed a clear outer boundary and 1–5 sinuses (0.5–7 μm in diameter). These sinuses formed a continuous series in the sections of ~1 cm distance along the vessel. Figure 8.2b shows a low-magnification TEM image of the cross section. We can also see a clear outermost boundary and a sinus. We observed 6–10 cell nuclei (2–5 μm in diameter) with small amounts of cytoplasm scattered in the collagenous ECMs. The collagenous fibers were relatively homogenous in shape and dimension (about 30 nm in diameter) and were aligned mostly in the longitudinal direction of the primo vessel. Figure 8.2c shows an enlarged TEM image of a sinus in the primo vessel. The boundary of the sinus was clear and regularly surrounded by mainly collagenous fibers and partly by 1–2 cells. Parts of the sinus were surrounded by the cell membrane (or membrane-like materials). The nuclei of the primo vessel had mostly euchromatin.

4 Discussion

For the in situ Trypan blue staining of the primo vessels and nodes, we conjecture that the dye first has been deposited on the vessels and nodes, and then bound to a certain kind of fibers in the ECMs of the vessels and nodes. The dye deposited

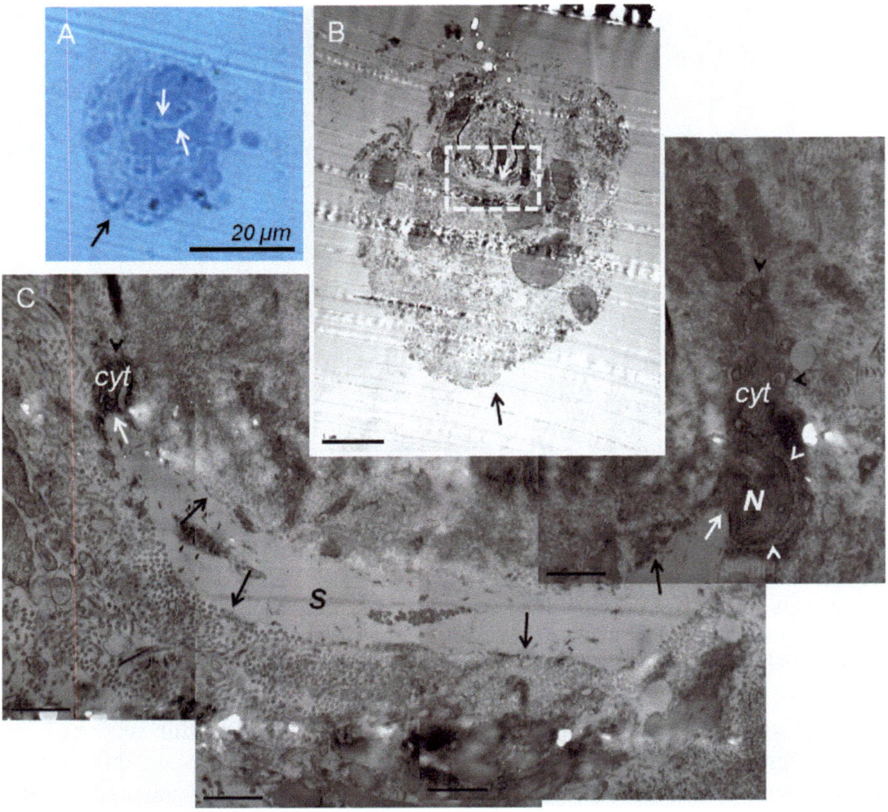

Fig. 8.2 Optical and transmission electron microscopic images of a cross section of the primo vessel. (**a**) Phase-contrast microscopic image of a cross section (semi-thin section; 1 μm in thickness) of the primo vessel. The semi-thin section, stained with Toluidine blue, shows a clear outer boundary (*black arrow*) and a sinus (*white arrows*). (**b**) Low-magnification view of the cross section. We can see also a clear outermost boundary (*black arrow*) and a sinus (*white arrow*). Eight cell nuclei with small amount of cytoplasm are scattered in the collagenous extracellular matrices. The *dotted rectangle* is magnified in (**c**). Scale bar = 5 μm. (**c**) An enlarged view of a sinus (S) in the primo vessel. Its boundary is clear and regularly surrounded by collagenous fibers (*black arrows*). Parts of the sinus (S) and the cytoplasm (cyt) of the cells are surrounded by the cell membrane (*white arrows* and *black arrow heads*, respectively). The nucleus (N) has mostly euchromatin. Scale bar = 0.5 μm

inside the sinus (lumen) of the vessel might be washed out during the specimen fixation and dehydration processes for TEM.

The TEM images of the cross section of the bovine heart primo vessel showed that, inside the vessel, there exist sinus(es) with a smooth inner surface composed of ECMs (longitudinally aligned collagenous fibers) and surrounded partly by cell membrane-like materials. This does not correspond to the typical vessel wall structure in which a single layer of endothelial cells surrounds the lumen. However, it has characteristics similar to the wall structure of the canine hepatic portal venule

(10–30 μm in diameter) in which prominent connective tissue components (mainly collagenous fibers), endothelium and smooth muscle cells constitute a smooth inner surface of the vessel wall [13]. The collagen fibers, woven in the hepatic portal venule wall, formed a plane sheath in the wall surface. One major difference is that the collagen fibers showed a randomly networked architecture, unlike the regular alignment of the collagenous fibers in the primo vessel sinus wall. We may postulate that this longitudinal regular alignment of the fibers could contribute to lower the hydrodynamic resistance in the liquid flow through the sinus (lumen). Since abundant collagens are found in the environment of the cells in the primo vessel, we can speculate that the cells may belong to the group of myofibroblasts, which deposits and aligns extracellular collagen fibers, especially in viscoelastic vessel walls [14]. For a future investigation, the exact types of the cells and the fibers in the wall of the primo vessel sinus should be identified by using immunohistochemical and advanced bioimaging methods.

Acknowledgments This work was supported by a Systems Biology Infrastructure Establishment grant provided by the Gwangju Institute of Science and Technology in 2010 and by the Association of Korean Oriental Medicine. Electron microscopy was supported by staff scientists in the National Instrumentation Center for Environmental Management. We appreciate Mr. Jin-Gyoum Kim at the National Agricultural Cooperative Federation of Korea in Bucheon for kindly providing the bovine hearts.

References

1. Lee BC, Park ES, Baik KY et al (2004) Acridine orange staining method to reveal characteristic features of intravascular threadlike structure. Anat Rec 278B:27–30
2. Baik KY, Lee J, Lee BC et al (2005) Acupuncture meridian and intravascular Bonghan duct. Key Eng Mater 277–279:125–9
3. Lee BC, Yoo JS, Baik KY et al (2005) Novel threadlike structures (Bonghan ducts) inside lymphatic vessels of rabbits visualized with a Janus Green B staining method. Anat Rec 286B:1–7
4. Johng HM, Yoo JS, Yoon TJ et al (2007) Use of nanoparticles to visualize threadlike structures inside lymphatic vessels of rats. Evid Based Complement Alternat Med 4(1):77–82
5. Lee BC, Soh KS (2008) Contrast-enhancing optical method to observe a Bonghan duct floating inside a lymph vessel of a rabbit. Lymphology 41:178–85
6. Lee BC, Kim S, Soh KS (2008) Novel anatomic structures in the brain and spinal cord of rabbit that may belong to the Bonghan system of potential acupuncture meridians. J Acupunct Meridian Stud 1(1):29–35
7. Shin HS, Johng HM, Lee BC et al (2005) Feulgen reaction study of novel threadlike structures (Bonghan ducts) on the surfaces of mammalian organs. Anat Rec 284B:35–40
8. Lee BC, Yoo JS, Ogay V et al (2007) Electron microscopic study of novel threadlike structures on the surfaces of mammalian organs. Microsc Res Tech 70:34–43
9. Yoo JS, Kim MS, Sung B et al (2007) Cribriform structure with channels in the acupuncture meridian-like system on the organ surfaces of rabbits. Acupunct Electrother Res 32(1/2): 130–2
10. Sung B, Kim MS, Lee BC et al (2008) Measurement of flow speed in the channels of novel threadlike structures on the surfaces of mammalian organs. Naturwissenschaften 95(2): 117–24

11. Sung B, Kim MS, Corrigan A et al (2009) In situ microextraction method to determine the viscosity of biofluid in threadlike structures on the surfaces of mammalian organs. Phys Rev E Stat Nonlin Soft Matter Phys 79:022901
12. Ogay V, Bae KH, Kim KW et al (2009) Comparison of the characteristic features of Bonghan ducts, blood and lymphatic capillaries. J Acupunct Meridian Stud 2(2):107–17
13. Ninomiya H, Inomata T, Ogihara K (1999) Collagen fiber arrangement in canine hepatic venules. J Vet Med Sci 61(1):21–5
14. Wagenseil JE, Mecham RP (2009) Vascular extracellular matrix and arterial mechanics. Physiol Rev 89:957–89

Chapter 9
Finding a Novel Threadlike Structure on the Intra-abdominal Organ Surface of Small Pigs by Using In Vivo Trypan Blue Staining

Ayati M. Hossein, Tian Yu-Ying, Huang Tao, Zhang Yu-Qing, Che Yong-Zhe, and Zhang Wei-Bo

Abstract Recently K.S. Soh and his colleagues in Seoul National University discovered novel threadlike structures named primo-vessels in organ surfaces, blood vessels, lymph vessels, and spinal cord in rabbits, rats, and mice. In this work, we tried to investigate whether the same structures can be found on the internal organ surface of small pigs. After being anesthetized with phenobarbital sodium and the abdomen opened, Trypan blue was applied to visualize the primo-vessels and primo-nodes on the internal organ surface of seven small pigs. Threadlike, movable, elastic structures could be observed on five small pigs. Morphological study on the obtained samples revealed that there are lots of rod-shaped nucleuses longitudinally along the structure. The structure is similar to the structures observed by K.S. Soh and his team in rat, rabbit, and mice. Further study is needed to investigate the function of these structures.

1 Introduction

Recently, a novel threadlike structure was found on the surface of various organs, such as stomach, liver, large and small intestine, and bladder in rabbits, mice, and rats [1–4]. Later on, it was named primo-vessel (PV) by Prof. K.S. Soh after more experiments. These studies reminded us of the earlier work of Kim Bonghan from North Korea who claimed to find the structure of acupuncture meridians [5] and the follow-up work of Fujiwara from Japan [6]. The findings from Korea were obtained by using Alcian blue on the surface of internal organs of anesthetized rats or rabbits and observing the surface with a stereomicroscope. Their finding was 50–100 μm

Z. Wei-Bo (✉)
Institute of Acupuncture and Moxibustion, China Academy of Chinese Medical Science, Beijing, China
e-mail: zhangweibo@hotmail.com

K.-S. Soh et al. (eds.), *The Primo Vascular System: Its Role in Cancer and Regeneration*, DOI 10.1007/978-1-4614-0601-3_9, © Springer Science+Business Media, LLC 2012

thick semitransparent threadlike structures which do not adhere to the surface, but move freely and are sparsely and irregularly fixed to the peritonea. Using the Feulgen reaction, which specifically stains DNA, an endothelial structure was found which had rod-shaped nuclei, 10–20 μm long, and aligned in a broken-line striped fashion. The PV consists of a bundle of several subducts. This structure was claimed to have a potential relationship with acupuncture meridians. In the most recent work [7], Alcian blue (AB) dye was injected at the rat acupoint BL23 and after 2 h, AB-stained primo-vessels were observed on the right abdominal cavity distributed on the surfaces of the duodenum, colon, and rectum. The work implies the possibility of one of meridian functions which connects peripheral tissue to the internal organs. In recent work, a new visualizing dye, Trypan blue, that stains the PV but not other known structures such as lymphatics, blood vessels, and adipose tissues was found by Lee et al. [8], which makes the PV visible even inside adipose tissues.

In the other aspect of meridian research in China, Zhang et al. found a low hydraulic resistance channel (LHRC) on the peripheral subcutaneous tissue along meridians [9]. It has a low impedance feature [10] and can be changed by acupuncture [11]. These two structures appear in different areas of the body and may have a certain connection between each other to form the whole network through peripheral tissue to internal organs as in the description of ancient meridian theory.

As the LHRC was found in small pigs and is hard to do in small animals like rats and rabbits, the first step is to find and repeat the finding of the Korean's work on pigs and the repeated work is also a significant step to confirm the previous results which were found in rats and rabbits.

2 Material and Methods

2.1 *Experimental Animals*

The experiment was conducted on seven anesthetized healthy small pigs weighing 10–19 kg, which were obtained from the Beijing Animal Husbandry Training and Demonstration Centre of the Ministry of Agriculture. All animals were kept according to guidelines issued by Beijing Municipal Administrative Committee for Experimental Animals. All animals were observed at least once daily.

2.2 *Preparation*

1.5–2 mg/kg of 2% phenobarbital sodium solution was injected intraperitoneally. The pig was anesthetized after several minutes. The animal was carefully carried to the lab room. The vital signs of the animal were observed carefully. The temperature of the lab room was kept 24–26°C. The four limbs were fixed to the surgery bed while the subject was in supine position.

2.3 Operation

Under deep anesthesia, a midline incision was carefully performed and we passed through the following structures: skin, linea alba, transversalis fascia, extraperitoneal fat, and peritoneum. The incision was extended by cutting around the umbilicus, while avoiding the falciform ligament above the umbilicus. Special care was also taken for the urinary bladder. Intra-abdominal organs were exposed carefully.

Hemorrhage was strictly avoided and minimal bleedings were immediately stopped while we tried not to let the blood enter the abdominal cavity.

2.4 Intraoperative Visualization and Imaging

Trypan blue solution (Sigma) 0.4% was obtained from the market and was carefully diluted to 0.1%. We filtered this 0.1% Trypan blue through 0.22 μm pore-sized filter paper just before the experiment. After exposure of the internal organs of the pig, Trypan blue was applied on the exposed organs such as small intestine, large intestine, and stomach.

After about 1 min the dye was washed away with warm saline, and threadlike structures were identified by searching through a surgical microscope (YSX, 30 times, made in China) or directly examining the surface of organs. Finally, the images were captured with a high resolution(8.0 MP)digital camera (Sony828, made in Japan).

2.5 Histological Analyses

After a threadlike structure was found, it was isolated from the organ carefully with microforceps to get the samples. One piece of sample was put on a glass and was observed under a microscope (DP71, Olympus). The other samples were fixed immediately with 4% paraformaldehyde (4°C, 48 h) prior to washing with PBS (0.01 M, 5′, 3 times). Washed samples were then stained with DAPI (1:1,000, room temperature) for 20 min in the dark and washed again with PBS (0.01 M, 5’, 3 times). Then they were picked on the slide and detected with a fluorescence microscope (Leica RXA2) with 359 nm. The images were then captured by a CCD camera (DF300F, Leica).

3 Results

Without much experience, the first pig had a serious bleed when opening the abdomen. This made the observation hard and no threadlike structure was found. The fourth pig stopped breathing during the operation because of very deep anesthesia and the threadlike structures were not found in this case. Apart from the two special cases, threadlike structures were found on the other five pigs with an appearance rate of 70%. On the second pig, a threadlike structure was found between the liver and stomach which was 100–300 μm in diameter (Fig. 9.1a). A thicker part, which may be a capsule, could be seen which connected three threadlike structures. The threads seemed thicker than the one found in rats which could often be seen clearly without using a microscope. Another threadlike structure was found between the small intestine and stomach (Fig. 9.1b) and on the surface of the liver (Fig. 9.1c).

On the third pig, a threadlike structure was found between the abdominal wall which was thinner and even. The other threadlike structure was found between the liver and stomach. On the fifth pig, threadlike structures could also be found on the surface of liver and small intestine. On the sixth pig, threadlike structures were found between the gall-bladder and stomach and between hepatic leaves. On the seventh pig, a threadlike structure was found between gall-bladder and liver.

Two samples from the surface of the liver (Fig. 9.1c) and intestine (Fig. 9.1b) were obtained in two pigs. Both images by fluorescence microscope showed broken lines in striped fashion in linear alignment parallel to the samples which represent rod-shaped nuclei distributed along the threadlike structure (Fig. 9.2).

4 Discussion

The finding of an unknown vessel on the surface of organs was first reported by Kim Bonghan in early 1965 and it was named Bonghan ducts. It was not repeatable by most scientists in their several attempts until 2005 when the similar structure was found in Soh's lab in Seoul National University, using Alcian blue and later using Trypan blue. As the experimental animals were small, a stereomicroscope with about 40 times magnification should be used to capture the small-sized PV and this structure has not been found in large animals like dogs and pigs. Our result showed that in small pigs which were 10–19 kg in weight, the novel threadlike structures which may represent PV could also be found if a good physiological condition is maintained. The threadlike structure was larger than the one found in rabbits and rats and could also be seen by the naked eye, sometimes even without staining. The images can easily be taken by a camera under the bright yellow light. Although the structure has not been studied in the human body, it should exist also in human as it seems to be a common structure in mammals. Some doubt arose as to why surgeons or anatomists have not noticed it during operations. Our study suggested that bleeding can seriously influence the appearance of such a structure and if the circulation stops, it may also disappear. The structure is also often hidden beneath the deep organs like the liver and can only be found by the person who intensively looks for it.

In our experiment, the threadlike structures appeared on several organs such as liver, stomach, intestine, and gall-bladder. We have not tried to observe such structure on heart, lung, and kidney where it is difficult to do the operation. The connection of the threadlike structure is not very certain and more cases should be studied to obtain statistical results about the distribution of the structure.

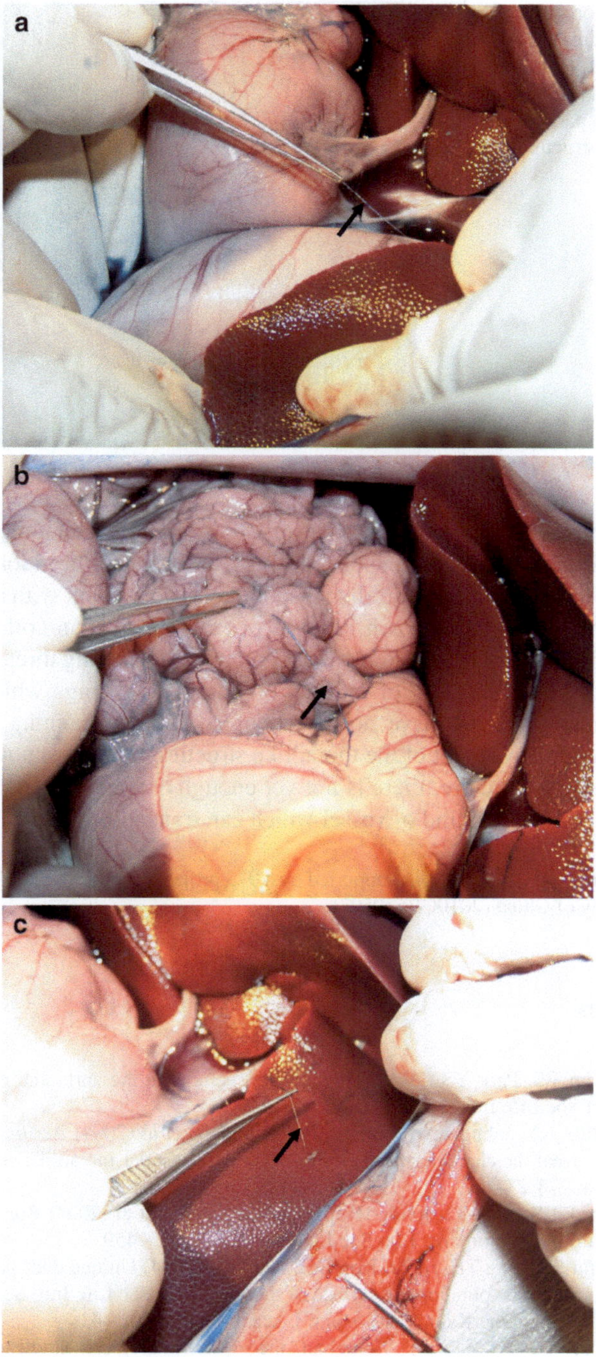

Fig. 9.1 (**a**) Stereomicroscope image of a corpuscle with three threadlike structures was found on the surface between the liver and stomach stained by Trypan blue. (**b**) A threadlike structure was stained by Trypan blue on the surface between small intestine and stomach. (**c**) A threadlike structure was stained by Trypan blue on the surface of liver

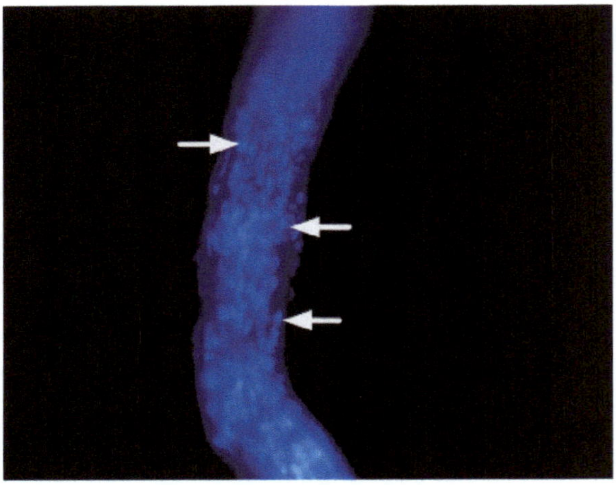

Fig. 9.2 A fluorescence image of the threadlike structure which showed many rod-shaped dots along in a striped fashion parallel to the threadlike structure

The DAPI-stained picture showed several rod-shaped nuclei along the surface of the threadlike structure which appear to be blue because the fluorescence wavelength is 359 nm. The image of rod-shaped nuclei is coincident with the result found by Shin in rabbit which means it consists of cells rather than collagen fibers and coagulated fibrins. Our results imply a phenomenon of existing threadlike structures between and on the surfaces of internal organs in small pigs which may have a cellular structure. As we have not used confocal laser scanning microscope to observe the cellular content of the threadlike structures and have not yet observed a liquid flow in the structure, there is not yet enough evidence for us to say that this threadlike structure is a kind of vessel which can transport a liquid.

Acknowledgments This work was supported by a 973 project grant from the Ministry of Science and Technology of China (2010CB530507).

References

1. Cho SJ, Kim BS, Park YS (2004) Threadlike structures in the aorta and coronary artery of swine. J Int Soc Life Info Sci 22:609–611
2. Lee BC, Baik KY, Johng HM, Nam TJ, Lee J, Sung B et al (2004) Acridine orange staining method to reveal the characteristic features of an intravascular threadlike structure. Anat Rec B New Anat 278:27–30
3. Lee BC, Park ES, Nam TJ, Johng HM, Baik KY, Sung B et al (2004) Bonghan ducts on the surface of rat internal organs. J Int Soc Life Info Sci 22:455–459
4. Lee KJ, Kim SY, Jung TE, Jin D, Kim DH, Kin HW (2004) Unique duct system and the corpusclelike structures found on the surface of the liver. J Int Soc Life Info Sci 22:460–462
5. Kim BH (1963) On the Kyungrak system. J Acad Med Sci 10:1–41
6. Fujiwara S, Yu SB (1967) "Bonghan theory"morphological studies. Igakuno Aumi 60:567–577
7. Han HJ, Sung B, Ogay V, Soh KS (2009) The flow path of Alcian blue from the acupoint BL23 to the surface of abdominal organs. J Acupunct Meridian Stud 3:182–189

8. Lee BC, Kim KW, Soh KS (2009) Visualizing the network of Bonghan ducts in the omentum and peritoneum by using Trypan blue. J Acupunct Meridian Stud 2:66–70
9. Zhang WB, Tian YY, Li H, Tian JH, Luo MF et al (2008) A discovery of low hydraulic resistance channel along meridians. J Acupunct Meridian Stud 1:20–28
10. Zhang WB, Zhuang FY, Tian YY, Li H (2001) A simulating study of biophysical features along meridians on a gel model. J Biomed Eng 18(3):357–361
11. Zhang W, Ruiming Xu, Zhu Z (1999) The influence of acupuncture on the impedance measured by four electrodes. Acupunct Electrother Res 24:181–183

Chapter 10
Observation of the Primo Vascular System on the Fascia of Dogs

Zhaofeng Jia, Kwang-Sup Soh, Qiang Zhou, Bo Dong, and Wenhui Yu

Abstract Trypan blue is known to be preferentially effective for a primo vascular system (PVS) compared to blood vessels, lymphatic vessels, or adipose tissues. We observed the PVS of the dog in various membrane structures, such as the mesentery, peritoneum, and omentum, by using the Trypan blue technique. The freely movable PVS did not adhere to the surfaces or wrap membranes of internal organs. The features of the PVS, such as the distribution of rod-shaped nuclei and the bundle structure, were observed, as expected from previous reports on rabbits, rats, and mice.

1 Introduction

In order to reveal the hardly visible primo vascular system (PVS), the scientists at the Seoul National University found a simple but effective method to visualize the novel threadlike structures on the organ surfaces of rabbits, rats, and nude mice. As they reported, Trypan blue was successfully applied as an effective staining dye to visualize the PVS [1]. It was a great step forward in PVS research because no one except a Japanese anatomist Fujiwara [2] was able to reproduce Bong-Han Kim's discovery [3] for a long time due to the lack of a proper visualizing technique for the PVS.

As far as the authors are aware, there have been no reports on the observation of the PVS in dogs. Only small animals such as rabbits, rats, and mice were studied in the previous works of Kim [3], Fujiwara and Yu [2], and Soh [4] even though Kim had hinted at the existence of PVS in the human body. In this article, we present for the first time the observation of the PVS on the surfaces of intestinal organs in the abdominal cavities of dogs. We used the Trypan blue technique and our observational results were in good agreement with the previous reports [4]. We were able to confirm the distinctive characteristic feature of the PVS in dogs.

W. Yu (✉)
Department of the Traditional Chinese Veterinary Medicine, School of Veterinary Medicine, Northeast Agricultural University, Harbin, China
e-mail: yuwh666666@yahoo.com.cn

K.-S. Soh et al. (eds.), *The Primo Vascular System: Its Role in Cancer and Regeneration*,
DOI 10.1007/978-1-4614-0601-3_10, © Springer Science+Business Media, LLC 2012

2 Materials and Methods

2.1 Animal Preparation

The dogs (Papillan, both sexes, 2.8–5.0 kg) used for this study were obtained from the Aijian Dog Company. The animals were housed in a constant-temperature controlled environment (26°C) with 60% relative humidity under a 12-h light/dark cycle. All the dogs had ad libitum access to food and water. The procedures were in accordance with international laws and policies (Guide for the Care and Use of Laboratory Animals, National Academy Press, 1996). The dogs were anesthetized with Zolertil (1.5 mL/kg) administered intraperitoneally, and all surgical procedures were performed under systemic anesthesia. Under deep anesthesia, we cut the medial alba of the abdomen carefully to avoid contaminating the organs with blood; then we performed the Trypan blue staining procedure.

2.2 Trypan Blue Staining

We used a 0.4% Trypan blue solution (Sigma, USA) for staining the PVS on the internal organs of the dogs. After exposure of the internal organs of the dog, we splashed Trypan blue on internal organs such as the small intestine, the liver, and the stomach. Then, we washed the internal organs several times with physiological saline for about 15 seconds. We could observe the web net structure of the PVS stained with Trypan blue and take images with a digital camera.

2.3 Ethidium Bromide Staining

In order to characterize the nuclei in the PVS, we stained the primo vascular specimens by using ethidium bromide (EB). The specimens were examined with a phase-contrast microscope (Axiophot, Zeiss, Germany).

3 Results

Threadlike primo vessels were observed on the surfaces of the internal organs. The threadlike primo vessels had 2–3 branches and were joined to an oval-shaped primo node. A piece of the mesentery containing the web-like network of primo vessels was isolated and stained using EB. The features of the nuclei of the threadlike structure are in good agreement with Bong-Han Kim's original description [3] and with those of recent works [4].

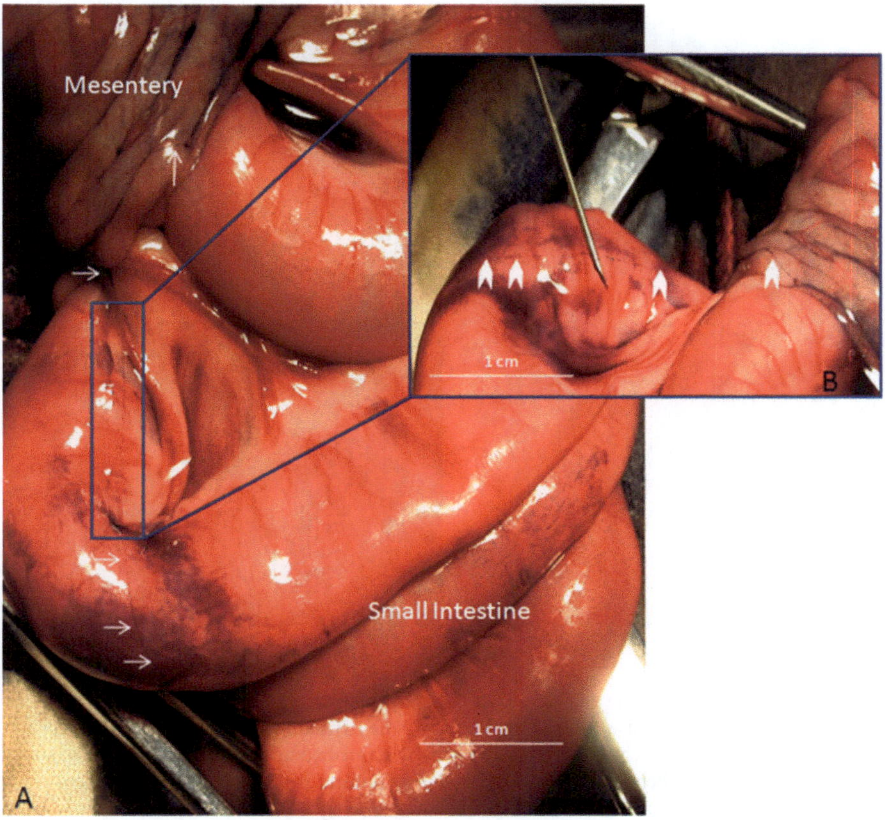

Fig. 10.1 The primo vessel of the dog extended above the small intestine from the mesentery to the bladder area. (**a**) The primo vessel weakly stained by EB was movable above the surface of the small intestine. (**b**) The primo vessel (*arrow heads*) was easily stirred up by using a syringe needle

A threadlike structure above the small intestine starting from the stomach and going to the bladder area was found, as shown in Fig. 10.1. The threadlike structure did not adhere to but floated on the surface of the intestine and could be lifted up by using syringe needles. After the primo vessel had been taken and stained with EB, we observed that corpuscular nodes were sparsely located along the threadlike structures.

Figure 10.2a shows a node that had been weakly stained with the EB and whose diameter and length were respectively about 0.3 and 1 mm. The node was more heavily stained compared with the threadlike structure. This suggests the existence of more nuclear content inside the node than in the threadlike structure. Moreover, we could clearly observe the primo vessel connecting with the node. In Fig. 10.2b, if one looks at the threadlike structure very carefully, one can see the dim image of two sub ducts (marked with ∘1, ∘2) that contain many rod-shaped nuclei [5] along the primo vessel.

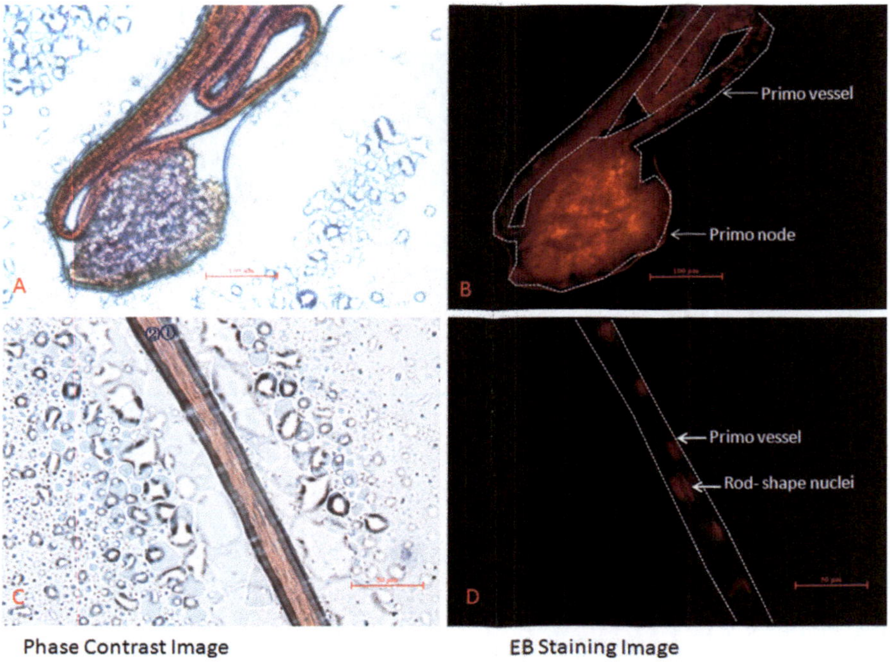

Fig. 10.2 Both the primo node and the primo vessel taken from the Fig. 10.1 were stained by EB. The up panel showed the phase contrast image (a) and the EB staining image (b) of both the primo vessel and the primo node. The node was more heavily stained compared with the threadlike structure. This suggests that there is much higher number of nuclear inside the node than in the threadlike structure. The down panel showed the phase contrast image (c) and the EB staining image (d) of the higher magnification of the primo vessel. The rod- shape nuclei distributed on the layers of the subducts, and along several broken lines in a striped fashion are marked with 1 and 2 which are marked in (c)

4 Discussion

Trypan blue is usually used for staining dead cells to distinguish them from live ones, but it was used here as a visualizing agent in vivo and in situ. By using it, we could observe the PVS on the surfaces of the internal organs of dogs. As far as the authors are aware, this is the first observation of the PVS in dogs. Previous works dealt only with small animals like rabbits, rats, and mice [6–8].

The significance of our work is the suggestion that the PVS is a system that exists in most mammals and most importantly in the human body. Another important point is that the length of the primo vessel sample in dogs is much larger than those of small animals. This will make further analysis of the PVS for various physiological characteristics easier. Finally, clinical applications in veterinary hospitals will be based upon this basic research. More details of this work will be reported elsewhere.

Acknowledgments This work was supported by the Association of Korean Oriental Medicine.

References

1. Lee BC, Woo Kim Ki, Soh KS et al (2009) Visualizing the network of Bonghan ducts in the omentum and peritoneum by using Trypan blue. J Acupunct Meridian Stud 2(1):66–70
2. Fujiwara S, Yu SB (1967) 'Bonghan theory' morphological studies. Igaku no Ayumi 60:567–577
3. Kim BH (1963) On the Kyungrak system. J Acad Med Sci DPR Kor 90:1–35
4. Soh KS (2009) Bonghan circulatory system as an extension of acupuncture meridians. Journal of Acupuncture and Meridian Studies 2(2):93–106
5. Ogay V, Bae KH, Kim KW et al (2009) Comparison of the characteristic features of Bonghan ducts, blood and lymphatic capillaries. Journal of Acupuncture and Meridian Studies 2(2):107–117
6. Yi SS, Hwang IK, Kim MS et al (2009) The origin of endothelial cells in novel structures, Bonghan ducts and Bonghan corpuscles determined using immunofluorescence. Journal of Acupuncture and Meridian Studies 2(3):190–196
7. Lee KJ, Kim S, Jung TE et al (2004) Unique duct system and the node-like structures found on the surface of the liver. J Int Soc Life Info Sci 22:460–462
8. Yoo JS, Hossein Ayati M, Kim HB et al (2010) Characterization of the primo-vascular system in the abdominal cavity of lung cancer mouse model and its differences from the lymphatic system. PloS One 5(4):e9940

Chapter 11
Development of the Putative Primo Vascular System Before the Formation of Vitelline Vessels in Chick Embryos

Seung-Yoon Lee, Byung-Cheon Lee, Kwang-Sup Soh, and Gil-Ja Jhon

Abstract We found threadlike structures in the vitelline membrane and its surrounding dense albumen of a chick embryo by using the Trypan blue staining technique. The threadlike structure, a putative primo vessel, had DNA bodies as revealed by DAPI staining and was developed before the formation of blood vessels.

1 Introduction

Chick embryos are widely used as laboratory material for studying vertebrate embryology, so it is desirable to investigate the development of the primo vascular system (PVS) in the early stages of their development as the PVS has been studied only in the adult anatomy of animals [1]. Every animal starts life as a single cell, the zygote. Two questions when and where the PVS appear in the course of an animal's development from a single fertilized ovum into a highly complex living being with beautifully functioning systems like cardiovascular and nervous systems. There are interesting and important questions.

We were particularly motivated by the claim of Bong-Han Kim who allegedly observed that a PVS had already formed at between 10 and 15 h of incubation of chick embryos [2]. He also claimed that major blood vessels developed later around the primo vessels so that the PVS was surrounded by blood vessels and became an intravascular primo vessel (or intravascular Bonghan duct). His remarkable observation, however, was not reproduced by others until the present time.

In this chapter, we planned experiments to determine whether the PVS develops before blood vessels formation. For this purpose, we searched the PVS around the

G.-J. Jhon (✉)
Department of Chemistry and Nano Science, College of Natural Sciences,
Ewha Woman's University, Seoul, Korea
e-mail: gjjhon@ewha.ac.kr

K.-S. Soh et al. (eds.), *The Primo Vascular System: Its Role in Cancer and Regeneration*,
DOI 10.1007/978-1-4614-0601-3_11,

vitelline area in the period of 16–24 h of incubation, which is earlier than the formation of extraembryonic blood vessels where it requires 33–35 h of incubation for the establishment of afferent vitelline channels [3]. Thus, our observation period was after the formation of the PVS and before the formation of extraembryonic blood systems around the vitelline.

For finding and identifying the PVS, we used the Trypan blue visualization technique developed to stain preferentially the PVS rather than blood, nerve, muscle, lymph, or adipose tissues [4, 5]. We applied this technique to the chick embryo in the vitelline and its surrounding albumin area, and found threadlike structures with associated corpuscular bodies well stained with Trypan blue, which might be primo vessels associated with primo nodes. Further study with 4′, 6-diamidino-2-phenylindole (DAPI) and acridine orange revealed the presence of DNA in these tissues, which is an essential requirement for the identification of the PVS. However, more conclusive histological examinations of this putative PVS have not yet been performed. Nevertheless, the observation of the Trypan blue-specific network of threadlike structures with associated corpuscular bodies around the vitelline membrane before the formation of blood vessels, nerves, or any other important structural systems of a chicken is a significant finding and is worth further study in connection with the PVS.

2 Materials and Methods

2.1 Egg Preparation

The fertilized fresh eggs (Hyline Brown, each about 60 g) needed in this study were obtained from Pulmuone Company (Seoul, Korea). The eggs were incubated in an automatic electric incubator (Alcom Mini, Autoelecs, Korea) which holds six eggs simultaneously in a dark, constant temperature (37°C), controlled environment with 45% relative humidity. The structure of an egg is schematically illustrated in Fig. 11.1a. Inside the outermost shell lies the thin and dense albumen with a yolk toward the center and latebra at the center.

2.2 Development Stages

The very early stages of a chick embryo are well established by the characteristic generation of indicative tissues. For example, Hamburger and Hamilton [6] proposed the following development stages: Stage III: primitive streak reaching the pellucid area (12–18 h of incubation), Stage IV: Hensen's node (18–19 h), Stage V: Beginning of head bending (19–22 h), Stage VI: completion of head bending (23–25 h), Stage VII: Beginning of somite formation (1st somite) (23–26 h),

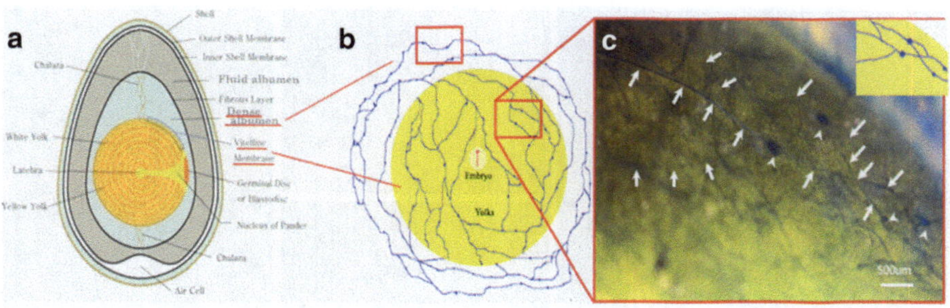

Fig. 11.1 (**a**) Chicken egg in a longitudinal section. (**b**) Top view of primo vessels on the vitelline membrane and in dense albumen by Trypan blue staining. (**c**) Stereoscopic image of primo vessels (*arrows*) and primo nodes (*arrowheads*) stained by Trypan blue on the vitelline membrane of a chicken egg for Stages IV–VII. The *inset* illustrates the network of a primo vascular system on the vitelline membrane of an egg

Stage VIII: 4th somite formation (26–29 h). In this study, we observed the PVS in Stages IV–VII (18–24 h of incubation), and we checked the appearance of Hensen's node or the somites. Kim reported the observation of the pre-PVS in the first 15 h of incubation [2] while the extraembryonic vitelline blood vessel formation did not begin before 24 h of incubation [3].

2.3 Trypan Blue Staining and Observation of the PVS

We used a 0.1% Trypan blue solution (0.4% Trypan blue solution, Sigma, USA with phosphate-buffered saline, pH 7.4) to visualize networks on the vitelline membrane and the albumin of chicken egg. After exposure of the yolks of the eggs, we removed albumin on the yolks with scissors; then, we dropped several milliliters of 0.1% Trypan blue on the vitelline membrane and the albumin. About one and half minutes after dropping of the Trypan blue, we washed the vitelline membrane several times with phosphate-buffered saline (PBS), pH 7.4. Using a stereomicroscope (STZ 10, Olympus) with a CCD camera (DP 70, Olympus), we observed and took pictures of novel networks consisting of threadlike and corpuscular structures stained by Trypan blue [7].

3 Results

Figure 11.1a is a diagram of the hen's egg in a longitudinal section where the relations of the various parts of the egg at the time of lying are indicated schematically [7]. The relevant part in this experiment is the vitelline membrane between the yellow yolk and the dense albumen, as indicated by a thin box. This boxed area is illustrated in Fig. 11.1b, where the weblike network of primo vessels is schematically drawn.

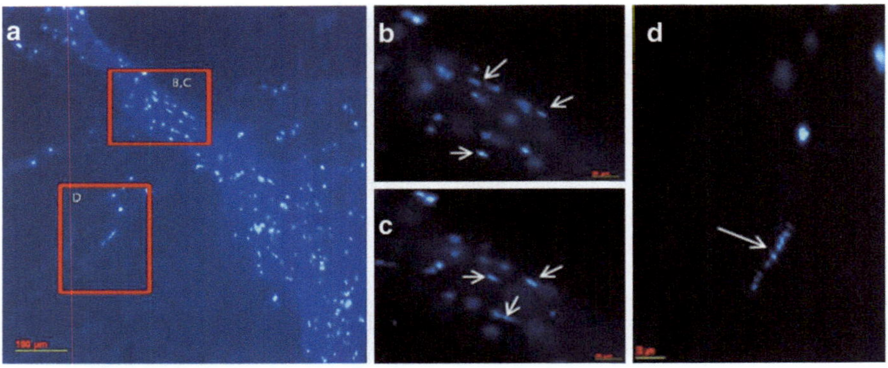

Fig. 11.2 (**a**) Primo vessels, as shown in Fig. 11.1c, on the vitelline membrane of a chicken egg for Stages IV–VII have many DAPI-stained DNA particles. (**b**) and (**c**) are magnified view of the *rectangle* labeled as *B*, *C* in (**a**). It is noticeable that the *B* and the *C* images show DNA particles that are longitudinally arranged. (**d**) DNA particles (*arrow*) magnified from the *rectangle* labeled as *D* in (**a**) clearly show that the DNA particles are arranged in longitudinal order

The primo vessels were present in the vitelline membrane and in the dense albumen as one connected system.

The visualization of the transparent and apparently not visible threadlike primo vessels was achieved by using Trypan blue as shown in Fig. 11.1c. These putative primo vessels were observed on the vitelline membrane of the hen's egg in Stage VII, that is, after about 24 h of incubation. As indicated by the arrowheads, corpuscular bodies, which are the putative primo nodes, were observed in the network of the primo vessels.

The diameter of the primo vessel was about 40–90 μm. The shape of primo nodes was oval and their size was about 80–200 μm in diameter. As shown in Fig. 11.2 further in vitro analysis by using DAPI showed that the putative PVS contained DNA granules, but it was not clearly determined whether they were parts of cellular nuclei.

4 Discussion

Weblike networks of putative primo vessels were observed in the vitelline membrane and in the surrounding dense albumen of a hen's egg after 18–24 h of incubation by employing the method of Trypan blue visualization of PVSs in animals [4]. The primo vessels branched to form a complex net covering the vitelline membrane and extended out into the dense albumen-like tree branches.

We were particularly interested in the question of whether the PVS develops before blood vessel formation as Bong-Han Kim claimed [2]. It should be noted that there are two distinct circulatory arcs of which the heart is the center. One complete circulatory arc is established entirely within the body of the embryo. A second arc is formed of the vitelline vessels whose terminals are located in the extraembryonic membranes enveloping the yolk. The main distribution of the vitelline circulation is extraembryonic even though the vitelline vessels communicate with the heart over the main vessels that traverse the embryonic body. Neither the intraembryonic nor the vitelline circulatory channels are completed in younger embryos before 33 h of incubation [3].

The formation of extraembryonic blood vessels is presaged by the appearance of blood islands in the vascular area of 24-h chicks. By 33–35 h of incubation, the extraembryonic vascular plexus has extended inward and has made a connection with the omphalo mesenteric veins, establishing the afferent vitelline channels [3].

The first definite structural indication of heart formation appears in a 25-h chick. Concurrently with the establishment of the heart, intraembryonic blood vessels are arising within the body of the embryo. The large vessels connecting the heart are the first of the intraembryonic channels to be established [3].

Since our observation of the PVS in the vitelline membrane was made after 16–24 h of incubation, the putative PVS was clearly developed earlier than the formation of the extraembryonic vessels, let alone the establishment of the heart and the intraembryonic vessels. Kim stated that a pre-primo vessel was formed at 15 h of incubation, and a blood vessel was formed, surrounding the primo vessel in 20–27 h of incubation [2]. He did not mention his experimental methods, nor did he describe explicitly whether he worked with the intra- or the extraembryonic blood vessels. Therefore, comparing our findings with his claims is difficult, yet both results seem to be consistent and not to be contradictory in the least.

One surprising discovery in the current work was the outward extension of the putative PVS into the dense albumen of the egg. This observation raises questions on the function of the PVS in the dense albumen. An interesting conjecture was proposed by one of us (G.J. Jhon) that the PVS is an oxygen transporting path from air outside the shell through the albumin layer to the yolk. Mere diffusion of oxygen through the albumin may not be sufficient for the generation of primordial structures in the embryo when blood vessels have not yet been formed. The PVS may work as a primitive oxygen conduit, in addition to its own major functions, i.e., generation of tissues and regeneration of damaged tissues, as claimed by Kim [8].

The current work is only a first step in a developmental study of the PVS. More detailed histological work is needed, some of which will be reported elsewhere. In order to elucidate the full developmental features of the PVS, we need more elaborate techniques to observe and identify the PVS from the very beginning of incubation to the fully developed chick stage. The close relation between the PVS and the cardiovascular system is worth further investigation.

Acknowledgments This work was supported by a "Systems Biology Infrastructure Establishment Grant" from the Gwangju Institute of Science and Technology. K. Soh thanks the Association of Korean Oriental Medicine for support.

References

1. Soh KS (2009) Bonghan circulatory system as an extension of acupuncture meridians. J Acupunct Meridian Stud 2(2):93–106
2. Kim BH (1965) The Kyungrak system. J Jo Sun Med 108:1–38 [In Korean]
3. Pattern BM (1957) Early embryology of the chick, 4th edn. McGraw-Hill, New York
4. Lee BC, Kim KW, Soh KS (2009) Visualizing the network of Bonghan ducts in the omentum and peritoneum by using Trypan blue. J Acupunct Meridian Stud 2(1):66–70
5. Lee BC, Bae KH, Jhon GJ et al (2009) Bonghan system as mesenchymal stem cell niches and pathways of macrophages in adipose tissues. J Acupunct Meridian Stud 2(1):79–82
6. Hamburger V, Hamilton HL (1951) A series of normal stages in the development of the chick embryo. J Morph 88:49–92
7. Pattern BM (1957) Early embryology of the chick, 4th edn. McGraw-Hill, New York, p 35
8. Kim BH (1965) Sanal theory. J Jo Sun Med 108:39–62 [In Korean]

Chapter 12
Characterization of Primo Nodes and Vessels by High Resolution Light Microscopy

Vitaly Vodyanoy

Abstract Morphological properties of primo vascular system were characterized by a high resolution light microscopy that makes structural features with size of ~100 nm clearly visible. High visibility and advanced contrast of the smallest features in the image are due to enhancement of high spatial frequencies in the optical transfer function. We have isolated primo nodes (PN), vessels (PV), and capillaries from the surface of organs in the peritoneal cavity of rats. These structures correspond to Bonghan corpuscle, ducts, and ductules, discovered by B.H. Kim in the early 1960s. The non-fixed samples stained with acridine orange and Bouin fixed and H&E stained slides were observed, photographed, and analyzed. The primo vessel is composed of 1–20 primo-capillaries of 3–25 μm in diameter. A thin external envelope of primo-capillary is composed of muscle-like endothelial cells with rod-shape 15–20 μm nuclei directed along the capillary axis. The primo-capillaries carry a liquid that contains granules and cell-like structures rich in nucleic acids. The bundle of primo-capillaries of the primo vessel is laid into an external jacket composed of endothelial cells with 6–12 μm spindle-like or oval nuclei. The primo nodes are oval-shape of 0.1–0.5 mm along the short and 0.5–1 mm along the long axis, on both ends connected to primo vessels of 3–6 cm in length and 40–100 μm in diameter. The primo node is essentially the interlacement of broadened and branched primo-capillaries covered with a 5–40 μm thick capsule. A single capillary bundle (B) of the incoming (afferent) vessel enters the node, branches into additional bundles, and fills the node interior by tightly spun and folded bundles. Capillaries converge, narrow, and come out from the node as a single bundle of the efferent primo vessel. The structural features of primo nodes, vessels, and capillaries observed with a high resolution light microscope are very different from those observed in blood and lymphatic vascular samples. Specifically, the jacketed architecture of primo vessels has no known parallel in other vascular systems.

V. Vodyanoy (✉)
Department of Anatomy, Physiology and Pharmacology, College of Veterinary Medicine, 109 Greene Hall, Auburn University, Auburn, AL 36849, USA
e-mail: vodyavi@auburn.edu

K.-S. Soh et al. (eds.), *The Primo Vascular System: Its Role in Cancer and Regeneration*, DOI 10.1007/978-1-4614-0601-3_12, © Springer Science+Business Media, LLC 2012

1 Introduction

Acupuncture is the stimulation of specific points (acupoints) that occur along virtual lines (meridians) on humans or other animals. According to western interpretation, there is no known anatomical foundation for the meridians and acupuncture effects are mediated by unknown nervous, circulatory, endocrine, and/or immune mechanisms.

Acupuncture is ancient but is still an important part of traditional Chinese medicine. According to contemporary interpretation, traditional meridians are the superficial representation of a channel network, the nodes of which underlie the acupoints on the body's surface. The material (*Qi*) that flows through the channels is essential for health [1]. Acupuncture is increasingly practiced in the United States by physicians and veterinarians and is comparable in effectiveness to conventional treatments for certain clinical conditions such as nausea and pain, and it promises to be useful in other areas [2].

Intuitively, the most logical and direct theory of acupuncture meridians is likely associated with anatomical structures. In the early 1960s, Bong-Han Kim, a North Korean scientist and professor of Pyongyang Medical College, reported that the superficial acupuncture meridian system embodied an underlying microcapillary system. He injected radioactive phosphorus (^{32}P) into a rabbit acupoint and reported that the ^{32}P followed the acupuncture meridians [3]. Kim reported that he isolated DNA granules (sanals) from the microcapillaries and induced their proliferation under artificial conditions. He stated that mature sanals had cell-like structure, contained chromosomes, and participated in the tissue regeneration [4]. This is what we know now, stem cell can do. Kim did not disclose his methods, and subsequent investigators failed to reproduce many of his results.

Crucially, radioactive visualization of acupuncture meridians was reported again [5, 6]. Both research groups injected the radioisotope technetium (^{99}Tc) at acupoints and reported that the preferential radiotracer pathways coincided with acupuncture meridians. Most recently, Hobbs [7] reported that infrared light introduced at acupoints travels in tracks detectible on the skin and that these tracks correspond to classical acupuncture meridians.

These studies support observations reported by a group of South Korean scientists led by Professor Kwang-Sup Soh of Seoul National University. Dr. Soh and colleagues developed tracing techniques including fluorescent stains and nanoparticles and have therewith described a vascular system, distinct from the blood and lymphatic systems that may be a good candidate for the physical entity underlying acupuncture meridians. Dr. Soh and associates have described properties – never reported before – that are related to the mechanisms mediating obesity and cancer [8–10].

The above experimental data synopsis allows us to pose questions, for example: "What is being stimulated in acupuncture? Are there anatomical (physical) structures that represent meridians and acupoints? How do they function?" A simple answer to these questions is a hypothesis: "There are anatomical (physical) entities that underpin the acupuncture meridian system."

There are challenges to this concept. The physical challenges include the unknown nature of the required structures and the absence of physical, chemical, or biological markers. It is also entirely possible that the required structures are similar to known ones, but have different functions. There are also perceptual challenges to overcome. These originate in different groups of scientists and acupuncture practitioners. One group comprises skeptics who find faults in the practice of, or inadequacy in, classical teaching of acupuncture. The second group is composed of enthusiasts who "explain" the acupuncture mechanisms with little or no experimental support [11]. A third group is composed of practitioners who shroud acupuncture in mystery. Lastly, a special group of doubters derives from anatomists and histologists who feel offended by the idea that a, hitherto unknown, anatomical system exists.

Even if this hypothesis is overturned, it will still bring a better understanding of the acupuncture meridian system's fundamental properties and focus future acupuncture research on the combined actions of the nervous, circulatory, endocrine, and immune systems [12]. However, if this hypothesis is not overturned, it will confirm the existence of a new, distinct vascular system and lay the foundation for a new paradigm in biology and medicine. Confirmation of the hypothesis could serve to bring together Western and Eastern medical philosophies, provide an unlimited source of individualized stem cells, bring new diagnostic and therapeutic methods, and much more. The highest potential impact is expected in pain management, developmental biology, tissue regeneration, organ reconstruction, diabetes, and cancer prevention and treatment.

In this work, high resolution optical microscopy was used to visualize structures that may be related to those described by Dr. Kim in 1960s and recently rediscovered and analyzed by Dr. Soh's group [9].

2 Methods

2.1 High Resolution Light Microscopy

The system produces the highly oblique hollow cone of light (NA 1.2–1.4). Coupled with a high aperture microscope objective with iris, this system provides two different regimes of illumination. When the iris is closed so that no direct light enters the objective after passing through the object, only refracted, scattered, or diffracted light goes in the objective. If the iris is open in such a way to allow the direct entrance of light into objective, the front lens of the objective is illuminated by the annular light produced by the empty cone of light entering the objective. In this case the mixed illumination is produced that combines the darkfield and oblique hollow cone brightfield illuminations. The cardioid condenser is an integral part of the illumination system so that the system comprises a collimation lenses and a first surface mirror that focus light onto the annular entrance slit of the condenser. As a part of the illumination system, the condenser is pre-aligned and therefore additional alignment is unnecessary. The illumination system is positioned in Olympus BX51 microscope by replacing a regular brightfield condenser. The illumination system is connected with a light source (EXFO120, Photonic Solution) by a liquid light guide. The objective used for this work is infinity corrected objective HCX PL APO 100×/1.40–0.70, oil, iris from Leica. The image is magnified with a zoom intermediate lens (2×-U-CA, Olympus), a homebuilt 40× relay lens, and captured by a Peltier-cooled camera (AxioCam HRc, Zeiss) and Dimension 8200 Dell computer. The microscope is placed onto a vibration isolation platform (TMS, Peabody, MA). Live images were recorded with Sony DXC-33 Video camera and Mac OS X Computer [13]. Test images were examined using Richardson slide [14].

2.2 Sample Preparation

Sprague–Dawley rats (~300 g) were deeply anesthetized with Nembutal (40 mg/kg) administered intraperitoneally. All isolation procedures were performed under deep anesthesia. The lateral abdominal wall was incised, and the large vessels in the skin of the abdomen and the thorax were held by hemostats so that blood flow over organ surfaces was minimized. The search for threadlike structures was carried out under stereomicroscope. The primo node and vessel (Bonghan corpuscle and duct) samples were collected from the surfaces of organs in the peritoneal cavity of rats. The careful exploratory search along the organ surfaces with forceps helped in finding the primo vessels and nodes. If undisturbed, the node and vessels blend into the background. When they are touched, they change color and become more noticeable. From each rat we collected on average 4–5 samples suspected of belonging to primo vascular system. After a close microscopic examination of the harvested samples, 2–3 were confirmed to be part of the system. Tissues harvested from rats were processed in two ways. For observation of unfixed samples under high resolution light microscope, the primo nodes and vessels were extracted and transferred into a phosphate buffer solution (PBS, pH=7.4) stained with 0.01% acridine orange. Other samples were fixed in Bouin's fluid, embedded in paraffin, sectioned at 6 μm, slide mounted, hematoxylin and eosin (H&E) stained, and cover-slipped. Surgical and staining procedures were adopted from Prof. Soh laboratory [15, 16].

3 Results and Discussion

3.1 Images of Light Microscope

The design of the light microscopy system was optimized for very thin samples of biological tissue, which typically have low optical density and some details of submicron size. The challenging goal was the correct imaging showing the shape and

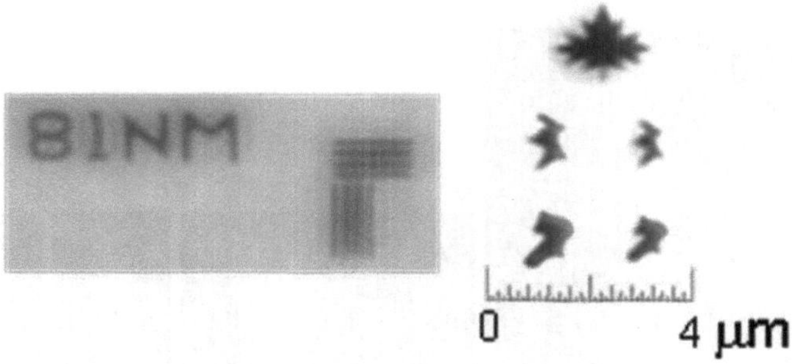

Fig. 12.1 Images of vertical/horizontal bar set (*left*) at $\lambda = 405$ nm and various shapes (*right*) at $\lambda = 546$ nm recorded using the home-built annular illumination system

internal features of fine structures. Obviously, this type of microscopy provides enormous additional information on structure of submicron length scale. It demands both the extreme resolving power and high contrast for visibility of the smallest details. This was achieved with the microscope illumination system by using the brightfield or mixed (comparable intensities of dark- and brightfield contributions to the image) illumination [13]. The examples of test pattern images using this system are shown in Fig. 12.1. The vertical/horizontal bar set consists of opaque lines with the width of 81 nm spaced at 81 nm and thus the period of 162 nm. The image was taken at $\lambda = 405$ nm. For the imaging system with NA 1.4 the Abbe limit for the pitch of resolved periodic structure is $L_{\text{Abbe}} = \lambda/(2\ \text{NA}) = 405/2.8 = 145$ nm. Therefore, the imaged period of 162 nm is in complete accord with the Abbe theory. For non-periodic shapes in Fig. 12.1 (right part) imaged at $\lambda = 546$ nm, the features with the size of ~100 nm are clearly visible. The two-point Rayleigh resolution criterion predicts $l_{\text{Rayleigh}} = 1.22\ \lambda/(\text{NA}_{\text{objective}} + \text{NA}_{\text{condenser}}) = 238$ nm. Thus with the annular condenser we observed the features about twice less than l_{Rayleigh}. The result can be explained by the standard diffraction theory of microscope [17, 18]. The theoretical analysis showed up to two times advantage in two-point resolution of the annular compared to the circular illumination system with the same numerical aperture [19, 20].

3.2 *Primo Node, Vessels, and Capillaries Structure*

The results presented in this work represent typical samples obtained from a total of ten rats, 936 slides, and 2,925 sections. According to Bong-Han Kim's classification, the primo vascular system includes intravascular (inside the blood and lymphatic vessels), intra-extravascular (along the surface of the internal organs), and extravascular (alongside the outer surface of the walls of blood and lymphatic vessels and under the layers of skin) primo nodes and vessels. The extravascular primo structures are the most likely to correspond to the traditional acupuncture meridian lines. Additionally, the neural primo vessels in the central and peripheral nervous systems and the intestinal primo vessels inside organs were also identified [4].

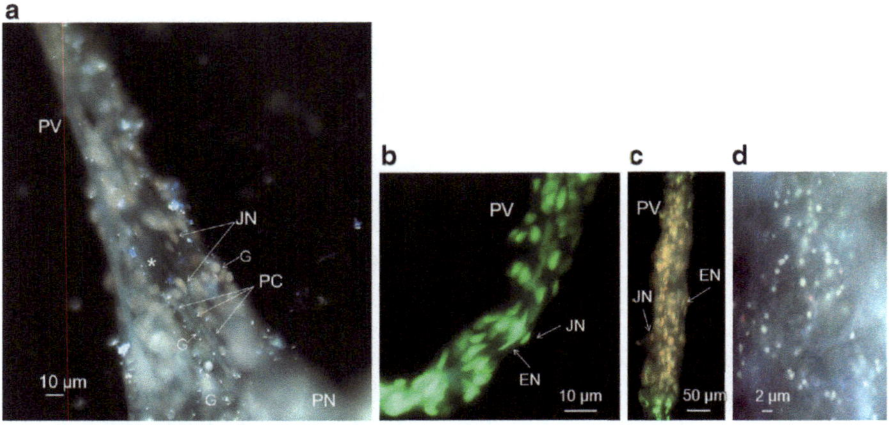

Fig. 12.2 Non-fixed primo-vascular samples. (**a**) A fragment of the primo vessel (PV) coming out of the primo node (PN). The vessel is formed by several primo-capillaries (PC). An external jacket of the vessel is composed of endothelial cells with spindle-like or oval nuclei (JN). Submicron and micron-size granules (G) escaped from the damaged capillaries are scattered along the vessel. The damaged vessel area is labeled with the *asterisk*. (**b**, **c**) Fragments of a primo vessel (PV) harvested a few centimeters of the primo node. The spindle-like or oval nuclei of the external vessel jacket (JN) and the rod-shape of the capillary envelope nuclei (EN) are clearly seen. (**d**) Micron and submicron size basophilic granules are observed inside a primo node

The primo nodes and vessels collected in this work are harvested from the surfaces of small intestine, kidney, and liver. The primo nodes are oval-shape of 0.1–0.5 mm along the short and 0.5–1 mm along the long axis. On both ends, the nodes are connected to vessels of 3–6 cm in length and 40–100 μm in diameter. When undisturbed, the nodes and vessels are translucent with coloration that blends into the background. When carefully touched with forceps, they become a more noticeable milky-yellowish color. The node and vessel can be lifted from the organ surface without a noticeable resistance. The vessels that come out of primo node are connected to another node or can go inside organs. All primo nodes are connected to primo vessels and all primo vessels are linked with primo nodes. If the primo vessel is cut, it immediately shrinks and loses the internal liquid.

The primo vessel is composed of 1–20 primo-capillaries (ductules) of 3–25 μm in diameter (Fig. 12.2a). An external envelope of primo-capillary is composed of endothelial cells and thin fibers. The envelope endothelial cells have very specific elongated rod-shaped nuclei that are clearly visible in non-fixed samples (Fig. 12.2b, c) and they are strongly stained by H&E in fixed samples (Fig. 12.3). The envelope nuclei of 15–20 μm are directed along the capillary axis and look similar to the nuclei of smooth muscle. The bundle of primo-capillaries of the primo vessel is surrounded by external jacket that is composed of cells with spindle-like or oval nuclei of 6–12 μm length. These nuclei are intensely stained with acridine orange in non-fixed primo vessels (Fig. 12.2a–c). Describing the external envelope of primo-capillary, Bong-Han Kim noticed that cells could be stretched and contracted. Additionally, he found that some of the primo-capillaries had an appearance of

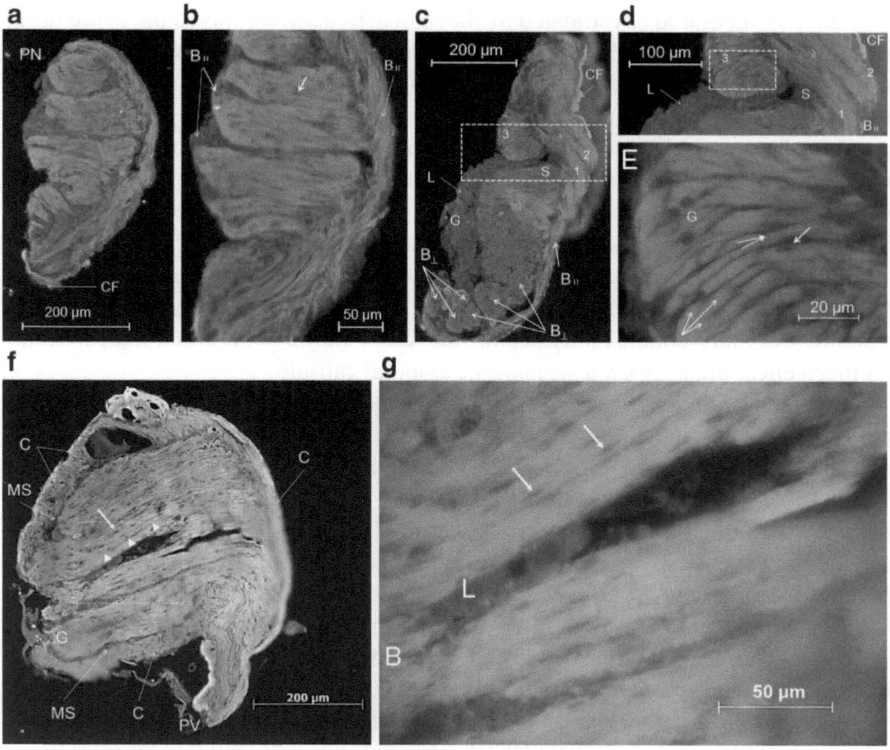

Fig. 12.3 Longitudinal sections of primo node. Primo node from the surface of small intestine was cut into 25 slides and 75 sections of 6 μm thick. (**a**) The section represents the top slide of this series showing that primo node (PN) is a complex interlacement of capillaries. *CF* capsule fragment. (**b**) Another section from the top primo node slide. The internal laying of capillaries conserves the bundle-like structure. *B||* a bundle of primo-capillaries in the interior of primo node. A *thick arrow* shows a typical rod-shape nucleus of the primo-capillary envelope cells. (**c**) The section cut at ~100 μm from the top surface of the ~500-μm node. *CF* capsule fragment; *B||* longitudinal bundle of primo-capillaries; *B⊥* cross sections of transversal capillary bundles. A primo-liquid stream (L) that came out open sinus (S) carrying small granules (G) covers the left side of the section. In *rectangle*: bundles (1) and (2) merge to create bundle (3). (**d**) Magnified view of the *rectangle* in Fig. 12.3c. *B||* the longitudinally arranged capillary bundle; *S* sinus; *L* primo liquid; *CF* capsule fragment. (**e**) Magnified view of the *rectangle* in Fig. 12.3d. Broadened and branched capillaries. Capillaries are converging or diverging (labeled by converging or diverging *arrows*, respectively). *G* basophilic granule in the lumens of cut capillaries. A *thick arrow* shows a typical rod-shape nucleus of the primo-capillary envelope cells. (**f**) The bottom section at ~450 μm from the top surface of the node. *Arrowheads* demarcate interface between two capillaries. A *thick arrow* shows a rod-shape nucleus of the primo-capillary envelope cells. *C* capsule; *MS* marginal sinus; *PV* primo vessel; *G* granules. (**g**) A magnified view of the *dashed rectangle* in Fig. 12.3f. *Thick arrows* show nuclei of a primo-capillary envelope cells. The broken capillary (B) shrank and spilled liquid (L) in the freed space

cross-striation that suggests their role in propelling the intro-capillary liquid by peristalsis. The observations of longitudinal, transversal, and wave mechanical motions of isolated primo vessels had revealed a quite large velocity of mechanical displacement of 100–600 μm/s [4]. The liquid flow velocity in primo vessels that was directly measured or estimated by the radioactive tracers was found in the range of 100–800 μm/s [5, 6, 21]. This velocity is impressively high compared to one observed in the lymphatic vessels that averages at 10 μm/s in small and does not exceed 100 μm/s in large vessels [22].

The primo node is essentially the junction of broadened and branched primo-capillaries. Figure 12.3 shows cross sections of primo node collected from the surface of the small intestine. Figure. 12.3a represents a top slide of this series that is closed to outmost top node surface. During preparation, the node sheds a large portion of its outside capsule exposing a complex interior. The node is heterogeneous in structure, consisting of tightly spun capillary bundles that fill practically the entire node volume. Capillary bundles interlace in such close-fitting pattern that node sustains its mechanical integrity even without the external capsule.

Another section from the top slide of the primo node (Fig. 12.3b) also shows that internal pattern of capillaries conserves the bundle-like structure. The round bundle of primo-capillaries (B//) on the right side of the section goes along entire node body. The capillary diameter in this bundle is ~3 μm. The bundles (B//) economically fill the node middle space by serpentine-like structures. The serpentine goes around the node long axis in the spiral fashion and produce ~6 coils (or folds). The bundles are in plane with the cross section. The middle bundles (B//) are ~50 μm width composed of ~7 μm capillaries. Although capillaries are laid in bundles, individual capillaries are clearly visible as parallel tubules with distinct rod-shape envelope nuclei directed along the capillary axis.

Figure 12.3c shows the node section that is further down from the top surface. The external capsule of this section is also stripped except the small capsule fragment (CF) in the upper right side of the node. The section shows the single longitudinal bundle of primo-capillaries (B//) on the right side of the section that goes up along the node body and gradually increases in size. In the upper part of the section (dashed rectangle), this capillary bundle (1) unites with another bundle (2) coming from the sample lower plane. The bundle (3) looks like a merger of (1) and (2) and has significantly wider capillaries. The middle and lower part of this sample shows the transversal cross sections of capillary bundles (B⊥), which are similar in appearance to patches of straw stubble. These patches resulted from cutting capillary strands that are perpendicular to the section plane. Damaged capillaries loose liquid (L) that become very visible when it comes out of large sinuses (S).

Capillaries inside the node that are far from vessels (Fig. 12.3c, d) are broadened and branched. The capillaries are converging or diverging (labeled by converging or diverging arrows, respectively). Basophilic granules (G) are seen at the lumens of cut capillaries. The average diameter of capillaries is ~7 μm that is nearly twice as large as those of vessel capillaries.

Figure 12.3f shows the section cut near the bottom at ~450 μm from the node top surface. This section of the node is surrounded by the capsule (C) of connective tissue composed of the endothelial cells and fibers. The cells look similar to those of the primo vessel external walls with the spindle-like or oval nuclei shown in Fig. 12.2. The capsule wall thickness varies between 5 and 40 μm. The space between envelop and node (marginal sinus, MS) is filled with a thin layer of liquid. This section shows

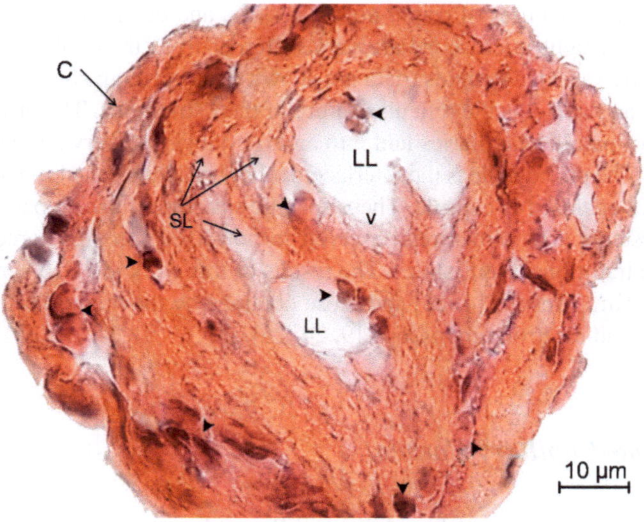

Fig. 12.4 Transversal section of a small primo node. Brightfield mode. The node was collected from the surface of small intestine. *LL* large lumens; *SL* small lumens of primo-capillaries; *v* vacuoles; *C* capsule. Basophilic granules are labeled with *arrowhead*

connection of the primo vessel (PV) with the primo node. The primo-capillaries are gradually broadened as they enter the primo node. Some of the primo-capillaries do not branch and maintain the same diameter along the node length and become narrow again at the other end of the node. Many capillaries are arranged in parallel rows. Primo-capillaries inside the node are filled with liquid carrying basophilic granules (G) spilled from broken capillaries. A magnified view of the dashed rectangle in Fig. 12.3f is shown in Fig. 12.3g. The parallel lines of capillaries with envelope rod-shape nuclei (thick arrows) show a strong resemblance of smooth muscle surface. The broken capillary (B) shrunk and spilled liquid (L) in the free space.

Figure 12.3 shows a set of primo node cross sections from the top to bottom surfaces with the internal arrangement of primo-capillaries. Combining these entire images one can imagine the primo node three-dimensional structure. The primo node is composed of broadened and branched primo-capillaries enclosed in capsule. A single capillary bundle of the incoming (afferent) vessel enters the node, branches into additional bundles, which in turn branch and fill the node interior by tightly spun and folded bundles. The internal pattern of capillaries conserves the bundle-like structure. Capillaries converge, narrow, and come out from the node as a single bundle of the efferent primo vessel.

Additional structural information can be obtained from the transversal section of a small primo node (Fig. 12.4). The section shows two large (LL) and a few small (SL) lumens of primo-capillaries that are perpendicular to the section plane. Lumens have smooth regular uninterrupted surface with visible vacuoles (v). Basophilic granules and cell-like structures of various sizes and texture are clearly visible in section image.

The structural features of primo nodes, vessels, and capillaries observed with a high resolution light microscope are very different from those observed in blood and lymphatic vascular samples (data not shown). The striking characteristic of the primo node and vessel is the reticular nature of their structures. The primo vessel is the bundle of parallel capillaries laid into an external jacket, very similar to a fiber-optic cable. The primo node is the interlacement of broadened and branched bundles of capillaries enclosed in capsule. These specific structures have no parallel in other vascular systems. The structures observed in non-fixed samples agree well with those of the traditional H&E slides. The results of this work are also generally consistent with Kim's results [4, 23] and those obtained by the method of immunohistochemistry and electron microscopy [9, 24, 25].

3.3 Primo-Capillary Content

The primo-capillaries carry a liquid that is different from the blood serum and lymphatic fluid.

The liquid was found to be richly supplied with basophilic granules. They have been observed as individual granules, and also have been seen as groups of two, three, and granular clusters. They are observed in the non-fixed samples stained with acridine orange (Fig. 12.2d) and the fixed H&E primo node slides (Figs. 12.3c, e, f and 12.4). Bong-Han Kim described similar granules (sanals) and found they contain large concentration of nucleic acids. He also found the abundance of proteins and free amino acids. Sugars, lipids, hyaluronic acid, hormones, mononucleotides, and many minerals were additionally discovered in the liquid taken from primo-capillaries. It was found that nucleic acid granules play a crucial role in physiological functions of primo system, especially in the regeneration of dying cells [4, 23]. More recently, the structure of the granules and liquid was thoroughly characterized by a variety of physical and biochemical methods [26–29]. Proteomic analysis of primo vessel and primo liquid (L) has revealed 70 different proteins in the liquid and 270 proteins in the vessels. Many of those proteins were not normally found in blood, lymph, or blood vessels [30]. These results also suggest the presence of stem cells in primo vascular system [8].

4 Conclusions

Structural properties of primo nodes, vessels, and capillaries were characterized by a high resolution light microscopy. Due to high spatial resolution and contrast, the method allowed seeing in detail small structural features of the chosen components of primo vascular system. Galen described a blood circulatory system in the second century AD. Fifteen hundred years passed before the lymphatic system was discovered. We presume that knowledge of the third, primo vascular system, that may represent acupuncture meridians, offers advances in medical science comparable to the progress generated by the lymphatic system's discovery. We should not wait hundreds of years or more for this to happen. For many, the acupuncture meridian system is nothing more than a network of lines drawn on a body map and labeled with hieroglyphs. We all need to make serious efforts to test and verify the hypothesis that there are real anatomical structures underpinning these dots and lines. We need to overcome the common prejudice and preconception against acupuncture. If this hypothesis is not overturned, our efforts will create a new paradigm in biology and medicine: discover an unlimited source of individualized stem cells and provide a natural way of a stem cell delivery; suggest new and efficient methods of organ regeneration and longevity; and bring new diagnostic and therapeutic methods.

Acknowledgments This work was supported by John R. Fetzer Memorial Trust, Inc. The author is grateful to John C. Dennis, Ludmila Globa, Oleg Pustovyy, Christopher Moore, Iryna Sorokulova, and Edward Morrison for help and encouragement, and Kwang-Sup Soh, Byung-Cheon Lee, and Min Su Kim for assistance and advice.

References

1. Hoizey D, Hoizey M (1988) A history of Chinese medicine. Edinburgh University Press, Edinburgh
2. NIH (1998) NIH consensus conference. Acupuncture. JAMA 280:1518–1524
3. Kim BH (1964) On the Kyungrak System. Foreign Language Publishing House, Pyongyang, DPRK (available at the Duke University Libraries)
4. Kim BH (1965) Kyungrak system and Sanal theory. Medical Literature, Pyongyang, DPRK (Russian)
5. Darras J-C, Albaredo P, de Vernejoul P (1993) Nuclear medicine investigation of transmission of acupuncture information. Acupunct Med 11:22–28
6. Zhang WB, Tian YY, Li H, Tian JH, Luo MF, Xu FL, Wang GJ, Huang T, Xu YH, Wang RH (2008) A discovery of low hydraulic resistance channel along meridians. J Acupunct Meridian Stud 1:20–28
7. Richard F, Hobbs MD III (2009) The biophysics of acupuncture: emerging patterns. From selected studies. Med Acupunct 21(4):251–256
8. Lee B-C, Bae K-H, Jhon G-J, Soh K-S (2009) Bonghan system as mesenchymal stem cell niches and pathways of macrophages in adipose tissues. J Acupunct Meridian Stud 2:79–82
9. Soh K-S (2009) Bonghan circulatory system as an extension of acupuncture meridians. J Acupunct Meridian Stud 2:93–106
10. Yoo JS, Kim HB, Ogay V, Lee B-C, Ahn S, Soh K-S (2009) Bonghan ducts as possible pathways for cancer metastasis. J Acupunct Meridian Stud 2:118–123
11. Chan SHH (1984) What is being stimulated in acupuncture – evaluation of the existence of specific substrate. Neurosci Biobehav Rev 8:25–33
12. Zhao Z-Q (2008) Neural mechanism underlying acupuncture analgesia. Prog Neurobiol 85:355–375
13. Vainrub A, Pustovyy O, Vodyanoy V (2006) Resolution of 90 nm (lambda/5) in an optical transmission microscope with an annular condenser. Opt Lett 31:2855–2857
14. Richardson TM (1988) Test slides: diatoms to divisions – what are you looking at? Proc Roy Microsc Soc 22:3–9

15. Shin H-S, Johng H-M, Lee B-C, Cho S-I, Soh K-S, Baik K-Y, Yoo J-S, Soh K-S (2005) Feulgen reaction study of novel threadlike structures (Bonghan ducts) on the surfaces of mammalian organs. Anat Rec B New Anat 284:35–40
16. Vodyanoy V (2009) Demonstration of Bonghan corpuscles and ducts in rabbits and rats by Korean scientists. J Acupunct Meridian Stud 2:169
17. Born M, Wolf E (1999) Principles of optics, 7th edn. Cambridge University Press, Cambridge
18. Martin LC (1966) The theory of the microscope. Elsevier, New York
19. Vodyanoy V, Pustovyy O, Vainrub A (2007) High resolution light microscopy of nanoforms. vol 6694. Proceedings of SPIE – The International Society for Optical Engineering, pp 1–12
20. Vodyanoy, V, Vainrub, A, Pustovyy, O (2009) High resolution optical microscope with cardioid condenser for brightfield and darkfield illumination. US Patent 7,542,203
21. Sung B, Kim MS, Lee BC, Yoo JS, Lee SH, Kim YJ, Kim KW, Soh KS (2008) Measurement of flow speed in the channels of novel threadlike structures on the surfaces of mammalian organs. Naturwissenschaften 95:117–124
22. Shayan R, Achen MG, Stacker SA (2006) Lymphatic vessels in cancer metastasis: bridging the gaps. Carcinogenesis 27:1729–1738
23. Kim BH (1965) The Sanal theory. J Acad Med Sci (DPR Korea) 108:39–62
24. Lee BC, Yoo JS, Ogay V, Kim KW, Dobberstein H, Soh KS, Chang BS (2007) Electron microscopic study of novel threadlike structures on the surfaces of mammalian organs. Microsc Res Tech 70:34–43
25. Ogay V, Bae KH, Kim KW, Soh K-S (2009) Comparison of the characteristic features of Bonghan ducts, blood and lymphatic capillaries. J Acupunct Meridian Stud 2:107–117
26. Baik KY, Ogay V, Jeoung SC, Soh K-S (2009) Visualization of Bonghan microcells by electron and atomic force microscopy. J Acupunct Meridian Stud 2:124–129
27. Baik KY, Sung B-K, Lee B-C, Johng H-M, Ogay V, Nam TJ, Shin H-S, Soh K-S (2004) Bonghan ducts and corpuscles with DNA-contained granules on the internal organs-surfaces of rabbits. J Int Soc Life Inform Sci 22:598–601
28. Ogay V, Baik Ku Y, Lee B-C, Soh K-S (2006) Characterization of DNA-containing granules flowing through the meridian-like system on the internal organs of rabbits. Acupunct Electrother Res 31:13–31
29. Sung B, Kim MS, Corrigan A, Donald AM, Soh KS (2009) In situ microextraction method to determine the viscosity of biofluid in threadlike structures on the surfaces of mammalian organs. Phys Rev E Stat Nonlin Soft Matter Phys 79(2 Pt 1):022901
30. Lee SJ, Lee B-C, Nam CH, Lee W-C, Jhang S-U, Park HS, Soh K-S (2008) Proteomic analysis for tissues and liquid from Bonghan ducts on rabbit intestinal surfaces. J Acupunct Meridian Stud 1:97–109

Chapter 13
Distribution of Primo Vessels in the Mesentery of a Mouse

Zhendong Su, Ping An, Jeong-No Lee, and Kwang-Sup Soh

Abstract We found that a combination of the phalloidin (specific to F-actin) and the DAPI staining methods was a reliable method to distinguish primo vessels from vascular and lymphatic vessels. Also, we traced primo vessels in the mesentery of a mouse with a combination of the phalloidin and the DAPI staining methods and a whole mounting technique. We showed that primo vessels with a threadlike structure were distributed like a network in the mesentery between the colon and the root of the mesentery and were extended into the colon walls or fat tissues

1 Introduction

The primo vessel is a new circulatory system distributed throughout an animal's body and has been assumed to be the anatomical structure of the acupuncture meridians in traditional Oriental medicine. This was first proposed by Bong-Han Kim of North Korea in the early 1960s [1], and recently extensive reinvestigations on his work with various modern techniques have been done [2].

Until now, the primo vessel has been observed in blood vessels of various animals [3–5], inside bovine hearts [6] and lymph vessels of rabbits [7] and rats [8], in brain ventricles and the central canal of the spinal cord of rabbits and rats [9], and on the surfaces of various internal organs of rabbits, rats, and mice [10–12]. In the case of cancer-bearing mice, the primo vessel was seen as an additional cancer metastasis path [13–15]. These works have been done by using various methods like modern bio-imaging techniques [16], in vivo staining with Trypan blue [3], and histological and immunofluorescent methods [14], etc.

J.-N. Lee (✉)
Department of Physics and Astronomy, Seoul National University, Seoul, Korea
e-mail: leejeongno2002@yahoo.co.kr

K.-S. Soh et al. (eds.), *The Primo Vascular System: Its Role in Cancer and Regeneration*,
DOI 10.1007/978-1-4614-0601-3_13, © Springer Science+Business Media, LLC 2012

In this work, we used a combined staining method with phalloidin and DAPI and a whole mounting technique to trace primo vessels in the mesentery of a mouse. As known, phalloidin is specific to F-actin of cells while Trypan blue just stains dying or dead cells. This means the double staining with phallodin and DAPI is more evidential because it can show more apparent direction and arrangement of cells. We used this method to investigate the distribution of primo vessels in the mesentery of a mouse.

2 Methods

The mice (Hsd:ICR (CD-1), both sexes, 6–8 weeks, 25–28 g) were obtained from Jung-Ang Laboratory Animal Company (Seoul, Korea). The animals were housed in a constant-temperature, controlled environment (24°C) with 50% relative humidity under a 12-h day and night cycle and could have ad libitum access to palatable, uncontaminated, and nutritionally adequate food and water. All the care and studies of the animals abided by international laws and policies according to the Guide for the Care and Use of Laboratory Animals, National Academy Press [17].

Mice were put under general anesthetic with Zoletil 50 (Virbac, France), i.m. Then, we cut off the skin along the medial alba of the abdomen and opened it carefully. At first, we used forceps to push the small intestine out of the abdominal cavity to expose the colon while dripping 39°C phosphate buffered saline (PBS) to keep the tissues fresh and alive. Next, we separated carefully part of the colon, the root of the mesentery, and the connective tissue between them under a stereo microscope (STZ10, Olympus, Japan). It was very important to keep the connective tissue complete. The experiments were repeated with five mice. The fresh samples were washed three times with PBS, and we placed them in 4% PFA/PBS (ph = 7.2) at 4°C for 1 h. After the fixation, we washed them three times with PBS. We put the sample on a plastic board, spread it carefully, and fasten it to the board with some needles. A group of photos of the samples was taken under a stereo fluorescence microscope (MVX10, Olympus, Japan). Then, we added phalloidin (Alexa 488, Invitrogen, USA) to immerse the sample completely and kept it in a wet box at 4°C. Every 12 h, we checked the sample and continued to add more phalloidin to keep it wet. After 48 h, the sample was washed three times with PBS. Finally, we observed it under a stereo fluorescence microscope (MVX10, Olympus, Japan). When we found a fluorescent signal, we could take out the parts with the signals from the original sample, put them on slides, and stain them with DAPI (Invitrogen, USA) for 30 min. After DAPI staining, we used PBS to wash them several times and mounted them through a coverslip with antifade reagent (Invitrogen, USA). When the slides became dry, we checked the slides under a fluorescence phase contrast microscope (BX51, Olympus, Japan).

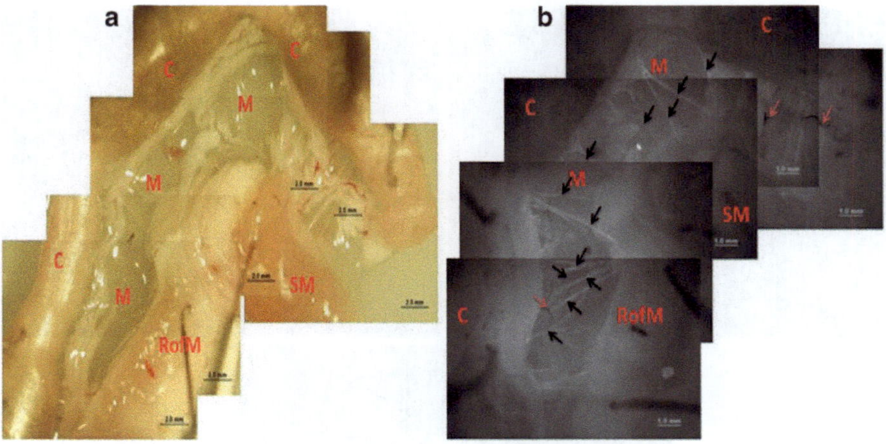

Fig. 13.1 Fluorescence threadlike structures in the mesentery between the colon and the root of the mesentery of a white mouse. (*C* colon; *M* mesentery; *SM* small intestine; *RofM* root of mesentery) (**a**) Parts of the mesentery that were observed are shown. (**b**) Phalloidin staining shows many fluorescence threadlike structures (*black arrows*), and some vascular vessels (*red arrows*) appear in the mesentery

3 Results

We ascertained that primo-vessels existed in the mesentery of a mouse. In Fig. 13.1a, b, we can observe that there are several fluorescent threadlike structures in the mesentery between the colon and the root of the mesentery and that the primo vessels are extended into the colon walls or fat tissues. Phalloidin was very sensitive to primo vessels, and it clearly showed the cell shape and arrangement of the primo vessels.

Primo vessels, vascular vessels, and lymphatic vessels were stained with phalloidin and constituted separate networks in the mesentery. The fluorescence threadlike structures could be primo vessels, vascular vessels, and lymphatic vessels, but after phalloidin and DAPI staining together, the three kinds of vessels show their own typical characters. Primo vessels have typical rod-shaped nuclei with a parallel linear arrangement. Vascular vessels have a characteristic transverse arrangement surrounded by smooth muscular cells. The remaining vessels must be lymphatic vessels. In most case, they could be observed separately in different slides but sometimes we were able to find both or all of them together in the same ones (Fig. 13.2).

4 Discussion

Because the primo vessels have prominent histological features such as rod-shaped nuclei with linear alignments, we could find primo vessels in the mesentery of a mouse more easily through the combination of phalloidin (specific to F-actin) and

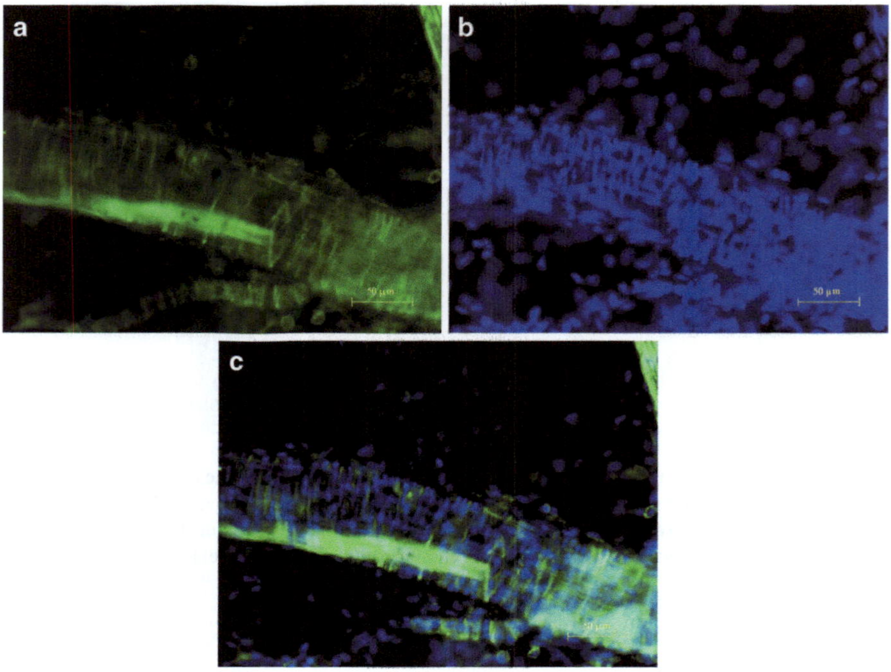

Fig. 13.2 Vascular vessels stained by phalloidin and DAPI in the mesentery between the colon and the root of mesentery. (**a**) Phalloidin is specific to F-actin in smooth muscular cells of vascular vessels. (**b**) DAPI shows two kinds of vascular cells: the longitudinal one is endothelial cells and the transverse one is smooth muscular nuclei. (**c**) Image merged by image (**a**) and image (**b**)

DAPI staining and a whole mounting method. In this work, even though we showed the distributions of the three kinds of vessels in the mesentery between the colon and the root of the mesentery of a mouse, we can use a similar method to find the distributions of the vessels in rabbits, rats, etc. The distributions and appearances of the three kinds of vessels will provide us with new support for the idea that primo vessels belong to a new system different from vascular and lymphatic vessels and will provide a basis for further study.

Acknowledgments This work is supported in part by the Association of Korean Oriental Medicine

References

1. Kim BH (1963) On the Kyungrak system. J Acad Med Sci (DPR Korea) 90:1–41
2. Soh KS (2009) Bonghan circulatory system as an extension of acupuncture meridians. J Acupunct Meridian Stud 2(2):93–106
3. Jiang X, Kim HK, Shin HS et al (2002) Method for observing intravascular Bonghan ducts. Korean J Orient Prevent Med 6:162–166

4. Baik KY, Lee J, Lee BC et al (2005) Acupuncture meridian and the intravascular Bonghan duct. Key Eng Mater 277:125–129
5. Yoo JS, Kim MS, Ogay V et al (2008) In vivo visualization of Bonghan ducts inside blood vessels of mice by using an Alcian blue staining method. Indian J Exp Biol 46:336–339
6. Sung B(2010) Bioimaging of primo-vascular system with fluorescent nanoparticles and cryo-electron microscopy of self-assembled spermine-DNA condensates. Ph.D. Dissertation, Seoul National University, Seoul
7. Lee BC, Yoo JS, Baik KY et al (2005) Novel threadlike structures (Bonghan ducts) inside lymphatic vessels of rabbits visualized with a Janus Green B staining method. Anat Rec B New Anat 286:1–7
8. Johng HM, Yoo JS, Yoon TJ et al (2007) Use of magnetic nanoparticles to visualize threadlike structures inside lymphatic vessels of rats. Evid Based Complement Alternat Med 4:77–82
9. Lee BC, Kim SK, Soh KS (2008) Novel anatomic structures in the brain and spinal cord of rabbit that may belong to the Bonghan system of potential acupuncture meridians. J Acupunct Meridian Stud 1:29–35
10. Shin HS, Johng H, Lee BC et al (2005) Feulgen reaction study of novel threadlike structures on the surfaces of rabbit livers. Anat Rec B New Anat 284:35–40
11. Lee BC, Yoo JS, Ogay V et al (2007) Electron microscopic study of novel threadlike structures on the surfaces of mammalian organs. Microsc Res Tech 70:34–43
12. Sung B, Kim MS, Lee BC et al (2008) Measurement of flow speed in the channels of novel threadlike structures on the surfaces of mammalian organs. Naturwissenschaften 95:117–124
13. Yoo JS, Won N, Kim HB et al (2010) In vivo imaging of cancer cells with electroporation of quantum dots and multispectral imaging. Appl Phys. doi:0.1063/1.3447858
14. Yoo JS, Hossein Ayati M, Kim HB et al (2010) Characterization of the primo-vascular system in the abdominal cavity of the lung cancer mouse model and its differences from the lymphatic system. PLoS One 5(4):e9940
15. Yoo JS, Kim HB, Won N et al (2010) Evidence for an additional metastatic route: in-vivo imaging of cancer cells in the primo-vascular system around tumors and organs. Mol Imaging Biol 13(3):471–480. doi:10.1007/s11307-010-0366-1
16. Kim BH (1963) On the Kyungrak system. J Acad Med Sci (DPR Korea) 90:1–41
17. National Academy Press (1996) Guide for the care and Use of Laboratory Animals

Chapter 14
Primo Vessels in the Mesentery of Nude Mice

Ping An, Zhendong Su, Hesheng Luo, and Kwang-Sup Soh

Abstract Our previous studies have observed primo vessels in the mesentery of rats between the root of the mesentery and the colon or between the small intestine and the cecum. Some research revealed the involvement of the primo vascular system in cancer metastasis in cancer-bearing nude mice. For our future studies on the role of the primo vessels in cancer, we designed the current experiment to confirm the presence of primo vessels in the mesentery of nude mice.

Methods With Trypan blue, phalloidin, and Hoechst staining, we estimated the morphologic characters of the primo vessels in the mesentery.

Results Our results manifested that these structures in nude mice not only had white, semitransparent, threadlike appearances, but had specific rod-shaped nuclei with linear alignments, were positive to phalloidin, and were sensitive to Trypan blue. All of these data were consistent with our previous studies.

Conclusion We conclude that primo vessels exist of the mesentery of many animals and suggest that these specific structures might have specific and important functions.

1 Introduction

Recently, using modern bioimaging techniques, our group manifested the primo vascular system (PVS), a new kind of circulatory system, in animals. We observed the primo vessels on the surfaces of internal organs [1–3], in blood vessels [4–6],

K.-S. Soh (✉)
Nano Primo Research Center, Advanced Institute of Convergence Technology,
Seoul National University, Suwon, South Korea
e-mail: kssoh1@gmail.com

K.-S. Soh et al. (eds.), *The Primo Vascular System: Its Role in Cancer and Regeneration*,
DOI 10.1007/978-1-4614-0601-3_14, © Springer Science+Business Media, LLC 2012

in the brain ventricles and the central canal of the spinal cord [7], and on the surfaces of tumors [2, 8, 9]. In addition, we successfully observed the primo vessels in the mesentery between the cecum and the small intestine and between the colon and the root of the mesentery [10]. These primo vessels had white color, semitransparent threadlike appearances, strongly positive to phalloidin, which is specific to F-actin, and had classic rod-shaped nuclei with linear alignments. Further investigation demonstrated that these structures were significantly different from vascular vessels and lymphatic vessels.

In vivo studies, we also explored the possibilities of the PVS acting as an additional metastasis pathway in cancer-bearing nude mice [2, 8, 9]. Some of our recent studies (not published) indicated that the primo vessels in the mesentery might be involved in gastric cancer metastasis. If the role of these primo vessels in cancer is to be confirmed, it is critical to verify the existence of the primo vessels in nude mice. Therefore, we designed the current studies and manifested the presence of the primo vessels in the mesentery of nude mice. These primo vessels had classic morphological characters that were consistent with our previous studies. These current findings supplied a solid foundation for further investigation of the role of PVS in cancer.

2 Materials and Methods

2.1 Animals

Nude mice (both sexes, 7 weeks) were obtained from Jung-Ang Laboratory Animal Company (Seoul, Korea). The animals were housed in a constant-temperature controlled environment (23°C) with 60% relative humidity under a 12-h light/dark cycle and had ad-libitum access to pelleted chow and tap water. The animal care and studies abided by international laws and policies according to the Care and Use of Laboratory Animals, National Academy Press (1996).

2.2 Reagents

Zoletil was obtained from Virbac (Virbac, France). Alexa Fluor 488 phalloidin, Hoechst 33342 and prolong gold antifade reagent were purchased from Invitrogen (Invitrogen, USA). Trypan blue (0.2%) and other reagents of highest purity were purchased from Sigma (Sigma, USA).

2.3 Surgical Operations and Trypan Blue Staining

Nude mice were anesthetized with Zoletil 50, i.m. Under a stereomicroscope, after opening the abdomen, we carefully searched for primo vessels by looking for white, semitransparent threadlike structures in the mesentery between the root of the mesentery and the colon or between the small intestine and the colon. We kept the internal organ surfaces humid by frequently dripping phosphate buffered saline (PBS) on to them. Then, we sprayed 0.2% Trypan blue on the primo vessels in the mesentery for 30 s and washed by PBS for 3 times. During the whole procedure, both PBS and Trypan blue were preincubated to 37°C and the experiments were repeated at least 5 times.

2.4 *Tissue Preparation*

The primo vessels were carefully taken out and fixed in 3.7% formaldehyde at room temperature (RT) for 10 min. After being washed with PBS 3 times, the samples were further treated with 0.2% Triton-X100 for 5 min. Then, the tissues were washed in PBS and placed on the slides for future examination.

2.5 *Phalloidin and Hoechst Staining*

After fixation and permeabilization as earlier, the samples were stained with Alexa Fluor 488 phalloidin (specific to F-actin) for 30 min at RT. Then, the tissues were washed in PBS 3 times and stained with Hoechst (specific combines to DNA) at 1 μg/mL for 10 min at RT. After PBS washing, we carefully flattened the samples on saline-covered slides under a stereomicroscope and mounted them with a prolong gold antifade reagent. After sealing, the specimens were analyzed with a fluorescence phase-contrast microscope.

3 Results

3.1 *The Primo Vessels in the Mesentery of Nude Mice Were Stained by Trypan Blue*

Our previous studies have demonstrated that primo vessels with a white color and asemitransparent threadlike appearance existed in the mesentery of rats. In this experiment, we further demonstrated that in nude mice, the primo vessels also existed in the mesentery between the root of the mesentery and the colon or between the small intestine and the colon. The primo vessels in nude mice had similar appearances similar to those of primo vessels in rats (Fig. 14.1). In addition, classic Trypan blue-staining studies revealed that these special structures were also stained blue, which were fairly consistent with our previous results in rats.

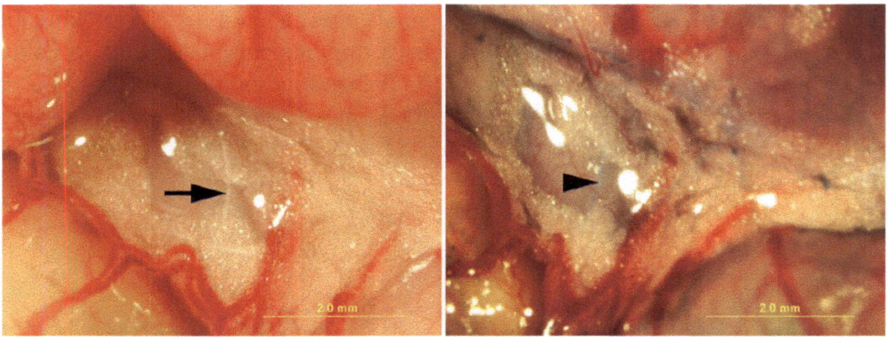

Fig. 14.1 Primo vessels in the mesentery between the root of the mesentery and the colon of nude mice. A white, semitransparent, threadlike structure was observed in the mesentery (*black arrow*). After spraying with Trypan blue and washing with PBS, this structure was stained to blue (*black arrowhead*)

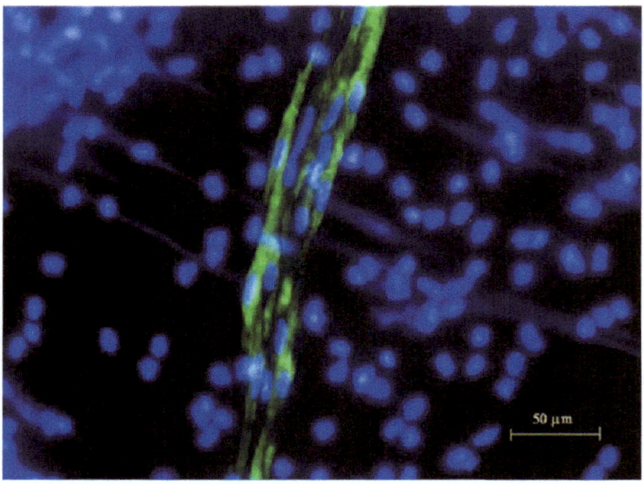

Fig. 14.2 Phalloidin and Hoechst images of the primo vessels in the mesentery. In the primo vessels, the nuclei stained by Hoechst (*blue*) showed a rod shape and were in a linear alignment. Palloidin staining (*green*) obviously improved the cells' shapes and the classic arrangement of the primo vessels

3.2 The Features of the Primo Vessels in Phalloidin and Hoechst Images

To further manifest that the structures we observed in nude mice belonged to the PVS, we investigated their morphologic characters by using phalloidin and Hoechst staining. As we shown in Fig. 14.2, in Hoechst images (blue fluorescence), the cells of the primo vessels in the mesentery had classic rod-shaped nuclei and linear alignments. Phalloidin staining (red fluorescence) allowed us to distinguish the arrangements of the primo vessels, which were obviously different from those in vascular vessels and lymphatic vessels. Taken together, our current data indicate that in nude mice, primo vessels exist in the mesentery.

4 Discussion

Our previous studies revealed primo vessels in the mesentery of rats [10]. They had classic rod-shaped nuclei with linear arrangements, a sinus, outermost membranes, and large amounts of collagen fibers, and were relatively sensitive to Trypan blue staining. Furthermore, we also found that a combination of phalloidin staining improved the results to greatly show the structure of the primo vessels. In the phalloidin images, the cell shapes and their arrangements are much more obvious. According to these results, we examined these structures in nude mice.

In the same locations where the primo vessels in the mesentery were detected in rats, we found similar structures in nude mice. In the mesentery between the root of the mesentery and the colon or between the small intestine and the colon, the primo vessels had white, semitransparent, threadlike appearances, and were sensitive to Trypan blue (Fig. 14.1). These results were in accordance with our previous studies.

As in cancer studies, we always employed green fluorescent protein (GFP) or red fluorescent protein (RFP) transfected tumor cells to built cancer models; we chose phalloidin with red fluorescence here. In addition, we employed Hoechst 33342 to stain nuclei for its higher ability to permeate into fresh tissues. Our results in Fig. 14.2 show that the primo vessels in nude mice had the same features as the primo vessels in rats. They had rod-shaped nuclei with linear arrangements and were positive to phalloidin.

In conclusion, our current study demonstrated that primo vessels also exist in the mesentery of nude mice. These observations enrich our concept that the primo vessels are specific structures exist in many animals, even in human beings. Here, we give some highlights of these widespread specific structures, which might be explored for specific and important functions in the future.

Acknowledgments K. Soh thanks the Association of Korean Oriental Medicine for support.

References

1. Yoo JS, Hossein Ayati M, Kim HB et al (2010) Characterization of the primo-vascular system in the abdominal cavity of the lung cancer mouse model and its differences from the lymphatic system. PLoS One 5(4):e9940
2. Yoo JS, Kim HB, Won N et al (2011) Evidence for an additional metastatic route: in-vivo Imaging of cancer cells in the primo-vascular system around tumors and organs. Mol Imaging Biol 13:471–480. doi:10.1007/s11307-010-0366-1

3. Lee BC, Bae KH, Jhon GJ et al (2009) Bonghan system as mesenchymal stem cell niches and pathways of macrophages in adipose tissues. J Acupunct Meridian Stud 2:79–82
4. Baik KY, Lee J, Lee BC et al (2005) Acupuncture meridian and the intravascular Bonghan duct. Key Eng Mater 277:125–129
5. Yoo JS, Kim MS, Ogay V et al (2008) In vivo visualization of Bonghan ducts inside blood vessels of mice by using an Alcian blue staining method. Indian J Exp Biol 46:336–339
6. Sung B (2010) Bioimaging of primo-vascular system with fluorescent nanoparticles and cryo-electron microscopy of self-assembled spermine-DNA condensates. Ph.D. dissertation, Seoul National University
7. Lee BC, Kim SK, Soh KS (2008) Novel anatomic structures in the brain and spinal cord of rabbit that may belong to the Bonghan system of potential acupuncture meridians. J Acupunct Meridian Stud 1:29–35
8. Yoo JS, Won N, Kim HB et al (2010) In vivo imaging of cancer cells with electroporation of quantum dots and multispectral imaging. J. Appl. Phys. 107, 124702, doi:10.1063/1.3447858
9. Yoo JS, Hossein Ayati M et al (2010) Characterization of the primo-vascular system in the abdominal cavity of thelung cancer mouse model and its differences from the lymphatic system. PLoS One 5(4):e9940
10. An P, Dai J, Zhendong Su et al (2010) Putative primo-vascular system in mesentery of rats. J Acupunct Meridian Stud 3(4):232–240

Chapter 15
Comparison of the Primo Vascular System with a Similar-Looking Structure

Cheon-Joo Choi and Chae-Hun Leem

Abstract The primo vascular system (PVS) is evaluated as one of the candidates for an anatomical structure of the acupuncture meridian system. There have been many reports on the visualization of the PVS on organ surfaces, inside blood and lymph vessels, and in the ventricle of the brain. However, in practice, it is not easy to find the PVS and it is difficult to identify the actual PVS. At this moment, four criteria are used to identify the PVS: the node and threadlike structure, the bundle structure, the rod-shaped nuclei, and trypan blue stainability. The aim of this study is to investigate whether the criteria are appropriate for identifying the PVS, what are the possible artifacts, and how to avoid them. Toward this end, this study searched for a threadlike structure under various conditions and investigated its characteristics. As a result, the major artifacts were coagulated strings and debris of the visceral peritoneum, caused by urethane damage. Both of these artifacts showed all characteristics of the PVS, as reported previously. In conclusion, in order to identify the PVS, stricter criteria for PVS discrimination are required so as not to confuse the PVS with the artifacts.

1 Introduction

In the 1960s, Bonghan Kim, a physiologist in North Korea, reported that he had discovered a new anatomical structure that looked like a thread connecting nodes. The node and the threadlike structure are called the Bonghan corpuscle and Bonghan duct, respectively [1].

Since 2002, this long forgotten theory has been studied by Professor Soh's research team in South Korea. Most of their work was done to confirm Kim's

C.-H. Leem (✉)
Department of Physiology, University of Ulsan College of Medicine, Seoul, South Korea
e-mail: leemch@gmail.com

K.-S. Soh et al. (eds.), *The Primo Vascular System: Its Role in Cancer and Regeneration*,
DOI 10.1007/978-1-4614-0601-3_15, © Springer Science+Business Media, LLC 2012

description. Because of extensive investigation, the existence of a threadlike structure was verified inside blood and lymph vessels, on the surface of organs, and in brain ventricles [2]. It was shown that the Bonghan system has rod-shaped nuclei distributed in a broken-line striped fashion [3]. The bundle-like structure of the Bonghan duct was observed on a phase contrast microscope [4]. This characteristic was used as evidence of the existence of ductules inside the Bonghan duct. It was reported that trypan blue can preferentially stain the Bonghan system [5]. Thus, this staining dye is used as the principal staining dye method for the Bonghan system.

In 2009, the Bonghan system, corpuscle, and duct were renamed the primo vascular system (PVS), primo node (PN), and primo vessel (PV), respectively, because they are known to be generated prior to vasculogenesis.

However, in practice it is very difficult to find the PVS, even on the surface of organs. Sometimes a threadlike structure can be found, but it is not known in what condition this is possible. To solve this problem, a threadlike structure was searched for on organ surfaces under various conditions.

We already knew that slight bleeding makes it easy to find a threadlike structure on the organ surface, while the structure was rarely found without bleeding. Therefore, we could not exclude the possibility of mistaking a coagulated string for the PVS. The previous report argued that the PVS was not a coagulated string because urokinase did not dissolve them [6]; however, urokinase is the only converter from plasminogen to plasmin that can dissolve fibrin. As a result, the possibility of the coagulation artifact was not excluded with confidence. The PVS on a visceral organ surface was frequently found in phenylhydrazine (PHZ)-induced anemic rats [2] and it is reported that PHZ can cause hypercoagulability [7]. Furthermore, urethane IP injection was used as an anesthesia, even though it causes damage to the visceral tissue [8]. Therefore, there was still a possibility that the artifacts were mistakenly regarded as the PVS. To confirm this, a threadlike structure was searched for when heparin was administered intravenously. Moreover, the threadlike structure was searched for in cases of PHZ-induced anemia, bleeding-induced anemia, and urethane anesthesia to investigate whether it is found more frequently than in normal cases. If a threadlike structure was found then characteristics, such as rod-shaped nuclei, trypan blue stainability, and bundle-like structures, were investigated. The characteristics of artificial threadlike structures formed by blood plasma were also investigated.

Therefore, the present report aims to present the possible PVS artifacts and to determine whether the criteria for identifying the PVS are appropriate.

2 Materials and Methods

2.1 Animals

For this study, Sprague Dawley rats (male, 6–8 weeks) were obtained from DooYeol Biotech. The animals were housed in a constant temperature-controlled environment (23°C) with 60% relative humidity under a 12-h light/dark cycle. All rats had ad libitum access to food and water. The procedures involving the animals and their care were in accordance with international laws and policies (Guide for the Care and Use of Laboratory Animals, National Academy Press, 1996).

2.2 *Anesthesia*

Generally, the rats were anesthetized with Zoletil (50 mg/kg) plus xylazine (10 mg/kg), administered intramuscularly. However, five rats were anesthetized by intraperitoneal injection of urethane (1.5 g/kg) to investigate whether threadlike structures are more easily found under this condition. Urethane anesthesia was generally used in the PVS studies.

2.3 *Administration of Heparin*

Some of the rats were injected with heparin (1,000 IU) inserted into their femoral veins at 5-min before dissection, to inhibit blood coagulation. This group was compared with rats who did not receive heparin injections.

2.4 *Anemia Model*

It is known that the PVS is easily found in anemic rats because of its hematogenesis function. The anemia model was induced by IM injection of PHZ (5 mg/100 g) 3 times a week [9]. In some rats, anemia was induced by extracting blood (1 mL) from the femoral vein.

2.5 *Artificial Threadlike Structure by Blood Plasma Application*

Blood plasma was prepared by centrifugation of fresh blood. The blood (0.5 mL) was extracted from the femoral vein of the rats under deep anesthesia with a syringe (31 G) including EDTA solution. After cutting open the abdomen, blood plasma was applied onto the organ's surface and, then, threadlike structures could be observed within minutes. Before applying the blood plasma, the trypan blue staining method was used to confirm that no threadlike structure was present. Thus, we could be sure that the threadlike structures were fully formed from the blood plasma coagulation.

2.6 *Procedure for Finding the Threadlike Structure*

The medial alba of the rat's abdomen was cut under deep anesthesia. The incision line was opened with forceps to expose the internal organs as fully as possible.

Slight bleeding from the incision line was stopped by forcipressure so as not to flood the abdominal cavity. Sometimes the bleeding was allowed to permeate to the internal organs in order to investigate its effect on finding a threadlike structure.

A stereomicroscope (Olympus, SZX12) was used to search for a threadlike structure on the surface of the organs; the surface of the organs was kept constantly wet using warm saline. If a threadlike structure exists, it is mostly found within about 5 min. In this case, a 0.4% trypan blue solution was dropped onto the threadlike structure to determine whether it could be stained or not. If the threadlike structure was not found after more than 10 min, a warm trypan blue solution was applied to the entire abdominal cavity for 30 s and washed out with warm saline before the search was conducted once more. A "No threadlike structure" result means the threadlike structure could not be found even using this process.

2.7 Observation of Nuclei and a Bundle-Like Structure

Nuclei were stained using 4′,6-diamidino-2-phenylindole (DAPI). A bundle-like structure was observed using a phase contrast microscope. For these two purposes, an upright microscope (Olympus, BX51) was used.

3 Results

Through experience, it was known that a threadlike structure is well developed when slight bleeding occurs during an operation. Thus, it is reasonable to think that a threadlike structure can be formed by blood coagulation. To confirm this, heparin was administered to inhibit blood coagulation. In this condition, no threadlike structures were observed regardless of bleeding. Moreover, the artificial threadlike structure resulting from blood plasma coagulation also has specific characteristics of the PVS, such as a threadlike structure connecting nodes, a bundle-like structure, a rod-shaped nuclei, and good trypan blue stainability (Fig. 15.1a, b).

In the case of the PHZ-induced anemia rat model, a threadlike structure was found easily and it had four kinds of PVS characteristics. However, a threadlike structure was not found when heparin was administered. After cutting the abdomen open, the peritoneal fluid was taken by a glass capillary tube and dropped on a slide glass for observation with a phase contrast microscope. As a result, formation of fiber on the slide glass was observed, and the cells in the peritoneal fluid gathered around it. However, this phenomenon was not observed in the heparin-administered PHZ-induced anemic rats, and in normal rats.

Anemia was also induced by extracting blood from the femoral vein. Two days after the extraction, a threadlike structure was searched for, but was not found.

Another possible similar threadlike structure could be the debris of the visceral peritoneum by urethane. Threadlike structures barely observed in the rats anesthetized by zoletil IM injection, were observed extensively on the organ surface. The end of the thread spreads to the visceral peritoneum. Rubefaction of the organ

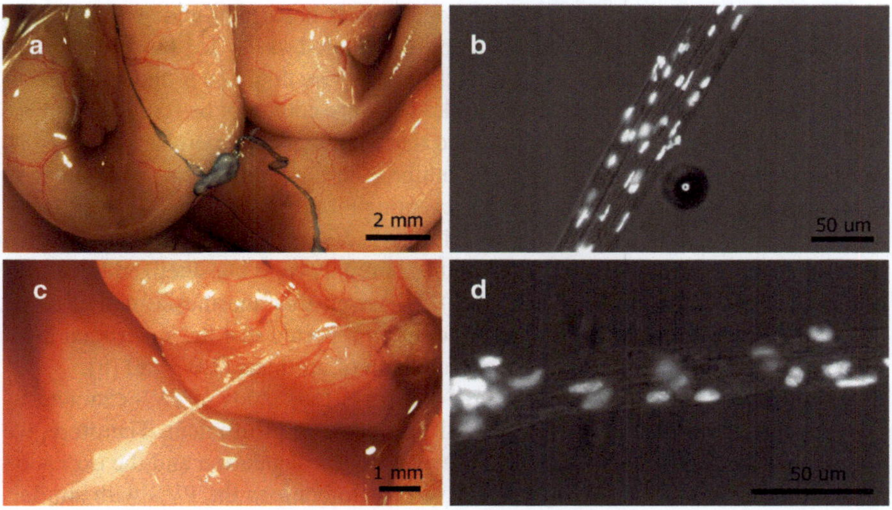

Fig. 15.1 Artificial coagulated string (**a**, **b**); debris of visceral peritoneum (**c**, **d**); node and thread-like structure (**a**, **c**); and rod-shaped nuclei and bundle-like structure (**b**, **d**)

Table 15.1 Observation of the threadlike structure in various conditions: the threadlike structure could not be found when heparin was administered. On the other hand, the structure was easily found in rats anesthetized by urethane IP injection. There was no exceptional case difference with this result (*N* is number of animals)

Anesthesia	Bleeding on operation	Heparin IV injection	Other conditions	Threadlike structure	*N*
Zoletil IM	X	X	None	Not found	>5
	O	X	None	Found	>5
	X	O	None	Not found	>5
	O	O	None	Not found	>5
	X	X	PHZ-induced anemia	Found	5
	X	O	PHZ-induced anemia	Not found	5
	X	X	Bleeding-induced anemia	Not found	2
	X	X	Blood plasma application	Found	4
Urethane IP	O or X	X	None	Found	5

surface was observed and the peritoneum was easily torn when it was pulled. In this case, the threadlike structure also had four kinds of PVS characteristics (Fig. 15.1c, d). These results are summarized in Table 15.1.

4 Discussion

The results, thus far, show that the structures similar to the PVS are coagulated string and debris of the visceral peritoneum caused by urethane damage. These artifacts also have four characteristics that are used for discriminating the PVS.

A threadlike structure could not be found in all heparin administration cases. This suggests the possibility of mistaking coagulated string for the PVS. Fibrin formation may be the reason why a threadlike structure could be found in the PHZ-induced anemia rats, because the threadlike structure could not be found when heparin was administered. One study reported that PHZ causes hypercoagulability [7], such as vascular thrombosis [10], and it inhibits the function of nitrogen monoxide [11], which plays an important role in preventing blood coagulation [12].

The other structure similar to the PVS is debris of the visceral peritoneum caused by urethane anesthesia. In most PVS studies, animals were anesthetized by intraperitoneal injection of urethane (1.5 g/kg). However, this anesthesia method is no longer used because of this problem. Meer et al. reported that the superficial layer of organs is damaged by intraperitoneal injection of urethane (1.0 g/kg) [8].

Since these similar-looking structures also have four kinds of PVS characteristics, the criteria for discrimination of the PVS appear to insufficient. Therefore, we recommend heparin administration to prevent blood coagulation and the exclusion of anesthesia by urethane IP injection, so as not to confuse the PVS with similar-looking structures.

The counterargument to this is that fibrin makes the PVS thicker, allowing it to be more easily found. Furthermore, the PVS is originally present at the remaining part of the membrane that has been damaged by urethane. To confirm this, additional tissue analysis is required. Most importantly, it is known that the PVS is a vascular structure for circulation. Thus, if a serial section of a sample, having the four characteristics mentioned previously, exhibits continuous lumen, it will then certainly be the PVS.

Acknowledgments This work was supported by the Association of Korean Oriental Medicine (AKOM) and partially supported by KOSEF (No, 20090084124).

References

1. Kim BH (1964) On the Kyungrak system. J Acad Med Sci DPR Korea 90:1–41
2. Soh KS (2009) Bonghan circulatory system as an extension of acupuncture meridians. J Acupunct Meridian Stud 2(2):93–106
3. Lee BC, Baik KY, Johng HM et al (2004) Acridine orange staining method to reveal the characteristic features of an intravascular threadlike structure. Anat Rec 278(1):27–30
4. Shin HS, Johng HM, Lee BC et al (2005) Feulgen reaction study of novel threadlike structures (Bonghan ducts) on the surfaces of mammalian organs. Anat Rec B New Anat 284(1):35–40
5. Lee BC, Kim KW, Soh KS (2009) Visualizing the network of Bonghan ducts in the omentum and peritoneum by using Trypan blue. J Acupunct Meridian Stud 2(1):66–70
6. Jiang X, Lee BC, Choi CH et al (2002) Threadlike bundle of tubules running inside blood vessels: new anatomical structure. Arxiv preprint physics/0211085
7. Sato H, Shinozuka J et al (2008) Acute thrombus formation in the lungs of phenylhydrazine-treated rats. J Toxicol Pathol 21(4):249–251
8. Van Der Meer C, Versluys-Broers J et al (1975) The effect of ethylurethane on hematocrit, blood pressure and plasma-glucose. Arch Int Pharmacodyn Ther 217(2):257

9. Jacob H, MacDonald R et al (1963) Regulation of spleen growth and sequestering function. J Clin Invest 42(9):1476
10. Ramot Y, Nyska A (2007) Drug-induced thrombosis – experimental, clinical, and mechanistic considerations. Toxicol Pathol 35(2):208
11. Luangaram S, Kukongviriyapan U et al (2007) Protective effects of quercetin against phenylhydrazine-induced vascular dysfunction and oxidative stress in rats. Food Chem Toxicol 45(3):448–455
12. Sasaki Y, Seki J et al (1996) Effects of NO-donors on thrombus formation and microcirculation in cerebral vessels of the rat. Thromb Haemost 76(1):111–117

Chapter 16
Effect of the Primo Vascular System on Liver Tissue Recovery After Irreversible Electroporation: A Preliminary Study

Hong-Bae Kim, Chang-Kyu Sung, and Saeyoung Ahn

Abstract Primo vessels have been found on various organs of the animals. The liquid in these vessels reveals distinct chemical compositions according to a proteomic analysis. A recent study showed the primo vessels to be functioning pathways of cancer metastasis. In this article, we report evidence of tissue healing in the liver via a primo vessel. In the hope that a sudden impact of irreversible electroporation (IRE) on liver tissue could provide a dynamic response by the primo vessel, we employed it to induce apoptosis in ablated lesions, which yielded positive signals in H&E staining and on a TUNEL assay. The ablated regions were observed to draw primo vessels in order to cure the damage. The primo vessels were observed with a Trypan blue technique. In addition, capillaries were observed to have been structurally preserved after applying IRE while another study showed extensive hepatocellular regeneration in the post-IRE stage. These results suggest that the PV appears to contribute to curing the ablated lesions in the liver.

1 Introduction

The primo vessel (PV) has been found in animals, and in humans, it was first observed by Bong-Han Kim in the early 1960s [1]. PVs have been discovered on the surfaces of abdominal organs [2–4] and in blood [5, 6] and lymph vessels [7, 8] of rats and rabbits. They were also found over the liver surfaces of rabbits [9]. Additionally, based on the result of a proteomic analysis, the PV was revealed to have chemical compositions in relation to stem cells [10, 11], cancer cells [12], and differentiated myeloid cells [13]. Recent studies also demonstrate that PVs might be

S. Ahn (✉)
Department of Physics and Astronomy, Seoul National University, Seoul, South Korea
e-mail: saeyounga@yahoo.com

K.-S. Soh et al. (eds.), *The Primo Vascular System: Its Role in Cancer and Regeneration*, DOI 10.1007/978-1-4614-0601-3_16, © Springer Science+Business Media, LLC 2012

pathways of cancer metastasis [14]. In spite of these findings, the function of the PV is not yet fully understood. In this study, we carefully demonstrated the function of PV in a limited way, especially in wound healings, by applying an external irreversible electroporation (IRE).

IRE is widely utilized in clinical cases [15, 16] and is minimally invasive and nonthermally ablational. Introduction of higher and shorter electric pulses brought these advantages in tissue ablation [15–23]. Application of IRE to the liver shows remarkable ablating results without thermal change, unlike established thermal ablation techniques [19–22]. In addition IRE allows a quick recovery after treatment [21, 22]. These are the merits of using IRE in correlation studies of the PVs' function, that is, the aspect of healing, in this preliminary work. The dynamic nature of IRE will be a valuable approach to better observe induced traffic in the PV for the discovery of its function and role.

2 Materials and Methods

Thirteen 6-week-old male Sprague-Dawley rats (from Doo-Yel Bio Co., Ltd.) were used and weighed 200–300 g, and all animals received care as approved by the Institutional Animal Care and Use Committee and in compliance with the institutional guidelines of Seoul National University and current international laws and policies (Guide for the Care and Use of Laboratory Animals, National Academy Press, 1966). Rats were maintained at room temperature (24°C) with a natural day/night light cycle in a conventional animal colony and were fed food and water ad libitum. They were adapted for 2 weeks to these conditions before testing.

The experimental setup has previously been described [22]. Briefly, two monopolar electrodes were used for all mice IRE procedures. The electrode was constructed using two needles made with tungsten (each 50 mm in length with a diameter of 0.6 mm) with the surface coated in pure platinum through unbalanced magnetic field sputter to a thickness of 10 μm. The rats were anesthetized with intraperitoneal injection of 10 mg/kg Zoletil (Virbac, Carros, France) solution. Under deep anesthesia, the medial alba of the rat abdomen was cut open, and the liver was exposed. Electrodes were then inserted into the liver with 2.5 mm spacing between electrodes. The electrode were enveloped with a plastic blocker for insulation, and eight rectangular direct current pulses of 100 μs width were applied at a frequency of 1 Hz with voltage amplitudes of 300, 700, and 1,000 V/cm at the middles of the electrodes by using a laboratory-built pulse generator. Two IRE lesions were created in two of three lobes of the liver; only one IRE lesion could be created in a lobe. After the IRE treatment, skin incisions were sutured, and the animals were kept alive to 10 h. Before 10 min of the 10 h, the rats were again surgically cut along the medial alba. After the liver had been exposed, we dropped several milliliters of Trypan blue of 0.4% (Sigma, USA) on the liver and then, after 1 m of the dropping, washed the liver with phosphate-buffered saline, pH 7.4 (PBS), in order to stain and investigate the PV on the surface of the liver.

Liver tissues were taken after observing the PV stained by Trypan blue with a stereomicroscope (STZ10, Olympus), that is, after 10 h of IRE. In the liver tissues, a peak of an apoptotic pathway, which agrees with previous studies [22], was observed, and the rats were euthanized. Liver samples were fixed in 5% (v/v) formalin solution and then sectioned through a plane normal to the inserted electrodes. H&E staining and TUNEL assay (S7100; Millopore, Temecula, CA) were performed to elucidate necrosis and cell apoptosis for all samples.

3 Results

Of all animals, 24 ablation zones were created (12 in the right lobe and 12 in the left lobe). After staining with Trypan blue, a PV was revealed on the surface of the liver (20 lobes), linking into the electrode position (Fig. 16.1b) and within the electroporated region (Fig. 16.1c). A PV was visualized within the injured area, as indicated by the arrows in Fig. 16.1c.

Microscopic images were obtained after H&E staining at 10 h after the IRE, as shown in Fig. 16.1. Well-delineated boundaries were always observed between untreated and ablated tissues in the H&E staining images, demonstrating eosinophilic

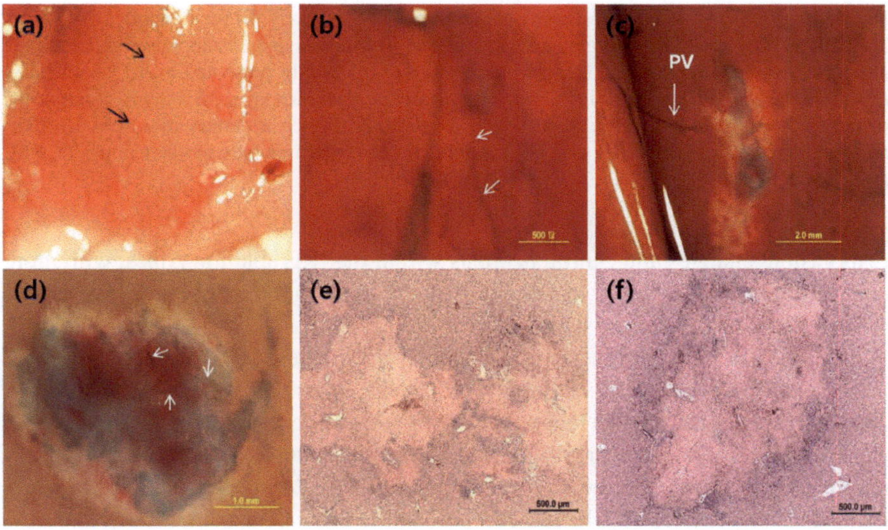

Fig. 16.1 A photograph showing the IRE electrode sites on liver tissue, which are indicated by *arrows*, with minimal lesion due to IRE application of 300 V/cm before staining with Trypan blue (**a**). A primo vessel (PV) (*arrow*) on the liver tissue was stained with Trypan blue; the PV was connected to the IRE lesion (**b**, **c**) and emerged around an inserted electrode (**d**), at 10 h after 300, 700 and 1,000 V/cm IRE procedures, respectively. IRE lesions with H&E staining corresponding to the tissue (**c–f**) (×10), show a well-delineated boundary between untreated and ablated tissues, with cellular necrosis in the ablation area

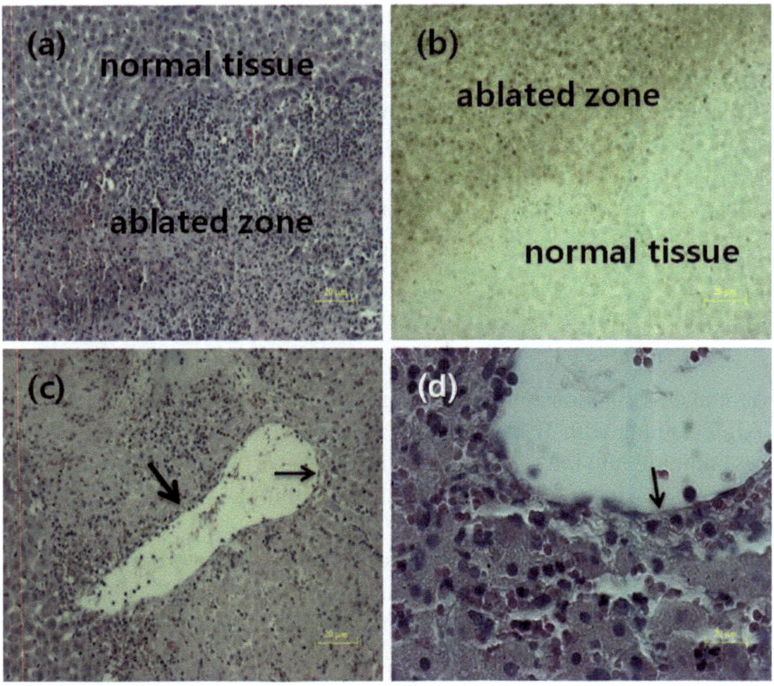

Fig. 16.2 IRE lesion with H&E staining (**a**, **c**, **d**) and TUNEL assay (**b**) for an applied 700 V/cm. The IRE-ablated zone has predominantly augmented neutrophils and eosinophils compared with normal liver tissue (**a**) (×10). It is possible to see the structurally preserved blood vessel (*thick arrow*) in the ablated area, showing a morphologically intact tunica layer (*thin arrow*) (composed of tunica media and tunica adventitia) (**c**) (×10), (**d**) (×40). A clear demarcation between IRE-ablated tissue and normal liver tissue is shown due to increased apoptotic markers in the TUNEL assay (**b**) (×10)

cytoplasmic changes and central necrosis within the IRE ablation zones, with more congestion in the interstitial spaces compared with normal liver tissue. However, the normal hepatic architecture was preserved.

In Fig. 16.2a, a focused image for an applied electric field 700 V/cm shows areas of acute, extensive, and severe cell death. The TUNEL assay that was performed brought positive results in the ablated region, indicating involvement of apoptotic cell death, as shown in Fig. 16.2b. Capillaries appeared in the ablated area, being structurally well-preserved, demonstrating sings of vaculitis. Figure 16.2d shows a morphologically intact tunica layer (arrow), which is composed of tunica media and tunica adventitia.

4 Discussion

PVs have been confirmed and verified to be floating on the surface of liver tissue [24]. The PVs were connected to surrounding organs, such as the omentum and the abdominal wall. However, the functions of the PVs over a liver have not yet been reported because indentifying these functions is not easy. Our limited study has, thus, suggested a correlation between a PV over the liver and IRE. IRE can effectively produce controlled cell death without thermal effects. In contrast to a thermal ablation by radiofrequency units, the IRE has definite advantages of ablating some undesirable tissues and of acquiring quickly treated tissues. The histological analysis of IRE treatment has shown neutrophil infiltrations, eosinophilic cytoplasmic change, and congestion in interstitial spaces in H&E staining while it shows severely positive signals (apoptosis) in the TUNEL assay within the ablated lesions of the post-IRE liver. Thus as in previous studies, the ablated lesion recovered while necrosis tissues were removed [22].

The PVs, which are likely to be coupled to lesions in the ablated area, were visualized with Trypan blue. The PVs were connected to surrounding organs or existed in the ablated lesion, which is an important finding to correlate a PV and IRE. PVs contained proteomically chemical compositions associated with stem cells [10, 11], cancer cells [12], and differentiated myeloid cells [13]. These points suggest strongly that the PV likely contributed curing ablated lesions. Another fact may back up the fact that capillaries within the ablated area were structurally intact (Fig. 16.2d); capillaries play a role in tissue recovery and the circulatory system to transport blood for life. The other may support the fact that some ablated areas show focally extensive hepatocellular regeneration after applying IRE [24]. However, not all samples showed PVs over livers. There were a few noises due to debris made in the process of cutting and suturing of the rat abdomen. Additional examinations are being planned to derive more concrete results.

References

1. Kim BH (1965) The Kyungrak system. J Jo Sun Med 108:1–38 [Korean]
2. Shin HS, Johng HM, Baik KY et al (2005) Feulgen reaction study of novel threadlike structures (Bonghan ducts) on the surfaces of mammalian organs. Anat Rec B New Anat 284:35–40
3. Sung B, Kim MS, Kim YJ et al (2008) Measurement of flow speed in the channels of novel threadlike structures on the surfaces of mammalian organs. Naturwissenschaften 95:117–124
4. Lee BC, Kim KW, Soh KS (2009) Visualizing the network of Bonghan ducts in the omentum and peritoneum by using Trypan blue. J Acupunct Meridian Stud 2:66–70
5. Lee BC, Baik KY, Sung B et al (2004) Acridine orange staining method to reveal the characteristic features of an intravascular threadlike structure. Anat Rec B New Anat 278:27–30
6. Lee BC, Yoo JS, Soh KS et al (2008) Development of a fluorescence stereomicroscope and observation of Bong-Han corpuscles inside blood vessels. Indian J Exp Biol 46:330–335
7. Lee BC, Yoo JS, Soh KS et al (2005) Novel threadlike structures (Bonghan ducts) inside lymphatic vessels of rabbits visualized with a Janus Green B staining method. Anat Rec B New Anat 286:1–7
8. Lee BC, Soh KS (2008) Contrast-enhancing optical method to observe a Bonghan duct floating inside a lymphatic vessel of a rabbit. Lymphology 41:178–185
9. Shin HS, Johng HM, Lee BC et al (2005) Feulgen reaction study of novel threadlike structures on the surface of rabbit livers. Anat Rec 284:35–40
10. Chen EI, Hewel J, Krueger JS et al (2007) Adaptation of energy metabolism in breast cancer brain metastasis. Cancer Res 67:1472–1486

11. Lian Z, Wang L, Yamaga S et al (2002) Genomic and proteomic analysis of the myeloid differentiation program. Blood 98:513–524
12. Forte G, Carotenuto F, Pagliari F et al (2006) Hepatocyte growth factor effects on mesenchymal stem cells: proliferation, migration, and differentiation. Stem Cells 24:23–33
13. Wuchter P, Boda-Heggemann J, Straub BK et al (2007) Processus and recessus adhaerentes: giant adherens cell junction systems connect and attract human mesenchymal stem cells. Cell Tissue Res 328:499–514
14. Yoo JS, Kim HB, Soh KS (2009) Bonghan ducts as possible pathways for cancer metastasis. J Acupunct Meridian Stud 2:118–123
15. Neal RE, Garcia PA, Rossmeisl JH et al (2010) A study using irreversible electroporation to treat large, irregular tumors in a canine patient. Conf Proc IEEE Eng Med Biol Soc 1:2747–2750
16. Pech M, Janitzky A, Wendler JJ et al (2010) Irreversible electroporation of renal cell carcinoma: a first-in-man phase I clinical study. Cardiovasc Intervent Radiol 34:132–138
17. Charpentier KP, Wolf F, Noble LB et al (2010) Irreversible electroporation of the pancreas in swine: a pilot study. HPB (Oxford) 12(5):348–351
18. Maor E, Ivorra A, Leor J et al (2007) The effect of irreversible electroporation on blood vessels. Technol Cancer Res Treat 6(4):307–312
19. Zhang Y, Guo Y, Ragin AB et al (2010) MR imaging to assess immediate response to irreversible electroporation for targeted ablation of liver tissues: preclinical feasibility studies in a rodent model. Radiology 256(2):424–432
20. Al-Sakere B, André F, Bernat C et al (2007) Tumor ablation with irreversible electroporation. PLoS One 2(11):e1135
21. Rubinsky B, Onik G, Mikus P (2007) Irreversible electroporation: a new ablation modality-clinical implications. Technol Cancer Res Treat 6(1):37–48
22. Choi YS, Kim HB, Chung J et al (2010) Preclinical analysis of irreversible electroporation on rat liver tissues using a microfabricated electroporator. Tissue Eng Part C Methods 16(6):1245–1253
23. Han HJ, Ogay V, Pak SJ et al (2010) Primo-vessels as new flow paths for intratesticular injected dye in rats. J Acupunct Meridian Stud 3(2):81–88
24. Edward WL, Christopher LL, Stephen TK (2007) Imaging guided percutaneous irreversible electroporation: ultrasound and immunohistological correlation. Technol Cancer Res Treat 6(4):287–294

Chapter 17
Detection of the Primo Vessels in the Rodent Thoracic Lymphatic Ducts

Inho Choi, Hee-Kyoung Chung, and Young-Kwon Hong

Abstract Since Hippocrates described the anatomy of the blood and the lymphatic systems for the first time at BC400, these two vascular systems have been believed to be the only circulatory systems existing and operating in vertebrates. However, the oriental medicine that originates much prior to the Hippocrates era in the Western world has suggested the presence of a third neurovascular network that serves as a conduit for acupuncture meridian. While the controversies on the therapeutic efficacies of acupuncture have not been entirely resolved, a novel primo-vascular system has been discovered to be present on the surface of the internal organs and within the large vessels in the rat, rabbit, and other animals. The morphological and anatomical properties of this new primo-vascular system were originally characterized by a North Korean scientist, Dr. Bonghan Kim, and thus this primo-vascular system was named after him as Bonghan thread/duct. In this chapter, we will report our endeavor to study this primo-vessel present in the thoracic ducts of rodents.

1 Introduction

A novel primo-vascular system has been reported to exist on the surface of the internal organs and within the large caliber blood and lymphatic vessels in rabbit and other animals by a North Korean scientist, Dr. Bonghan Kim [1]. Despite, the new primo-vessel has long been ignored or rejected by the mainstream medical and anatomical science due to its elusive morphological and functional properties. However, recent endeavor by Dr. Kwang-Sup Soh and colleagues have brought

Y.-K. Hong (✉)
Departments of Surgery and Department of Biochemistry and Molecular Biology,
Norris Comprehensive Cancer Center, University of Southern California,
Los Angeles, CA, USA
e-mail: Young.Hong@usc.edu

K.-S. Soh et al. (eds.), *The Primo Vascular System: Its Role in Cancer and Regeneration*,
DOI 10.1007/978-1-4614-0601-3_17, © Springer Science+Business Media, LLC 2012

new attention to this controversial neurovascular network, which was originally proposed to function as a conduit for acupuncture meridian. Dr. Soh and colleagues have rediscovered and extensively characterized the physicochemical properties of this primo-vascular system [2–6]. We are interested in characterizing the cellular nature of this primo-vessel present within the thoracic duct, the largest caliber lymphatic vessel. We have recently identified a lymphatic-specific fluorescent transgenic mouse that expresses green fluorescent protein (GFP) under the promoter of lymphatic-specific gene, Prox1 [7]. This Prox1-GFP BAC transgenic mouse model was initially created by the Gene Expression Nervous System Atlas (GENSAT) BAC-transgenic mouse project, a large-scale effort to investigate the expression patterns of the central nervous system (CNS)-related genes [8]. We detected that all lymphatic vessels, including the thoracic duct, in this Prox1-GFP mouse are labeled by GFP and that this transgenic mouse allows a convenient isolation of the thoracic duct, which is usually difficult to isolate due to its fragile property and elusive morphology. Here, we report primo-vessels within mouse and rat thoracic ducts.

2 Methods

Prox1-GFP BAC transgenic mouse (Tg(Prox1-EGFP)221Gsat/Mmcd), purchased from the Mutant Mouse Regional Resource Centers (MMRRC), was used to visualize the thoracic ducts. Prox1-GFP transgenic mice are of outbred background (FVB/N-Crl:CD1(ICR)). Normal wild-type Balb/C mice and Sprague Dawley rat were used to isolate thoracic ducts isolation. Animal-related works have been approved by the Institutional Animal Care and Use Committee of the University of Southern California (PI: YK Hong). Thoracic ducts were fixed with 4% paraformaldehyde (PFA) for 10 min and embedded in OCT solution to make 6 μm frozen sections, as previously described [9].

3 Results and Discussion

Development of a lymphatic-specific GFP fluorescent mouse model allowed us to easily locate and isolate the thoracic duct, the biggest caliber lymphatic vessel. Under a fluorescent stereomicroscope, the thoracic duct was clearly distinguished by its GFP signal and conveniently dissected out by using fine forceps (Fig. 17.1). Surgical skills obtained from the GFP mice enabled us to identify and isolate rat thoracic ducts. Consistent with previous studies from rabbits [1, 10–12], we were able to detect primo-vessels within the thoracic ducts of mouse and rat (Figs. 17.2 and 17.3). In both species, the primo-vessels appeared to be highly elastic and contain extracellular matrix materials with residing cells based on H&E staining. However, elastic nature of the primo-vessels and the thoracic ducts made it difficult

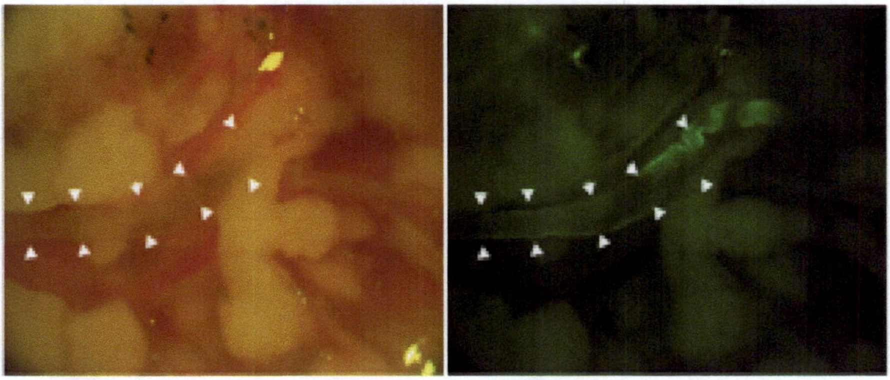

Fig. 17.1 Detection of mouse thoracic duct of mouse. GFP-positive thoracic duct of the lymphatic-GFP mouse was easily located in the thoracic cavity and pointed by *arrowheads*

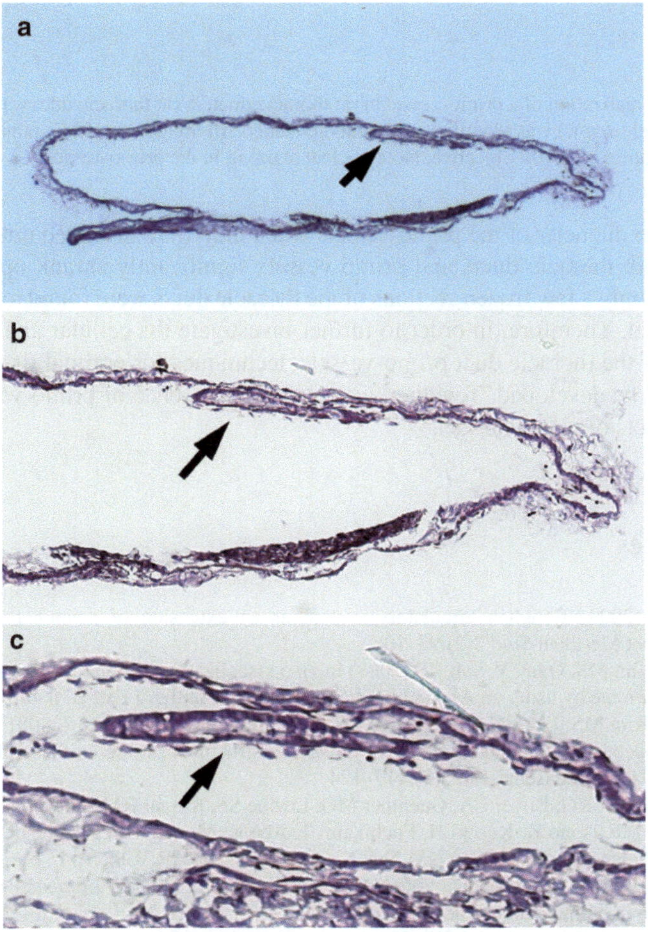

Fig. 17.2 Visualization of primo-vessels in the thoracic lymphatic duct of mouse. A primo-vessel (*arrow*) was found in the isolated thoracic duct of mouse. Images of a H&E stained frozen section were captured by using 10× (**a**), 20× (**b**), and 40× (**c**) objectives

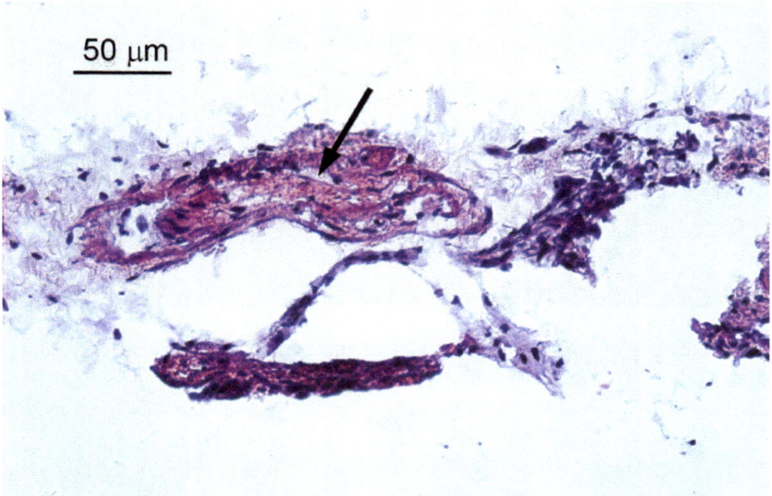

Fig. 17.3 Visualization of a primo-vessel in rat thoracic duct. A rat thoracic duct was isolated and a primo-vessel (*arrow*) was identified in the thoracic duct. An image of a H&E stained frozen section was captured by a 10× objective. Note nuclear staining in the primo-vessel

to assess the diameter of the primo-vessels when they were dissected out. Moreover, because both thoracic ducts and primo-vessels significantly shrank upon surgical dissection, only a few frozen sections of the thoracic ducts were found to contain the primo-vessel. Therefore, in order to further investigate the cellular and anatomical structure of the thoracic duct primo-vessels, techniques for optimal tissue preparation should be developed. Together, we report the presence of primo-vessels in the thoracic ducts of mice and rats.

References

1. Soh KS (2009) Bonghan circulatory system as an extension of acupuncture meridians. J Acupunct Meridian Stud 2(2):93–106
2. Yoo JS, Kim MS, Ogay V, Soh KS (2008) In vivo visualization of bonghan ducts inside blood vessels of mice by using an Alcian blue staining method. Indian J Exp Biol 46(5):336–339
3. Sung B, Kim MS, Lee BC, Yoo JS, Lee SH, Kim YJ, Kim KW, Soh KS (2008) Measurement of flow speed in the channels of novel threadlike structures on the surfaces of mammalian organs. Naturwissenschaften 95(2):117–124
4. Lee TI, Jenner RG, Boyer LA, Guenther MG, Levine SS, Kumar RM, Chevalier B, Johnstone SE, Cole MF, Isono K, Koseki H, Fuchikami T, Abe K, Murray HL, Zucker JP, Yuan B, Bell GW, Herbolsheimer E, Hannett NM, Sun K, Odom DT, Otte AP, Volkert TL, Bartel DP, Melton DA, Gifford DK, Jaenisch R, Young RA (2006) Control of developmental regulators by Polycomb in human embryonic stem cells. Cell 125(2):301–313
5. Ogay V, Baik KY, Lee BC, Soh KS (2006) Characterization of DNA-containing granules flowing through the meridian-like system on the internal organs of rabbits. Acupunct Electrother Res 31(1–2):13–31

6. Lee BC, Yoo JS, Baik KY, Kim KW, Soh KS (2005) Novel threadlike structures (Bonghan ducts) inside lymphatic vessels of rabbits visualized with a Janus Green B staining method. Anat Rec B New Anat 286(1):1–7
7. Choi I, Chung HK, Ramu S, Lee HN, Kim KE, Lee S, Yoo J, Choi D, Lee YS, Aguilar B, Hong YK (2011) Visualization of lymphatic vessels by Prox1-promoter directed GFP reporter in a bacterial artificial chromosome-based transgenic mouse. Blood 117(1):362–365
8. Gong S, Zheng C, Doughty ML, Losos K, Didkovsky N, Schambra UB, Nowak NJ, Joyner A, Leblanc G, Hatten ME, Heintz N (2003) A gene expression atlas of the central nervous system based on bacterial artificial chromosomes. Nature 425(6961):917–925
9. Hong YK, Lange-Asschenfeldt B, Velasco P, Hirakawa S, Kunstfeld R, Brown LF, Bohlen P, Senger DR, Detmar M (2004) VEGF-A promotes tissue repair-associated lymphatic vessel formation via VEGFR-2 and the alpha1beta1 and alpha2beta1 integrins. FASEB J 18(10): 1111–1113
10. Ogay V, Bae KH, Kim KW, Soh KS (2009) Comparison of the characteristic features of Bonghan ducts, blood and lymphatic capillaries. J Acupunct Meridian Stud 2(2):107–117
11. Lee BC, Kim KW, Soh KS (2009) Visualizing the network of Bonghan ducts in the omentum and peritoneum by using Trypan blue. J Acupunct Meridian Stud 2(1):66–70
12. Lee BC, Kim S, Soh KS (2008) Novel anatomic structures in the brain and spinal cord of rabbit that may belong to the Bonghan system of potential acupuncture meridians. J Acupunct Meridian Stud 1(1):29–35

Chapter 18
Histological Comparison of Primo Nodes in Abdominal Membrane and Lymph Nodes of Rat

Kyoung-Hee Bae, Zhendong Su, Kwang-Sup Soh, and Hee Min Kwon

Abstract We report a comparison of hematoxylin and eosin (H&E) stained images between primo nodes (PNs) in the abdominal membrane and lymph nodes (LNs) of a rat. Samples of PNs were taken from the region of acupuncture points in the conception vessel meridian (CV8-12) of rats which were aged between 7 and 9 weeks old. Samples of LNs were taken near the abdominal vein and the caudal vena cava. Both primo nodes and lymph nodes samples were fixed in a paraffin block and H&E stained. The images were taken using an optical microscopy equipped with a CCD camera. The observed features in LNs were absent in PNs, namely, capsule, subcapsular sinus, and follicles. In addition, there was a high density of blood capillaries in the primo nodes as Bong-Han Kim described on superficial primo nodes. The result showed that the primo nodes in the putative conception vessel meridian were different from lymph nodes, and further investigation of cell types and functions of the primo node is necessary to fully characterize the putative PN in the conception vessel.

1 Introduction

One of the most important acupuncture meridians is the conception vessel (CV) along the midline of ventral part of an animal. Even though the acupuncture points and meridians have been known for a long time, their anatomical positions and characteristics have never been observed despite various efforts by Heine [1], Langevin and Yandow [2], Ifrim and Ifrim [3], and Van wijk et al. [4]. Even Kim [5] and Fujiwara [6] who claimed discovery of anatomical structures corresponding to acupoints and meridians did not report any specific study on the CV.

H.M. Kwon (✉)
Biomedical Physics Laboratory, Department of Physics and Astronomy,
Seoul National University, Seoul 151-747, South Korea
e-mail: heemin.kwon@gmail.com

K.-S. Soh et al. (eds.), *The Primo Vascular System: Its Role in Cancer and Regeneration*, DOI 10.1007/978-1-4614-0601-3_18,

Only very recently did Eom [7] and Eom et al. [8] report on the observation of putative acupuncture points along the CV line of a rat. They were located not in the skin but in a deep muscle layer in the surface of the abdominal wall facing internal organs. Eom traced the Primo vessel from CV primo nodes while this chapter focuses on the difference between the CV primo node and lymph node. No one has yet proven whether these structures are indeed primo nodes corresponding to acupuncture points of CV line. More importantly, it was not clear whether they were simply variations of lymph nodes or novel tissue, i.e., primo nodes. Therefore, it was necessary to reproduce Eom's and Eom et al.'s work [7, 8], and independently examine the histological structure to find peculiar properties of the putative CV primo nodes. In this work we confirmed the existence of putative primo nodes (PN) along the CV line of a rat, and compared the primo nodes with lymph nodes by using hematoxylin and eosin (H&E) staining.

2 Methods

2.1 Animal Preparation

We used white rats (SD/male/about 7–9 weeks old) purchased from Dooyeol Biotech (Seoul, Korea). The rats were maintained at a constant temperature of 25°C, relative humidity 60%, and had access to food and water all the times. The use and maintenance of the rats were in accordance with the guidelines of Seoul National University and international policy (Guide for the Care and of Laboratory Animals, Nation Academy Press 1996).

The experiment was done under whole body anesthesia of a rat. When the rat was fully anesthetized by using Zoletile (10 mg/kg, Virbac Laboratories, France), the abdominal hair was removed.

2.2 Isolation of CV Primo Nodes and Lymph Nodes

We made marks along the conception vessel meridian between CV8 (umbilicus) and CV16 (Xiphosternum) by first marking CV8 and CV16, and then evenly dividing and marking in the order of CV12-CV10-CV14-CV9-CV11-CV13-CV15. We checked the CV8 point during surgery of the skin to see if the point was correct. This procedure ensured locating primo nodes along the CV line. Since the CV line is along the ventral midline of the body, an incision was made 4–5 mm next to the midline. After the abdominal cavity had been exposed, we searched for the primo nodes for the corresponding CV points. The CV primo nodes were located parallel to the arteries and veins. We isolated and washed the primo node with saline. All these operations were undertaken under a stereo microscope (SZX 12, Olympus). Lumbar lymph nodes with sizes of 4–5 mm were taken near the caudal vena cava by using a micro pincet.

2.3 *Hematoxylin and Eosin (H&E) Staining*

Samples of primo node and lymph nodes were fixed with paraformaldehyde (PFA 4%). Then the samples were fixed in a paraffin block and sliced in 3-μm slices, and a standard procedure of H&E staining was applied.

3 Results

Figure 18.1a shows the primo node before it was isolated, and Fig. 18.1b shows the isolated primo node. Generally, the primo node was covered by a band of fat tissue that usually ran the midline of the abdominal wall, and the width of the band was about 1–2 mm. The CV primo node of a rat was a size of 1–2 mm in diameter. Usually, a primo node was detected at each CV acupoint, but sometimes they were absent at some acupoints, or sometimes there were two or three nodes at one point, as shown in Fig. 18.1c. There were two large blood vessels in the fat band running

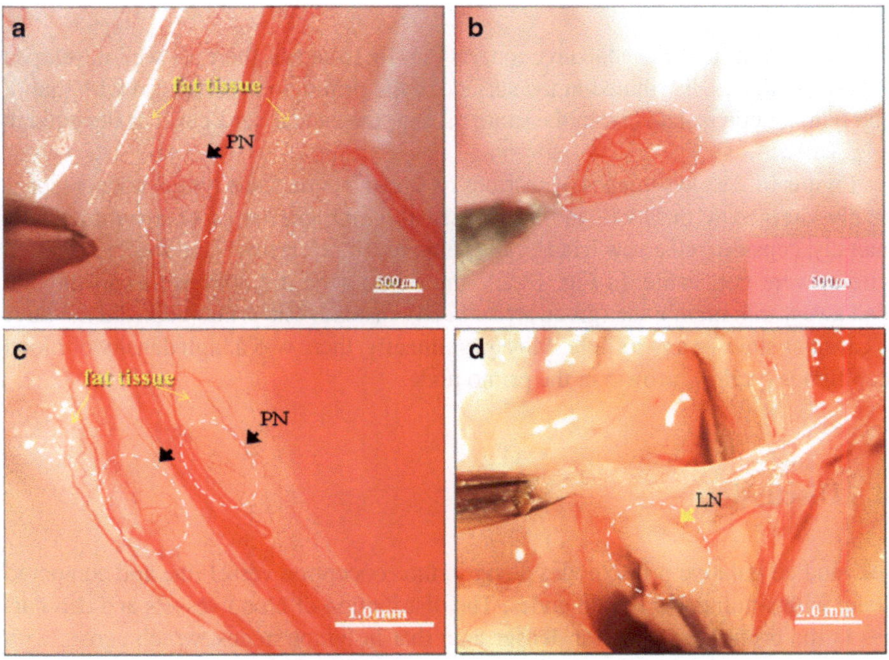

Fig. 18.1 Stereomicroscopic images of primo nodes in the CV and a lymph node from near the caudal vena cava in the abdominal cavity of a rat. The PNs are indicated with a *dotted oval circle*, and the lymph node with an *arrow*. (**a**) Is the PN in situ stretched by forceps and (**b**) is an isolated PN that demonstrates the well-developed capillaries in the surface. (**c**) Shows two PNs at one acupoint covered with fat tissue. Scale bars are 0.5 mm (**a, b**), 1.0 mm (**c**), and 2.0 mm (**d**). (**d**) Is a typical lymph node, which has only a low density of blood capillaries

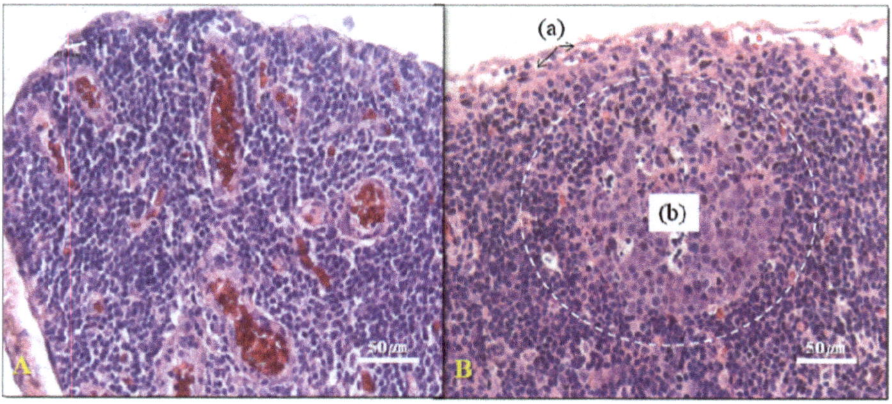

Fig. 18.2 H&E staining images of a (**A**) primo node and a (**B**) lymph node. The PN had no capsular sinus or follicles, the typical morphological signatures of LN (*black arrow* and *dotted circle* in (**b**)). The PN had a thin outermost membrane (*green arrow head*), in contrast to the capsule of a LN (*red arrow head*). The PN had a high density of blood capillaries (*yellow arrows* in (**a**)) compared with low density for the LN

from the cranial to the caudal direction. As Fig. 18.1b shows, there are well-developed blood capillaries distributed in the surface of the primo node compared with that of the lymph node isolated from the same rat near the caudal vena cava, as depicted in Fig. 18.1d.

Figure 18.2 shows images of sections of a primo node and a lymph node, respectively, stained with H&E. In Fig. 18.2B, the characteristics of a lymph node, namely, capsule (red arrow head), capsular sinus (black arrow), and follicles (dotted circle) were noticeable. In Fig. 18.2A, the primo node did not have a capsule, but had a very thin membrane (green arrow head), instead. The primo node did not have a capsular sinus or any follicle. Most prominently, there was a high density of blood capillaries (yellow arrows) in the primo node.

4 Discussion

We studied putative primo nodes along in the conception vessel line and compared them with similar size lymph nodes. The distribution of blood vessels and capillaries in the primo node was far denser than that in lymph nodes, which agrees with Kim's original claims [5] and with the recent work by Eom [7] and Eom et al. [8]. Furthermore, the morphological characteristics of the primo node revealed with H&E staining were also distinctively different from those of lymph nodes. A denser capillary distribution inside the primo node was again noticeable. The dense distribution of blood capillaries is consistent with the hematopoietic function of the primo node as claimed by Kim [5]. We observed that primo nodes of anemia-induced rats became larger than those of normal rats, which needs to be investigated further. More investigation and detailed reports will be presented elsewhere.

Our study suggests that the putative PN in the CV line may have an anatomical correspondence to classical acupuncture points. However, for the complete establishment, we need much more investigation. For example, we need to analyze and characterize the cells in the primo node and compare them with those in lymph nodes, and we are currently pursuing such a study. It is particularly important to observe the primo microcells (San Al) in the primo node because Kim claimed that a San Al was kind of small embryonic-like stem cell [9]. The present work is a preliminary stage for the investigation of primo microcells extracted from primo nodes. We are currently applying centrifuge method to separate and identify various cells and microcells from our PN samples.

Acknowledgments We would like to thank Drs. Byung-Cheon Lee and Ping An for helpful discussion. K. Soh thanks the Association of Korean Oriental Medicine for support.

References

1. Heine H (1988) Anatomical structure of acupoints. J Tradit Chin Med 8:207–212
2. Langevin HM, Yandow JA (2002) Relationship of acupuncture points and meridians to connective tissue planes. Anat Rec B New Anat 269:257–265
3. Ifrim-Chen F, Ifrim M (2005) Acupuncture and meridians: a histochemical study. Ital J Anat Embryol 110:51–57
4. Van Wijk R, Soh KS, Van Wijk EPA (2007) Anatomic characterization of acupuncture system and ultra-weak photon emission. Asian J Phys 16:443–474
5. Kim BH (1963) On the Kyungrak system. J Acad Med Sci DPR Korea 90:1–41
6. Fujiwara S, Yu SB (1967) 'Bonghan theory' morphological studies. Igaku no Ayumi 60:567–577 [in Japanese]
7. Eom K-H (2010) Imaging of the primo vascular system in the conception vessel using fluorescent nanoparticles. Department of Physics and Astronomy, The Graduate School, Seoul National University
8. Eom KH, Lee SW, Ahn SH, et al Fluorescent nanoparticle tracing from putative primo nodes in the conception vessel-like line to the bladder of a rat via primo vessel and ligament. unpublished
9. Baik KY, Ogay V, Jeoung SC et al (2009) Visualization of Bonghan microcells by electron and atomic force microscopy. J Acupunct Meridian Stud 2(2):124–129

Chapter 19
Visualization of the Primo Vascular System by Using Trypan Blue in the Subarachnoid Space of Rats

Inhyung Lee, Zhen-dong Su, Ki Woo Kim, Byung-Cheon Lee, and Kwang-Sup Soh

Abstract One of the most recognized theories for the acupuncture meridian has been the concept that acupuncture effects propagate through the nervous system. Beyond this general concept in Western medicine, there was one revolutionary idea for the acupuncture meridian, named "Bonghan theory." The Bonghan theory insisted that acupuncture effects transmit not via the nerve system but via specific channels, named Bonghan ducts (primo vessels, PVs). Bonghan theory insisted that the PVs also existed in the nerve system. In order to visualize the nerve primo vessels (NPVs) under arachnoid mater of rat, Trypan blue (TB) was intrathecally injected into the spinal cord of rats. TB worked to visualize NPVs as effectively as in previous works. The NPVs were investigated by using light microscopic and transmission electron microscopic (TEM) images. Those images of the NPVs showed that they had regular patterned rod-shaped nuclei and that many microparticles existed as in other NPVs inside rabbit brain ventricles and central canals. The new structure, the NPVs, under the arachnoid mater of the spinal cord of rats could encourage one to reconsider that there may be other potential channels in the nerve system corresponding to acupuncture meridian.

B.-C. Lee (✉)
Ki Primo Research Laboratory, KAIST Institute for Information Technology Convergence, Korea Advanced Institute of Science and Technology, Daejeon, 305-701, Korea

Pharmacopuncture Medical Research Center, Korean pharmacopuncture Institute, Seoul 157-200, Korea
e-mail: donboscolee@gmail.com

K.-S. Soh et al. (eds.), *The Primo Vascular System: Its Role in Cancer and Regeneration*,
DOI 10.1007/978-1-4614-0601-3_19,

1 Introduction

Beyond modern medical point of view, there has been only one theory, named "Bonghan theory," to understand acupuncture's effects by using a new anatomical system, called nerve primo vessels (NPVs, formerly nerve Bonghan ducts, 1960s) embedded in nerve bundles [1]. After visualization of primo vessels (PVs) inside rabbit brain ventricles and central canals [2, 3], new staining methods were developed to visualize the PVs and the primo nodes on the internal organs of rats [4] and were applied to visualize PVs in the brain, spinal cord, and sciatic nerve of rats in situ and in vivo [5].

In a previous study, Trypan blue (TB) was injected to visualize primo vascular networks in the spinal cord after decapitation of the rat brain. In order to overcome the drawbacks caused by decapitation, Dai and Lee et al. used the intraventricular injection method to visualize PVs in the rat brain [6]. In order to visualize PVs under the arachnoid mater of the spinal cord, TB was injected into the spinal cord in situ and in vivo. This chapter reports a new application of the intrathecal injection method to visualize primo vessels networks without decapitation and new findings on the networks of PVs under the arachnoid mater of a rat's spinal cord.

2 Materials and Methods

All experiments were performed in accordance with the Principles of Laboratory Animal Care prepared by the National Society of Medical Research and with the Guide for the Care and Use of Laboratory Animals prepared by the Institute of Laboratory Animal Resources and published by the National Institute of Health (NIH Publ. 85–23, Rev. 1985). The study was approved by the Ethical Committee of Seoul National University.

Rats were anesthetized by intramuscular injection of urethane (1.5 g/kg) into the right hind limb. Under deep anesthesia, 0.2% TB in artificial cerebrospinal fluid (CSF) was intrathecally injected into the spinal cord between the fifth lumbar (L5) and the sixth lumbar (L6) vertebrae. At 10–20 min after injection of TB, a laminectomy of the spine was performed. After exposing the spinal cord, the NPVs under arachnoid mater of the rat were examined under stereomicroscope (SZX12, Olympus, Japan). Then, the NPV specimens were isolated and stained with 4′, 6-diamidino-2-phenylindole (DAPI), and were observed with a fluorescence light microscope. For transmission electron microscopic (TEM) examination, general TEM processes were performed after the specimens had been fixed with 2.5% glutaraldehyde in a 0.1-M sodium-cacodylate buffer at 4°C for 4 h.

3 Results

Figure 19.1a shows an illustration of the brain and the spinal cord of a rabbit in which blood vessels and primo vessels are drawn by red and yellow lines, respectively. Especially, the magnified view in Fig. 19.1b of the dotted rectangle in Fig. 19.1a demonstrates clearly the blood vessels and primo vessels, respectively by using red and yellow lines. It is noticeable that the slightly blue-colored membrane in Fig. 19.1b is the arachnoid mater of the spinal cord of the rabbit. As shown in Fig. 19.1c and d, after the intrathecal injection of TB in rats, it was possible to visualize blue-colored primo vessels under the arachnoid mater. An isolated view of a primo vessel under the arachnoid mater showed that the DAPI-stained primo vessel had rod-shaped nuclei in a longitudinally regular pattern (Fig. 19.1e).

Figure 19.2a is a magnified view of the PVs from Fig. 19.1d and demonstrates that the PV consists of two primo vessels. A TEM image of primo vessel shows many microparticles and several collagen bundles in a cross-section view of a PV.

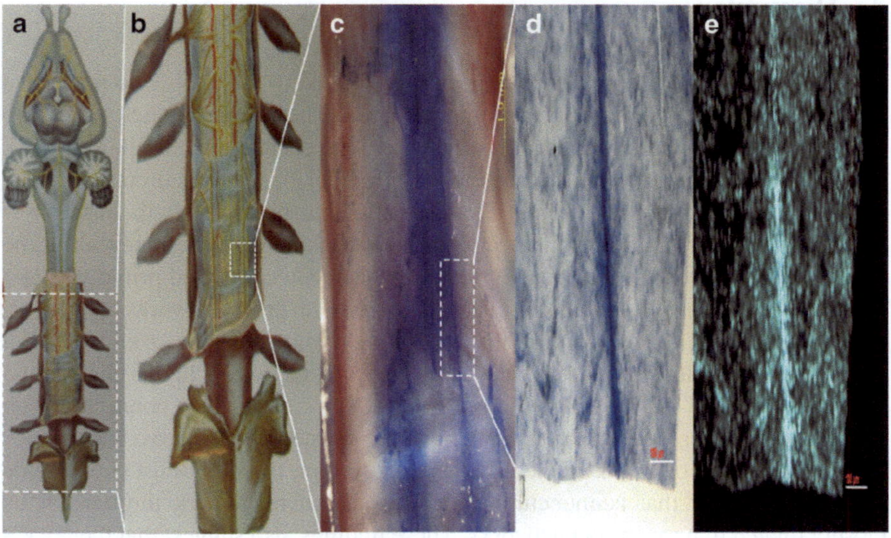

Fig. 19.1 Structures of primo vascular system (PVS) in the central nervous system. (**a**) An illustration of the brain and spinal cord of rabbit. (**b**) Magnified view of the dotted rectangle of (**a**) in which blood vessels and primo vessels are illustrated respectively by *red* and *yellow lines*. Slight blue-colored membrane means the arachnoid mater of the spinal cord of rabbit. (**c**) Stereoscopic image of primo vessels under the arachnoid mater stained by using Trypan blue with intrathecal injection in rat. (**d**) Isolated view of a primo vessel under the arachnoid mater. Scale bar is 100 μm. (**e**) DAPI-stained primo vessel showing rod-shaped nuclei. Scale bar is 100 μm

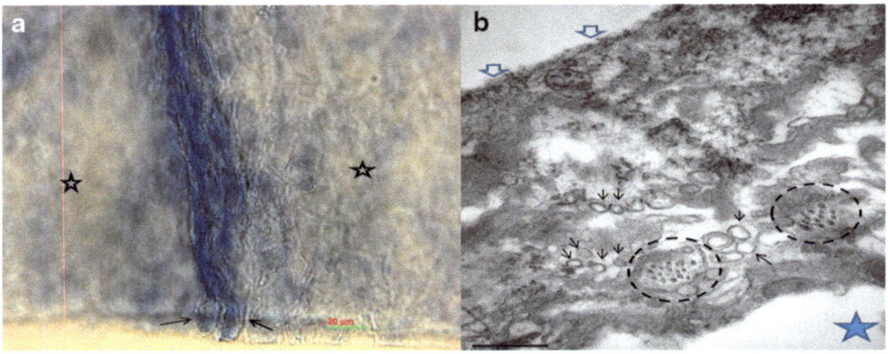

Fig. 19.2 Magnified view of a primo vessel (PV). (**a**) Isolated magnified view of a PV under the arachnoid mater in Fig. 19.1d. It is noticeable that PV consists of two primo vessels (*arrows*). Two *asterisks* indicate the arachnoid matter attached to the PV. (**b**) One of transmission electron microscopic images of a primo vessel in which there are many microparticles (*arrows*) and two collagen bundles (*dotted circles*). *Open arrows* indicate some of outermost membrane of PV. A *star* mark indicates the orientation to the arachnoid mater. Scale bar is 0.5 μm

4 Discussion

Recently PVs and primo nodes have been investigated by Soh [7], who made two observations. First, NPVs contain microparticles. These microparticles are also found in other NPV in rabbit brain ventricles and the central canal of the spine [3]. Second, NPVs are embedded in the arachnoid mater of the spinal cords of rats, which is a fascia system of a nerve. Thus, attention has been paid to the fascia system of a nerve in relation to the acupuncture meridian system. For revealing the relationship between the meridian system and the fascia connected primo vascular system (PVS), a new "Bonghan-Fascia Model" based on previous works on the PVS was hypothesized [8]. According to this model, the primo networks are embedded in the fascia system, loose connective tissues. Therefore, the results in this chapter and previous works [3–5] were considered that all NPVs are tightly connected with nerve fascia, such as the arachnoid mater and the perineurium of nerve bundles.

In future study on the acupuncture meridian system, these findings could guide one to reconsider that acupuncture and moxibustion may work through NPVs closely related to nerve fascia networks. These findings also suggest that the NPV's function is modulated by many DNA particles, named primo microcells (formerly Bonghan Sanals) [1, 3], and microparticles [3].

Acknowledgments This work was supported by the National Research Foundation of Korea (NRF) grant funded by the Korean government (MEST) in 2010 (No. 2010-0025289) and by the Korean Pharmacopuncture Foundation Grant funded by the Korean Pharmacopuncture Institute (KPI-2010-003).

References

1. Kim BH (1965) Kyungrak system. J Acad Sci DPR Korea 168:1–38
2. Lee BC, Kim S, Soh KS (2008) Novel anatomic structures in the brain and spinal cord of rabbit that may belong to the Bonghan system of potential acupuncture meridian. J Acupunct Meridian Stud 1:29–35
3. Lee BC, Kim KW, Soh KS (2010) Characteristic feature of a nerve primo-vessel suspended in rabbit brain ventricle and central canal. J Acupunct Meridian Stud 3:67–138
4. Lee BC, Kim KW, Soh KS (2009) Visualizing the network of Bonghan ducts in the omentum and peritoneum by using Trypan blue. J Acupunct Meridian Stud 2:66–70
5. Lee BC, Eom KH, Soh KS (2009) Primo-vessels and primo-nodes in rat brain, spine and sciatic nerve. J Acupunct Meridian Stud 3:111–115
6. Dai JX, Lee BC, An P et al. In vivo in situ staining of the primo vascular system in ventricles and subarachnoid space of brain by injecting Trypan blue into the lateral ventricle submitted to Anatomical Record
7. Soh KS (2009) Bonghan circulatory system as an extension of acupuncture meridians. J Acupunct Meridian Stud 2:93–106
8. Lee BC, Soh KS (2009) A Novel Model for Meridian: Bonghan Systems Combined with Fascia (Bonghan-Fascia Model). Proceedings, Fascia Research II: Basic Science and Implications for Conventional and Complementary Healthcare. Amsterdam, Elsevier, p 144

Chapter 20
Network of the Primo Vascular System in the Rat Hypodermis

Byung-Cheon Lee, Zhendong Su, Baeckkyoung Sung, Ki Woo Kim, Jin-Myung Cha, Jin-Kyu Lee, Byung-Joon Chang, and Kwang-Sup Soh

Abstract Anatomical and histological studies on acupuncture points (acupoints) and meridians have been one of the most important approaches to revealing the mechanism of acupuncture analgesia and therapeutics. However, researchers have not yet reached a conclusion on the exact anatomical structure of the acupuncture meridians. In this chapter, the network of primo-vessels and nodes in rat hypodermis is reported. Fluorescence imaging showed the network of the primo-vascular system on the layer of superficial fascia (stratum fibrosum) that existed around the location of the stomach meridian from the knee to the middle of the tibia about 24 h after the subcutaneous injection of fluorescent nanoparticles at the acupoint ST36. Its fine morphologies were analyzed by using confocal laser scanning microscopy and transmission electron microscopy. The primo-vessel had a sinus (3–4 μm in diameter) surrounded by a layer of cells, and had a bundle-like collagen architecture (covered by a cellular layer) in which collagen fibers were regularly aligned along the longitudinal direction of the primo-vessel. Nanoparticles were also found in the cellular or extracellular layer. Since it seems evident that the primo-vascular system mainly exists in the fascia, a new perspective is needed to investigate the whole mechanism of the propagation of acupoint stimulation through the primo-vascular system in relation to fascia physiology.

1 Introduction

Anatomical and histological studies on acupuncture points (acupoints) and meridians have been one of the most important approaches to revealing the mechanism of acupuncture analgesia and therapeutics since the early era of scientific investigation

B. Sung (✉)
Department of Physics and Astronomy, Biomedical Physics Laboratory,
Seoul National University, Seoul 151-747, South Korea
e-mail: baeckkyoung@gmail.com

K.-S. Soh et al. (eds.), *The Primo Vascular System: Its Role in Cancer and Regeneration*, DOI 10.1007/978-1-4614-0601-3_20,

of acupuncture. In more than 60 years, researchers have reached a general conclusion that, at the location of the acupoint (skin, subcutaneous, and muscular layers), (1) a rich supply of peripheral nerve receptors and blood capillaries [1] and (2) sharp perforations of loose connective tissue (superficial fascia) with nerve-capillary bundles [2] exist. For the anatomical structure of meridians, on the other hand, few general agreements exist among the researchers. At present, there is evidence that the meridians (1) are fluid pathways of interstitial spaces mainly following vascular and nervous packs [3, 4] or (2) are representations of body-wide loose connective tissue networks that have an ability of mechanotransduction [5–8].

It should be noted that in the 1960s Bong-Han Kim conducted a series of systematic experiments that integrated most of these points of views [9, 10], although at that time his results could not be confirmed by other research groups (except for Satoru Fujiwara [11]) due to lack of detailed information on the experiments reported in Kim's papers. Kim's observation can be summarized as follows: the acupuncture meridian system (primo-vascular system) is anatomically composed of primo-nodes (Bonghan corpuscles: acupoints particularly in skin), which are interconnected by primo-vessels (Bonghan ducts: meridians and collaterals). Recently, observations of primo-vessels in bovine [12], swine [13], rabbits [14–17], rats [18–20], and mice [21] were made by applying modern bio-imaging and histological techniques (fluorescence imaging with nanoparticles or organic dyes, scanning and transmission electron microscopy (SEM & TEM), atomic force microscopy (AFM), etc.), which have not only confirmed Kim's claims but also provided new findings and information, such as in vivo staining of the primo-vessels with Trypan blue [22], to the surfaces of subcutaneous tumor tissues [21]. Furthermore, some evidence for a fluid-conducting function of the primo-vascular system has been provided by measurements of the flow speed [23] and the fluid viscosity [24].

In this chapter, the network of primo-vessels and nodes in rat hypodermis are reported for the first time. Fluorescence imaging showed the presence of a network of primo-vascular system on the layer of superficial fascia (stratum fibrosum) around the location of the stomach meridian from the knee to the middle of tibia about 20 h after the subcutaneous injection of fluorescent nanoparticles at the acupoint ST36. Its fine morphologies were analyzed by using confocal laser scanning microscopy and transmission electron microscopy. The primo-vessel had a sinus (3–4 μm in diameter) surrounded by a layer of cells, and had a bundle-like collagen architecture covered by a cellular layer.

2 Materials and Methods

All experiments were performed in accordance with the Principles of Laboratory Animal Care prepared by the National Society of Medical Research and with the Guide for the Care and Use of laboratory Animals prepared by the Institute of Laboratory Animal Resources and published by the National Institute of Health (NIH Publ. 85–23, Rev. 1985). The study was approved by the Ethical Committee of Seoul National University.

Rats were anesthetized by intramuscular injection of urethane (1.5 g/kg) into the right leg. As a tracer, we used fluorescent nanoparticles containing rhodamine B isothiocyanate (RITC, orange, λmax = 555 nm) and biocompatible poly (ethylene glycol) (PEG) with a silica shell. Under deep anesthetization, 0.06 mL of nanoparticle was subcutaneously injected (injection depth = 2–3 mm) into the left Zusanli point (ST 36). The Zusanli point (ST 36) was determined based on the traditional "skeletal metric" procedure corresponding to that for humans. About 24 h after injection of fluorescent nanoparticles, we dissected the rat's skin and exposed its subcutaneous layer. The exposed subcutaneous layer was observed under a fluorescence stereomicroscope (MVX 10, Olympus). Significant fluorescence signals from the subcutaneous layer of rat were recorded by using the MVX 10 microscope equipped with a charge-coupled device (CCD). Then, the specimens were isolated and stained with 4′,6-diamidino-2-phenylindole (DAPI) and were observed with a confocal laser scanning microscope (LSM 510, Carl Zeiss, Germany).

For TEM examination, tissues were fixed with 2.5% glutaraldehyde in a 0.1-M sodium-cacodylate buffer at 4°C for 4 h, postfixed in 1% OsO4 in a 0.2-M sodium-cacodylate buffer for 1 h, dehydrated with ethanol and propylene oxide, and embedded in epoxy resin (Epon 812). Ultrathin sections were collected on large-scale copper grids, contrasted using 2% uranyl acetate and Reynolds' lead citrate, and examined in a field-emission transmission electron microscope (LIBRA 120, Carl Zeiss, Germany) at an accelerating voltage of 80 kV, with images being obtained using a digital CCD camera (ES1000W, Gatan, USA) and its software processing (Gatan, USA).

3 Results

On the layer of the stratum fibrosum in the superficial fascia, we could find networks of threadlike structures (5–10 μm in diameter) in which the fluorescent nanoparticles existed (Fig. 20.1a). These threadlike structures were found from the knee to the middle of tibia, around the location of the classical "stomach meridian." This region was a maximum of 1–2 cm away from the diffusion area. Magnified under the fluorescence stereomicroscope (Fig. 20.1b), we could observe that the network of threadlike structures was composed of primo-nodes and vessels. These threadlike structures were partly and irregularly fixed on the substrate (stratum fibrosum). With confocal laser scanning microscopy (Fig. 20.1c), we could observe that the nanoparticles existed nonuniformly inside the primo-node and vessel and that there were round and rod-shaped nuclei. The rod-shaped nuclei were aligned along the longitudinal direction of the primo-vessel. The nanoparticles also existed in the blood and the lymph capillaries, which were measured to be in similar dimensions to the primo-vessels. However, they could be discriminated from the primo-vessels by morphological characteristics: (1) The primo-vessels had rod-shaped nuclei aligned in a single (longitudinal) direction, although the blood and the lymph capillaries generally showed transversely directed nuclei together. (2) The blood and the lymph capillaries always exhibited a single uniform lumen, by contrast, the primo-vessels

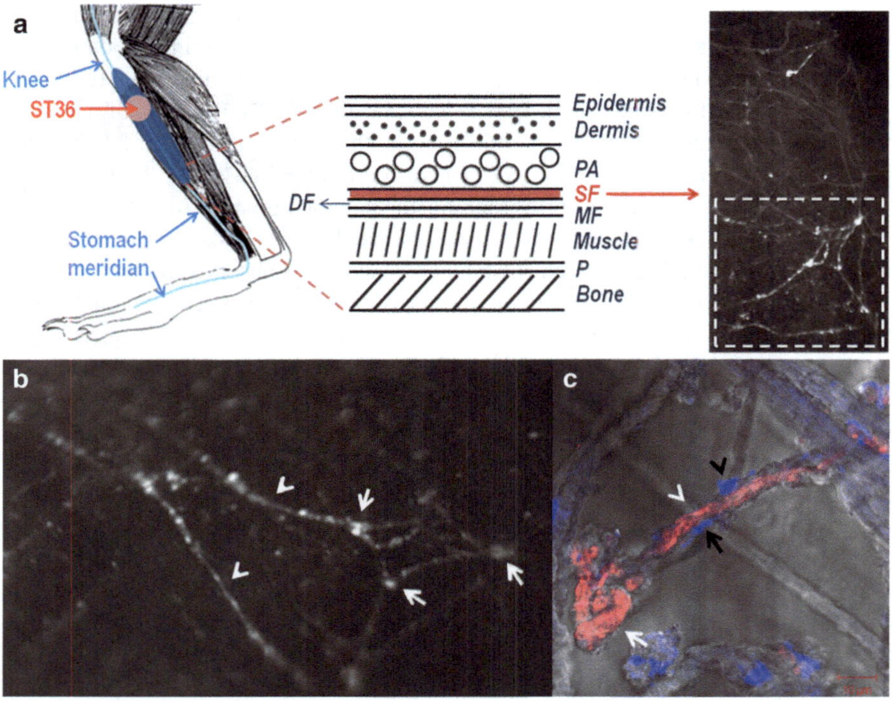

Fig. 20.1 (**a**) The anatomical location and distribution of the primo-vascular system in a subcutaneous layer. A suspension of fluorescent nanoparticles was subcutaneously injected at the acupoint ST36 (*red arrow*) which is located on the stomach meridian (*light-blue solid line*). About 24 h after the injection, the skin layer (epidermis plus dermis) was surgically removed and the subcutaneous layer (hypodermis: PA & SF) was observed with a fluorescence stereomicroscope. On the layer of stratum fibrosum in superficial fascia, we could find networks of primo-vessels (5–10 μm in thickness) in which the fluorescent nanoparticles flowed as shown in the stereoscopic image. The primo-vascular system was found from the knee to the middle of tibia, around the location of the stomach meridian (*blue* region). This region was a maximum of 1–2 cm apart from the diffusion area (*pink* region). *PA* panniculus adiposus; *SF* stratum fibrosum; *DF* deep fascia; *MF* muscle fascia; *P*: periosteum. (**b**) A magnified and oblique view of the *dotted box* in the fluorescence stereomicroscopic image shown in (**a**). The network of threadlike structures comprises primo-nodes (*white arrows*) and vessels (*white arrow heads*) is presented. These threadlike structures were partly and irregularly fixed on the substrate (stratum fibrosum). Part of these structures (not shown in the image) was observed by using a confocal laser scanning microscope, as shown in (**c**). (**c**) A confocal laser scanning microscopic image of a primo-node (*white arrow*) and vessel (*white arrow head*) shown in (**a**). This picture presents a merged view of the images of DIC (*gray*) and fluorescence of DAPI (*blue*) and nanoparticles (*red*). We can find that the nanoparticles exist nonuniformly inside the primo-node and vessel and that there are round (*black arrow head*) and rod-shaped nuclei (*black arrow*). Note that the rod-shaped nucleus is aligned along the longitudinal direction of the primo-vessel. Scale bar = 10 μm

showed the single and the multiple lumens together (even in the same vessel) depending on their dimensions and states. The nerve bundles were excluded from our considerations because we did not notice any nerve-like structures in the locations where the nanoparticles existed.

Field-emission transmission electron micrographic images of cross sections of the primo-vessel revealed that a bundle of collagen fibers was aligned in the longitudinal direction of the primo-vessel and that a cellular layer covered the extracellular matrix (Fig. 20.2a–d). The collagen fibers were longitudinally aligned along the primo-vessel in an ordered manner. All the collagen fibers seemed to be a single type of ~30 nm in diameter (Fig. 20.2d). In contrast to the case of the primo-vessel, the collagen fibers were randomly distributed in the substrate (stratum fibrosum) (Fig. 20.2a, d). In the cellular layer, an outer membrane was clearly seen in most of the regions of the cross section (Fig. 20.2d). A sinus (3–4 μm in diameter) existed with a bundle structure of extracellular matrices (Fig. 20.2a–c). The sinus was surrounded by a layer of cells. The inner and the outer boundaries of the sinus could be clearly seen (Fig. 20.2b). A continuous gap could be observed between the cells of the sinus and the other part of the primo-vessel (Fig. 20.2c). Nanoparticles were found in the cellular or the extracellular layer (Fig. 20.2c).

4 Discussion

The finding of primo-vascular networks in the loose connective tissue of the hypodermis, which corresponds to the location of the stomach meridian, may have an important implication for the mechanism of acupuncture therapeutics, especially in its anatomical and histological approaches. Our results present the exact anatomical structure (primo-vascular system) that is physically related to the acupoint (ST36) and is distributed around its corresponding meridian (stomach meridian). Furthermore, our results have clearly showed that acupoint-injected nanoparticles exist inside the primo-nodes and vessels (not on their surfaces), which implies that the hypodermic primo-vascular system may function as a microfluidic channel of the body. This makes evident how acupoint-injected radioactive agents showed clear trajectories along the meridians in previous reports [3, 25]. The interesting point is that such a microfluidic channel forms a network structure that is different from the plexus or the branching structures of blood or lymph capillaries. If the acupuncture meridian is a physical fluid channel, its clinical roles become evident in Western biomedicine. For example, it can be utilized as a migration pathway for drug delivery from the skin to organs: By injecting drugs transcutaneously, we may efficiently deliver them to internal organs.

This network of the hypodermic primo-vascular system is similar in shape to those of the primo-vascular system in the heart (endocardium) [12], in the omentum and peritoneum [22], on organ surfaces [26] and on tumor fasciae [21]. This is notable in the point that these primo-vascular systems are all located in/on fascia which is supposed to play a central role in the acupuncture mechanism [7].

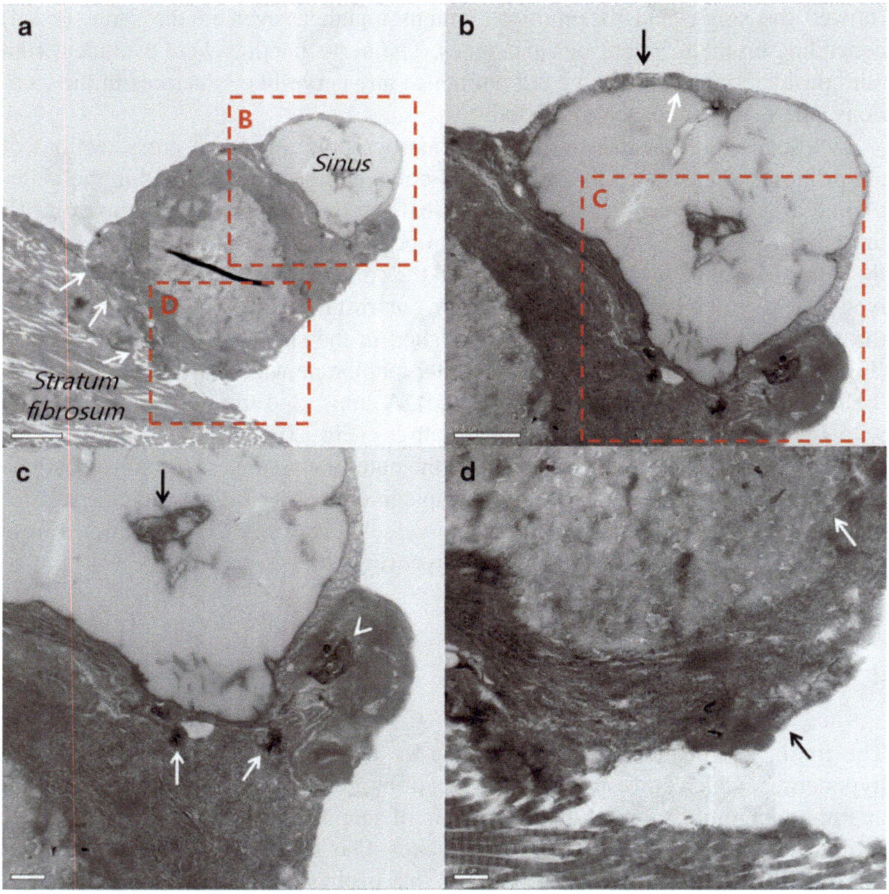

Fig. 20.2 Transmission electron microscopic images of a cross section of the primo-vessel. (**a**) Low-magnification view of the cross section of a primo-vessel and stratum fibrosum. A cellular layer covers the extracellular matrix, and a sinus can be seen. One can recognize a continuous gap between the primo-vessel and the stratum fibrosum (*arrows*). Scale bar = 1 μm. (**b**) The sinus is surrounded by a layer of cell(s). The inner and the outer boundaries of the sinus are clearly seen (*arrows*). Scale bar = 200 nm. (**c**) A continuous gap is clearly seen between the cells of the sinus and those of the other parts of the primo-vessel. Nanoparticles are seen in the cellular or extracellular layer (*white arrows*). We can see a part of cell-like structure inside the sinus (*black arrow*) and part of a nucleus (*arrow head*) of the cell surrounding the sinus. Scale bar = 200 nm. (**d**) Cross sections of densely packed collagenous fibers in a bundle structure are seen in the extracellular matrix. (A cross section of a fiber is indicated by the *white arrow*.) The *black arrow* shows the clear outer boundary of the primo-vessel. Scale bar = 200 nm

Especially, our result implies that the entity that mediates acupuncture stimuli may be not only the global linkage of fasciae in the body but also the network of certain channels (primo-vascular system) embedded in the fasciae. In addition, recalling the fact that all these fasciae (including the hypodermis) can be generally classified as connective tissues, much of the evidence for the role of loose connective tissues in acupuncture therapeutics can be directly related [6]. Especially, since the hypodermis is a representative of loose connective tissues, our finding of a hypodermic primo-vascular system may reveal an unknown mechanism of acupuncture in future investigations.

The regular alignment of collagen fibers along the longitudinal direction of the primo-vessel may provide us certain information not only on the structural aspects but also on the physiological functions of the primo-vessel. Our TEM images of the hypodermic primo-vessels and the substrate of stratum fibrosum indicate that there is a tendency for collagen fibers to exist in a more orderly manner in the primo-vessel than in the substrate. This may mean that the primo-vessels can be under more tension than the surrounding subcutaneous tissues [27]. This conjecture is justified by an observation of Kim [9], in which the primo-vessel can have a spontaneous mechanical motility in vital conditions. Since collagen fibers are thought to play a critical role in transmitting mechanical stimulation on the acupoint to the deep body through fascial networks [28], our images of the collagen architecture of hypodermic primo-vessels may provide an important cue for research on the acupuncture mechanism.

Although in this report, we have shown the structure of the hypodermic primo-vascular system for the first time, our method is limited in that it was not possible to visualize the primo-node in the dermis layer, which is thought to be directly related to the acupoint. In order to reveal the entire structure of the superficial primo-vascular system in the dermis and the hypodermis, more effective, noninvasive, and elaborate methods for bio-imaging and microfluidics need to be developed.

Acknowledgments This work was supported by a Systems Biology Infrastructure Establishment Grant provided by the Gwangju Institute of Science and Technology in 2009; by the National Research Foundation of Korea (NRF) grant funded by the Korea government (MEST) in 2010 (No. 2010-0025289); and by the Korean Pharmacopuncture Foundation Grant funded by the Koran Pharmacopuncture Institute (KPI-2010-003).

References

1. Dung HC (1984) Anatomical features contributing to the formation of acupuncture points. Am J Acupunct 12(2):139–143
2. Heine H (1988) Anatomical structure of acupoints. J Trad Chin Med 8(3):207–212
3. de Vernejoul P, Albarède P, Darras JC (1992) Nuclear medicine and acupuncture message transmission. J Nucl Med 33(3):409–412
4. Ma W, Tong H, Xu W et al (2003) Perivascular space: possible anatomical substrate for the meridian. J Altern Compl Med 9(6):851–859
5. Langenvin HM, Churchill DL, Cipolla MJ (2001) Mechanical signaling through connective tissue: a mechanism for the therapeutic effect of acupuncture. FASEB J 15:2275–2282
6. Langevin HM, Yandow JA (2002) Relationship of acupuncture points and meridians to connective tissue planes. Anat Rec (New Anat) 269:257–265
7. Schleip R, Klingler W, Lehmann-Horn F (2005) Active fascial contractility: Fascia may be able to contract in a smooth muscle-like manner and thereby influence musculoskeletal dynamics. Med Hyp 65:273–277

8. Langevin HM (2006) Connective tissue: a body-wide signaling network? Med Hyp 66:1074–1077
9. Kim BH (1963) On the kyungrak system. J Acad Med Sci DPR Kor 90:1–35
10. Chan SHH (1984) What is being stimulated in acupuncture: evaluation of the existence of a specific substrate. Neurosci Biobehav Rev 8:25–33
11. Fujiwara S (1967) Discovery of acupuncture meridians. Sogensha, Osaka [in Japanese]
12. Lee BC, Kim HB, Sung B et al (2011) Network of endocardial vessels. Cardiology 118:1–7
13. Cho SJ, Kim BS, Park YS (2004) Threadlike structures in the aorta and coronary artery of swine. J Intl Soc Life Info Sci 22:609–611
14. Lee BC, Yoo JS, Baik KY et al (2005) Novel threadlike structures (Bonghan ducts) inside lymphatic vessels of rabbits visualized with a Janus Green B staining method. Anat Rec 286B:1–7
15. Yoo JS, Kim MS, Sung B et al (2007) Cribriform structure with channels in the acupuncture meridian-like system on the organ surfaces of rabbits. Acupunct Electrother Res 32(1/2):130–132
16. Lee BC, Soh KS (2009) Contrast-enhancing optical method to observe a Bonghan duct floating inside a lymph vessel of a rabbit. Lymphology 41:178–185
17. Sung B, Kim MS, Lee BC et al (2010) A cytological observation of the fluid in the primonodes and vessels on the surfaces of mammalian internal organs. Biologia 65(5):914–918
18. Baik KY, Lee J, Lee BC et al (2005) Acupuncture meridian and intravascular Bonghan duct. Key Eng Mater 277–279:125–129
19. Lee BC, Park ES, Baik KY et al (2004) Acridine orange staining method to reveal characteristic features of intravascular threadlike structure. Anat Rec 278B:27–30
20. Lee BC, Yoo JS, Baik KY et al (2008) Development of a fluorescence stereomicroscope and observation of Bong-Han corpuscles inside blood vessels. Ind J Exp Biol 46(5):330–335
21. Yoo JS, Kim HB, Ogay V et al (2009) Bonghan ducts as possible pathways for cancer metastasis. J Acupunct Meridian Stud 2(2):118–123
22. Lee BC, Kim KW, Soh KS (2009) Visualizing the network of Bonghan ducts in the omentum and peritoneum by using Trypan blue. J Acupunct Meridian Stud 2(1):66–70
23. Sung B, Kim MS, Lee BC et al (2008) Measurement of flow speed in the channels of novel threadlike structures on the surfaces of mammalian organs. Naturwissenschaften 95(2):117–124
24. Sung B, Kim MS, Corrigan A et al (2009) In situ microextraction method to determine the viscosity of biofluid in threadlike structures on the surfaces of mammalian organs. Phys Rev E 79:022901
25. Kovacs FM, García A, Mufuraggi N et al (2000) Migration pathways of hypodermically injected technetium-99m in dogs. Eur Radiol 10:1019–1025
26. Lee BC, Yoo JS, Ogay V et al (2007) Electron microscopic study of novel threadlike structures on the surfaces of mammalian organs. Microsc Res Tech 70:34–43
27. Cooper JA Jr, Baily LAO, Carter JN et al (2006) Evaluation of the anterior cruciate ligament, medial collateral ligament, achilles tendon and patellar tendon as cell sources for tissue-engineered ligament. Biomater 27:2747–2754
28. Yu X, Ding G, Huang H et al (2009) Role of collagen fibers in acupuncture analgesia therapy on rats. Connect Tissue Res 50:110–120

Part III
Primo Microcell (SanAl) and Stem Cells

Chapter 21
Identification and Characterization of Small Stem-Like Cells in the Primo Vascular System of Adult Animals

Vyacheslav Ogay and Kwang-Sup Soh

Abstract Pluripotent stem cells have been recently identified in many tissues and intensively studied for their cell-therapeutic potential to renew and replace lost cells in any damaged tissue or organ. However, the true nature and origin of these cells is still unclear. Based on data obtained by Dr. Bonghan Kim (1965), we put forward a hypothesis that a rare population of small stem-like (SSL) cells found in the primo-vascular system (Bonghan system) possesses a broad tissue differentiation potential and that these cells are capable of differentiation into all specialized cells in the adult body. To examine the hypothesis in this study, we isolated SSL cells from primo-vascular system of rats and characterized their morphological and phenotypic features. Our microscopic examinations indicated that SSL cells are very small, approximately 3–4 μm in diameter. Hematoxyline-eosin staining revealed that SSL cells have an extremely dense eccentrically located round nucleus surrounded by basophilic cytoplasm. Immunofluorescence analysis showed that SSL cells expressed embryonic stem cell markers: CD133, Oct 4, and Nanog. Thus, based on our preliminary data, we suppose that SSL cells in the primo-vascular system might be variants of embryonic stem-like cells which can differentiate into tissue-specific cells and regenerate damaged tissue or organ.

1 Introduction

In the 1960s, for the first time, North Korean scientist Kim Bonghan described morphological and ultrastructural features of a new type of cells found in the Bonghan system (primo-vascular system) of adult animals [1]. These cells were small round

V. Ogay (✉)
Laboratory of Immunochemistry and Immunobiotechnology, National Center for Biotechnology of the Republic of Kazakhstan, Astana 010000, Kazakhstan
e-mail: ogay76@gmail.com

K.-S. Soh et al. (eds.), *The Primo Vascular System: Its Role in Cancer and Regeneration*, DOI 10.1007/978-1-4614-0601-3_21, © Springer Science+Business Media, LLC 2012

cells (1–3 μm) with basophilic stained nuclei and scant cytoplasm. According to Kim, these small cells were mostly located in the corpuscle-like structures (primo-nodes) where they actively proliferate and then migrate through Bonghan ducts (primo-vessels) into different tissues and organs to differentiate into adult stem cells. Kim suggested that these small cells regularly circulate in the primo-vascular system and are able to regenerate damaged tissues and organs. Although his data and suggested hypothesis met with scepticism at that time, recently there has been evidence of the presence of small cells with pluripotent/multipotent characters in different adult tissues and organs of human and animals [2]. Small stem cells have been found in bone marrow, brain, umbilical cord blood, cartilage, ovarian surface epithelium, kidney, and liver [3–7]. However, the most detailed study was performed for a rare population of very small embryonic-like stem cells (VSELs) that has been identified by Ratajczak et al. in bone marrow and other adult tissues [8, 9]. They revealed that VSELs measure ~3.6 μm and express several markers characteristic of pluripotent stem cells (PSCs) that are characteristic of epiblast/germ line-derived stem cells. It was suggested that some of these small stem cells are deposited in each organ during gastrulation and organogenesis, and they may serve as reserve pool of quiescent PSCs. Moreover, these cells may participate in cellular turnover and the rebuilding pool of the tissue-specific stem cells, especially under urgent circumstances such as tissue injury [10].

Recently, we reinvestigated Kim's data and confirmed the existence of small cells in the primo-vascular system of adult animals with Feulgen reaction, transmission electron and atomic force microscopy [11, 12]. Therefore, our next study was to examine whether these small cells possess the same features as pluripotent or multipotent stem cells which were found recently in adult tissues and organs. In the present study, we identified a rare population of small stem-like (SSL) cells in the primo-nodes of adult rats and characterized their basic morphological features. Moreover, we show here that SSL cells express stem cell marker CD133 and pluripotent transcription factors such as Oct4 and Nanog. It is suggested that a rare population of SSL cells found in the primo-vascular system have a broad tissue differentiation potential and that these cells are capable of differentiation into all specialized cells in the adult body.

2 Methods

2.1 Animal Preparation and Surgical Procedure

Male 10–12-week-old Wistar rats were housed in a temperature-controlled environment (23°C) with 60% relative humidity. All animals were exposed to a 12-h light–dark cycle and provided food and water *ad libitum*. Procedures involving animals and their care conformed to institutional guidelines and were in full compliance with current laws and policies. Animals were sacrificed by a high dose of ethyl ester. Primo-nodes and primo-vessels on the surfaces of internal organs were observed under a stereomicroscope (Olympus SZ51, Japan).

2.2 *Histology and Immunohistochemistry*

Primo-nodes were fixed in 10% neutral buffered formalin for 12 h at 4°C, rinsed in PBS, dehydrated in a graded ethanol series, clarified in xylene, and embedded in Paraplast (Sigma, USA). Paraplast sections, 7 μm in thickness, were next stained with hematoxylin-eosin (H&E) for general morphological observation, mounted with Neomount, and examined with a light microscope. To identify CD133 expression on small cells, tissue sections of primo-nodes were stained with CD133 antibody (dilution 1:200, Abcam, UK). After washing, the sections were incubated with the secondary antibodies Alexa 488-conjugated goat anti-rabbit IgG (H+L) (dilution 1:500, Invitrogen, USA). All antibodies were diluted in 10% normal goat serum. After each incubation with antibodies, the samples were washed with PBS containing 0.05% Tween 20. Immunostained samples were mounted with antifading reagent with DAPI (Invitrogen, USA) and covered by a coverslip. The slides were examined with fluorescent microscope.

2.3 *Isolation of SSL Cells*

The primo-nodes were finely minced with scissors and subjected to enzymatic digestion for 15 min at 37°C with 0.25% trypsin/0.02% EDTA solution. Then the resulting cell suspension was centrifuged at 300 × g for 10 min at room temperature. Afterward the cells were resuspended in DMEM supplemented with 10% fetal bovine serum and centrifuged at 300 × g for 10 min. The resulting cell pellet was resuspended in leftover volume of DMEM. For immunoselection of SSL cells from single-cell suspension, the cells were incubated with a rabbit polyclonal antibody to CD133 (1:200 dilution, Abcam, UK). The CD133-positive cells were purified by incubation with magnetic goat anti-rabbit IgG microbeads (1:500 dilution, Milteny Biotech, Germany) and selection on a Mini-MACS apparatus (Miltenyi Biotec, Germany) following the protocol recommended by the manufacturer.

2.4 *Immunocytochemistry*

SSL cells were fixed in 4% paraformaldehyde for 30 min and then washed in PBS containing 0.1% Saponin. After centrifuging at 300 × g for 10 min the cells were resuspended in leftover volume of PBS. Next, the cell suspension was incubated with mouse anti-Oct4 (1:200 dilution, Abcam, UK) and rabbit anti-Nanog (1:200 dilution, Abcam, UK). After washing, the cells were incubated with the appropriate secondary antibodies Alexa 555-conjugated goat anti-mouse IgG (H+L) and Alexa 488-conjugated goat anti-rabbit IgG (H+L) (Invitrogen, USA). All antibodies were diluted in 10% NGS in PBS containing 0.1% Saponin. After each incubation with antibodies, the cells were washed by PBS containing 0.1% Saponin.

Immunostained SSL cells were mounted with antifading reagent with DAPI and covered by a coverslip. The slides were examined with a confocal microscope.

3 Results

In order to identify SSL cells in the primo-nodes and describe their morphological features in situ, histochemical study was carried out. Our microscopic observation revealed that SSL cells have a size less than erythrocytes, approximately 3–4 μm in diameter (Fig. 21.1b). They have an extremely dense, eccentrically located round nucleus. The nucleus does not contain any nucleoli. SSL cells also exhibit medium nuclear to cytoplasmic ratio. The cytoplasm of SSL cells may contain a large number of ribosomes (basophilic staining). In addition, they form characteristic cytoplasmic bud-shaped protrusions. Immunohistochemistry revealed that these SSL cells were positive for a stem cell marker CD133 as shown in Fig. 21.1c–e. Next, to separate SSL cells, primo-nodes and primo-vessels were trypsinized and filtered through the

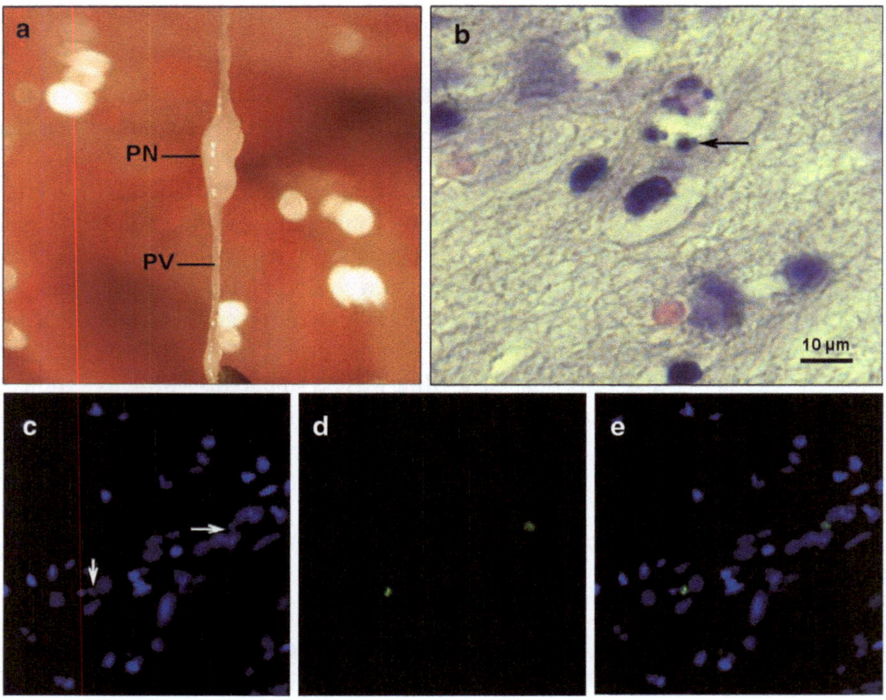

Fig. 21.1 SSL cells in the primo-node. (**a**) Stereomicroscopic image of typical primo-node (PN) and primo-vessel (PV) on the surface of internal organs, (**b**) H&E staining shows general morphology of SSL cells (*arrow*) in the primo-node, (**c**–**e**) immunofluorescence staining demonstrates that CD133 is expressed on SSL cells, (**c**) cell nuclei are stained with DAPI, (**d**) CD133 expression (*green*), (**e**) merge image of (**c**) and (**d**)

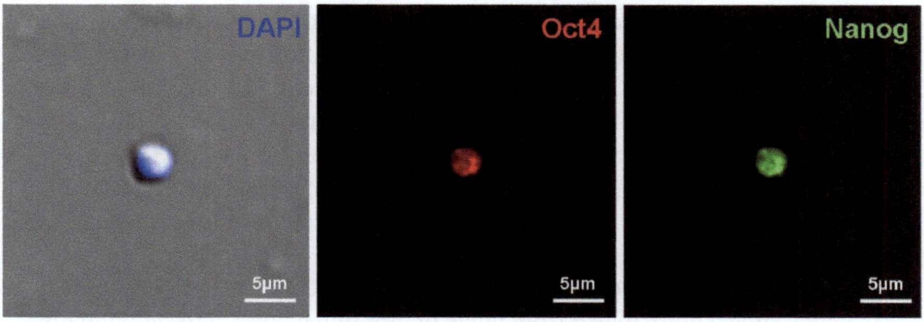

Fig. 21.2 Representative images showing expression of embryonic stem cell markers Oct4 (*red*) and Nanog (*green*) in SSL cells isolated from primo-nodes. Cell nucleus is stained with DAPI (*blue*). The scale bar indicates 5 μm

cell strainer to obtain a single-cell suspension containing cells of varying size and shape. Dissociated single cells were incubated with CD133 antibody and then immunoselected with magnetic microspheres. However, the number of separated CD133 positive cells was insignificant, about 0.05% of the total number of nucleated cells. As shown in Fig. 21.2, freshly isolated SSL cells measure about 4 μm. As soon as SSL cells were selected with CD133 antibody, we tested whether these cells express Nanog and Oct4. Nanog and Oct4 are transcription factors involved in stem cell self-renewal and are thought to be the intrinsic factors that mediate pluripotency in embryonic stem cells [13]. We found that most SSL cells were positive for these pluripotent transcription factors by immunocytochemistry (Fig. 21.2).

4 Discussion

In this study, we have identified and initially characterized a rare population of SSL cells found in the primo-vascular system of adult rats. They are very small, in the range less than 5 μm and have primitive morphology. These cells have relatively large round nuclei surrounded by a small proportion of cytoplasm. By immunohistochemistry, we revealed that SSL cells expressed CD133 which is well recognized as a stem cell marker for normal and cancerous tissues. In accordance with literature data, CD133 alone or in a combination with other markers is currently used for the detection and immunoselection of stem cells from numerous tissues, such as bone marrow, brain, kidney, liver, pancreas, and skin [14]. Therefore, to isolate SSL cells from primo-nodes, we used the method of indirect immunomagnetic separation with CD133 antibody.

SSL cells isolated from primo-nodes could be compared to the rare population of VSEL found in different human and animal adult tissues and organs as reported by Ratajczak et al. [7–9]. Ratajczak and co-workers isolated VSEL with a diameter from 3 to 5 μm, which is very comparable to the diameter of stem-like cells found in the primo-nodes. Similar to VSEL stem cells, our SSL cells were Nanog and

Oct-4 positive. These data could be confirmed by the findings of some other research groups, who published the existence of small stem cells with pluripotent characteristics in some adult tissues and organs. Virant-Klun identified small putative stem cells in human ovarian surface epithelium with embryonic character [6]. They measure 2–4 μm and express embryonic markers such as Oct4, Nanog, SSEA-4, Sox-2, and c-kit. McGuckin et al. found small PSCs (2–3 μm in diameter) in human umbilical cord blood [5]. These cells also exhibit embryonic stem cell markers including Oct4 and Nanog and have neural differential potential.

Interesting data were reported by Vacanti et al., who described a novel population of extremely small cells isolated from many organs of the body in the adult mammal [3]. They called these cells spore-like cells. Preliminary characterization of these cells utilizing basic and special stains, as well as scanning and transmission electron microscopy, revealed very small undifferentiated cells, which contain predominantly nucleus within a small amount of cytoplasm and a few mitochondria. In vitro, these cells had the capacity to enlarge, develop, and differentiate into cell types expressing characteristics appropriate to the tissue environment from which they were initially isolated. Taken together, data demonstrated by Vacanti are in accordance with the results of Bonghan Kim, which were described in his last paper on differentiation of SSL cells (Bonghan granules) into different types of blood cells [15]. Kim hypothesized that Bonghan granules found in the corpuscle-like structures may serve as pool of dormant PSCs, which can give rise to adult stem cells. Our data presented in this study partially confirm Kim's hypothesis on the existence of SSL cells with embryonic characteristics in the primo-vascular system. We suggest that SSL cells might be variants of embryonic stem-like cells that persist throughout development to reside in all tissue compartments. Apparently, they might generate diverse cell types during tissue renewal or repair in response to environmental cues. However to confirm this assumption, more research on pluripotent gene expression and differential potential of SSL cells should be carried out.

References

1. Kim BH (1965) Sanal theory. J Acad Med Sci DPRK 168:5–38
2. Zuba-Surma EK, Kucia M, Ratajczak J et al (2009) "Small stem cells" in adult tissues: very small embryonic stem cells stand up! Cytometry Part A 75A:4–13
3. Vacanti MP, Roy A, Cortiea J et al (2001) Identification and initial characterization of spore-like cells in adult mammals. J Cell Biochem 80(3):455–460
4. Chen MH, Lin S, Hsieh CH et al (2004) Identification and initial characterization of small cells in adult cartilage and bone marrow. J Formos Med Assoc 103(4):264–273
5. McGuckin C, Jurga M, Ali H et al (2008) Culture of embryonic-like stem cells from human umbilical cord blood and onward differentiation to neural cells in vitro. Nat Protoc 3(6):1046–1055
6. Virant-Klun I, Zech N, Rozman P et al (2008) Putative stem cells with embryonic character isolated from ovarian surface epithelium of women with no naturally present follicules and oocytes. Differentiation 76:843–856

7. Kucia M, Reca R, Campbell FR et al (2006) A population of very small embryonic-like (VSEL) CXCR4(+)SSEA-1(+)Oct-4+stem cells identified in adult bone marrow. Leukemia 20(5):857–869
8. Ratajczak MZ, Kucia M, Ratajczak J et al (2009) A multi-instrumental approach to identify and purify very small embryonic like stem cells (VSELs) from adult tissues. Micron 40:386–393
9. Ratajczak MZ, Zuba-Surma EK, Machalinski B et al (2008) Very small embryonic-like (VSEL) stem cells: purification from adult organs, characterization, and biological significance. Stem Cell Rev 4:89–99
10. Ratajczak MZ, Machalinski B, Wojakovski W et al (2007) A hypothesis for an embryonic origin of pluripotent Oct4+ stem cells in adult bone marrow and other tissues. Leukemia 21:860–867
11. Ogay V, Baik KY, Lee BC et al (2006) Characterization of DNA-containing granules flowing through the meridian-like system on the internal organs of rabbits. Acupunct Electrother Res 31:13–31
12. Baik KY, Ogay V, Jeoung SC et al (2009) Visualization of Bonghan microcells by electron and atomic force microscopy. J Acupunct Meridian Stud 2:124–129
13. Nichols J, Zevnik B, Anastassiadis K et al (1998) Formation of pluripotent stem cells in the mammalian embryo depends on the POU transcription factor Oct4. Cell 95:379–391
14. Yin AH, Miraglia S, Zanjani ED et al (1997) AC133, a novel marker for human hematopoietic stem and progenitor cells. Blood 90:5002–5012
15. Kim BH (1965) Sanal/cell cycle of blood cells. J Acad Med Sci DPRK 114:1–6

Chapter 22
Membrane Mechanical Property of Primo Microcells

Ku Youn Baik, Chang Ho Kim, Suk Yi Woo, Sae Chae Jeoung, and Kwang-Sup Soh

Abstract Primo microcells flowing in primo vessels are hypothesized to function like very small embryonic-like adult stem cells. We extracted them from organ surface primo nodes of rats, and studied their mechanical membrane properties with atomic force microscopy. Primo microcells have a higher elastic modulus than apoptotic bodies, which implies a difference in their origins and functions. This work suggests that primo microcells represent novel biological objects different from apoptotic bodies.

1 Introduction

Many kinds of intercellular micro-size particles in a body have attracted researchers' attention due to their transformative and communicative properties [1–5]. Representative micro-size particles are microparticles, apoptotic bodies, or platelets, and some apoptotic bodies contain DNA among them. Primo microcells were found in 1965 inside the primo-vascular system which had a role in tissue regeneration [6]. They had micro-scale size, round shape, and DNA as well as other intercellular micro-size particles. Since the study on primo microcells has not been continued by other researchers, a comparison to well-known micro-size particles is necessary. In previous works, we showed their morphological difference from apoptotic bodies by using confocal microscopy and electron microscopy [7–9]. In this proceeding, we report the distinct mechanical property of primo microcells in comparison with apoptotic bodies. Atomic force microscopy revealed the high elasticity of their outer membranes. This mechanical property can be used as a rough criterion for primo microcells without further molecular information.

K.Y. Baik (✉)
Plasma Bioscience Research Center,
Kwangwoon University, Seoul, South Korea
e-mail: a22joa@hanmail.net

K.-S. Soh et al. (eds.), *The Primo Vascular System: Its Role in Cancer and Regeneration*, 157
DOI 10.1007/978-1-4614-0601-3_22, © Springer Science+Business Media, LLC 2012

2 Methods

SD rats that were 8–10 weeks old (250–300 g) were used as experimental animals. Seoul National University Institutional Animal Care and Use Committee (SNUIACUC) oversaw this research and the approval number is SNU-070312-6.

The primo vascular corpuscles and ducts on organ surfaces were disrupted by using a glass homogenizer in 2 mL phosphate-buffered saline (PBS), and the homogenate containing tissue debris, whole cells, and primo microcells were sieved to remove big tissue debris. Then, the differential centrifugation method for microcell separation was applied [10]. The homogenate was centrifuged at 80×g for 10 min and at 1,000×g for 10 min in sequence. All steps of separation were performed with a fixed angled rotor (A1.5S-30) at 4°C in a Combi 514R centrifuge (Hanil Science Industrial, Korea).

CELLTAK (BD sciences, USA) mixed with 2 M sodium carbonate (4:1) was incubated on cellocate coverglass (55 μm square grids, sterilized, Eppendorf, Germany) at 37°C for 20 min and washed with distilled water. The separated primo microcells were loaded on the CELLTAK-coated cellocate and incubated at 37°C for 20 min in Hank's solution. After incubation, the unadhered granules were washed out using the same buffer. They were stained with acridine orange, and their position was specified on a cellocate by fluorescence from DNA. The DNA-containing microcells were imaged in tapping mode in air using AFM (Picoplus, Molecular Imaging, USA) with typical nitrite tapping cantilever, $k=0.16$ N/m.

To investigate the elastic value of the sample, the force distance curve was measured using AFM with high-resolution video microscopy. V-shaped silicon nitride cantilevers with a nominal spring constant $k=0.01$ N/m with a regular four-sided pyramidal tip of nominal semi-included angle $\theta=35°$ (MSCT-25) were used. A cantilever was immersed in PBS for 1 h to be stabilized before measuring. The photodiode was calibrated by recording the force curves of the bare coverglass in PBS using a deflection range of ~100 nm (F ~ 1 nN). One cycle took 3 s, and 2,000 points/cycle were recorded. A Petri dish with a sample immobilized coverslip was placed on the AFM plate, and a 1×1-μm area was scanned at a speed of 1 s/line, and the force curves were taken at 16 points per line. The Hertz model was used to obtain Young's modulus for the samples [11]. The value taken from each sample was plotted and log-normally fitted to get an expectation value.

In the case of apoptotic bodies, the neuroblastoma cell line (N2A) was cultured on coverglass in Dulbecco's Modified Eagle Medium media, and glutamate (80–120 mM) was added for 12 h to induce apoptosis [12]. The apoptotic bodies were gathered from 120-mM-treated N2A cells by using the same differential centrifugation and the same process followed for the primo microcells.

3 Results and Discussion

The force–distance curves from the highest positions of each sample (such as a nucleus for a cell) were chosen for calculation. N2A cells in normal media and N2A cells in 80, 100, and 120-mM glutamate added media were examined as controls (Fig. 22.1a and b).

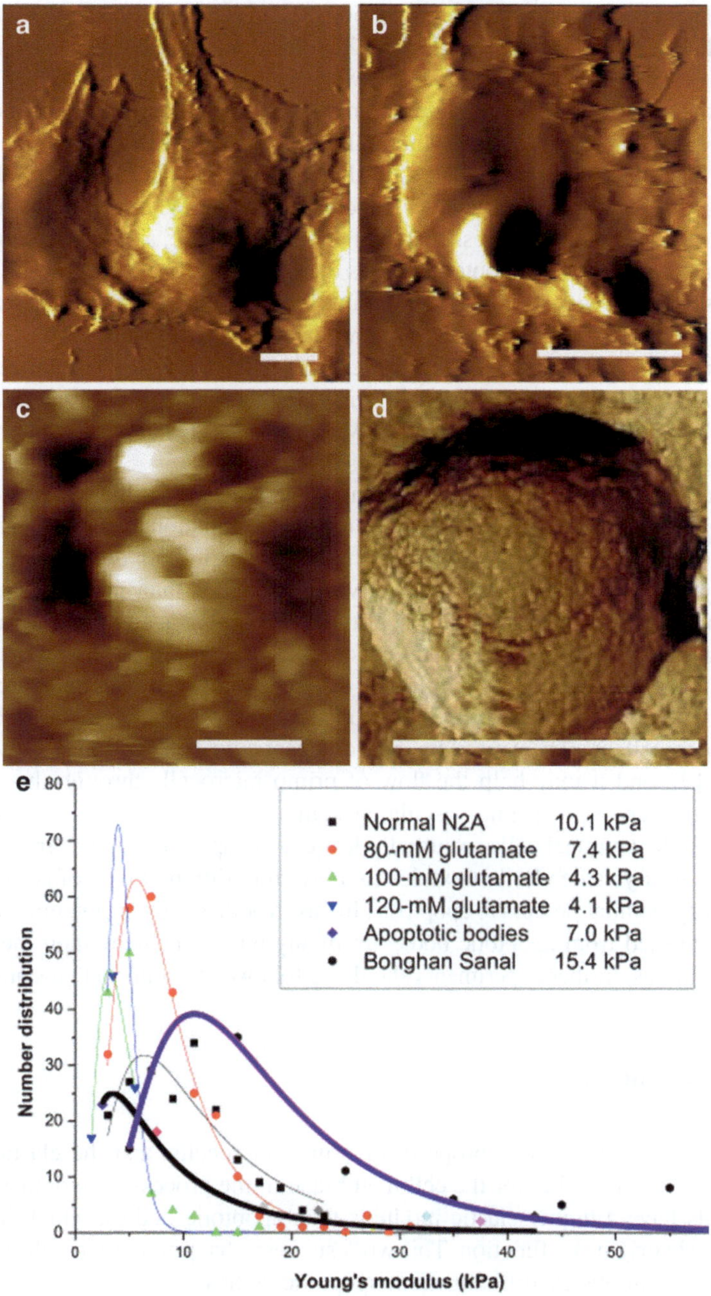

Fig. 22.1 Atomic force microscopy topography images (**a–d**) and force measurement (**e**) of N2A cells in normal condition (**a**), N2A cells under apoptosis (**b**), apoptotic bodies from N2A cells (**c**), and primo microcells from the primo-vascular system (**d**). The elastic modulus of cells decreased with increasing amount of apoptosis-inducing chemicals, and the elastic modulus of primo microcells is higher than that of apoptotic bodies by more than 2 times. All the scale bars are 1 μm

Cytoskeletal disruption and budding were observed in cells under apoptosis. The apoptotic bodies induced from N2A cells (Fig. 22.1c) and primo microcells taken from the primo-vascular system of rats (Fig. 22.1d) were examined together.

As expected, the Young's modulus from N2A cells was decreased with increasing amount of glutamate. The value of normal cells showed a wide distribution with an expectation value of 10.1 kPa, and the value of the cells under apoptosis showed a narrower distribution with a smaller expectation value. With 120-mM glutamate, the elasticity was reduced to less than half the normal value, as shown in Fig. 22.1e. The Young's modulus of apoptotic bodies detached from the cells showed a rather wide distribution with an expectation value of 6.96 kPa. On the other hand, the Young's modulus of primo microcells showed a much higher value. The expectation value was 15.4 kPa, which was even bigger than the value for normal cells. The elastic value of primo microcells is obviously smaller than those of bacteria or fungi whose elasticity is in the range of tens of MPa [13, 14]. This confirms that the primo microcells are not contaminated from outside.

The cell elasticity is known to be mainly determined by the cytoskeleton, an intricate polymer network [15]. Changes to cellular function during differentiation [16] or due to disease are mirrored in the cytoskeleton. For example, malignant hepatocytes (liver cells) have been shown to be stiffer than normal hepatocytes [17]. On the other hand, the elasticity of the environment is an important aspect of the niche and has been shown to influence differentiation pathways of stem cells [18]. In this point of view, the significant difference between primo microcells and apoptotic bodies reflects their different function in vivo. The high elasticity of the primo microcells may be caused by their compact cytoskeleton and closely packed vesicles [9], and it may help the flow of primo microcells through the complex fibrillar structure of the primo-vascular systems.

Though the neuronal cells are usually less elastic, apoptotic bodies from neurons may have nearly same elastic modulus to those from different type of cells because they undergo similar process to apoptosis. In this hypothesis, the primo microcells can be distinguished from apoptotic bodies from any type of cells by their mechanical property. In this, further experiments should be followed to confirm that statistically.

4 Conclusions

We showed the mechanical property of primo microcells, and the elasticity was compared with the value for the cell membrane in the process of apoptosis. Primo microcells have a higher elastic modulus than apoptotic bodies, which implies a difference in origin and function. This work suggests that primo microcells represent novel biological objects different from apoptotic bodies.

Acknowledgments This work was financially supported by Ministry of Knowledge Economy (SCJ). K. Soh thanks the Association of Korean Oriental Medicine for support. KY Baik acknowledges the support from the National Research Foundation of Korea Grant (No. 20100029418).

References

1. Hugel B, Martinez MC, Kunzelmann C et al (2005) Membrane microparticles: two sides of the coin. Physiology 20:22–27
2. Baj-Krzyworzeka M, Majka M, Pratico D et al (2002) Platelet-derived microparticles stimulate proliferation, survival, adhesion, and chemotaxis of hematopoietic cells. Exp Hematol 30:450–459
3. Distler JHW, Jüngel ALC, Seemayer CA et al (2005) The induction of matrix metalloproteinase and cytokine expression in synovial fibroblasts stimulated with immune cell microparticles. PNAS 102:2892–2897
4. Dini L, Coppola S, Ruzittu MT et al (1996) Multiple pathways for apoptotic nuclear fragmentation. Exp Cell Res 223:340–347
5. Buikis I, Harju L, Freivalds T (1999) Origin of microcells in the human sarcoma cell line HT-1080. Anal Cell Pathol 18:73–85
6. Kim BH (1965) The Sanal theory. J Acad Med Sci DPR Korea 108:39–62 [in Korean]
7. Ogay V, Baik KY, Lee BC et al (2006) Characterization of DNA-containing granules flowing through the acupuncture meridian system of rabbits. Acupunct Electrother Res 31:13–31
8. Kwon J, Baik KY, Lee NJ et al (2007) Scanning probe microscopy study of microcells from the organ surface Bonghan corpuscle. APL 90:173903
9. Baik KY, Ogay V, Jeoung SC et al (2009) Visualization of Bonghan Microcells by Electron and Atomic Force Microscopy. J Acupunct Meridian Stud 2:124–129
10. Stubblefield E, Pershouse M (1992) Direct formation of microcells from mitotic cells for use in chromosome transfer. Som Cell Mol Gen 18:485–491
11. Johnson KL (1985) Contact mechanics. Cambridge University Press, Cambridge
12. Zhang YM, Lu XF, Bhavnani BR (2003) Equine estrogens differentially inhibit DNA fragmentation induced by glutamate in neuronal cells by modulation of regulatory proteins involved in programmed cell death. BMC Neurosci 4:1–15
13. Boulbitch A, Quinn B, Pink D (2000) Elasticity of the rod-shaped gram-negative eubacteria. Phys Rev Lett 85:5246–5249
14. Zhao L, Schaefer D, Xu H et al (2005) Elastic properties of the cell wall of *Aspergillus nidulans* studied with atomic force microscopy. Biotechnol Prog 21:292–299
15. Titushkin I, Cho M (2007) Modulation of cellular mechanics during osteogenic differentiation of human mesenchymal stem cells. Biophys J 93:3693–702
16. Yourek G, Hussain MA, Mao JJ (2007) Cytoskeletal changes of mesenchymal stem cells during differentiation. ASAIO J 53:219–228
17. We ZZ, Zhang G, Long M et al (2000) Comparison of the viscoelastic properties of normal hepatocytes and hepatocellular carcinoma cells under cytoskeletal perturbation. Biorheology 37:272–290
18. Engler AJ, Sen S, Sweeney HL et al (2006) Matrix elasticity directs stem cell lineage specification. Cell 25:677–689

Chapter 23
Primo Microcell in a Primo Node as a Possible Origin of Adult Stem Cells

Seong-hun Ahn, Sung-won Lee, Sung-Yeoun Hwang, Jae-hyo Kim, and In-chul Sohn

Abstract Adult stem cells have been intensively studied for their cell-therapeutic potential to renew damaged tissues or organs because their use eliminates the ethical, legal, and political concerns with using embryonic stem cells. On the other hand, a microcell contains a micronucleus that has one or a few chromosomes with a small amount of cytoplasm. Recently, microcells have become a possible origin of adult stem cells because of their ability to induce undifferentiated cells. In the early 1960s, after its first observation by Kim, the primo vascular system (PVS) was thought to be an anatomical basis for traditional acupuncture systems (termed *kyungrak* theory or meridians). Kim also claimed that about 1.0-μm-sized DNA-containing granules, named primo microcells or *sanals*, flowed through the PVS consisting of primo vessels and primo nodes to internal organs to proliferate and differentiate into specific cells in the organs in a manner similar to that of adult stem cells. In this study, we harvested primo microcells from Sprague Dawley rats about 250–300 g. The primo microcells were observed to grow into cell-like body entities. These interesting results suggest that Kim's manual pictures from his article 1965 and *Sanal* theory may be true based on experimental studies similar to ours. Also, primo microcells in primo nodes might be one possible origin of adult stem cells. Understanding the primo microcell is thought to be one of the most fundamental studies for cell therapy.

1 Introduction

Stem cells having a self-renewal ability to proliferate in an undifferentiated or unspecialized state and the capability to differentiate or specialize along multiple lineages [1] are classified as mesenchymal stem cells, hematopoietic stem cells, and

I.-c. Sohn (✉)
Department of Meridian and Acupoint, College of Oriental Medicine,
Wonkwang University, Jeollabuk, South Korea

Research Center of Traditional Korean Medicine, Wonkwang University, Jeollabuk, South Korea
e-mail: ichsohn@wku.ac.kr

K.-S. Soh et al. (eds.), *The Primo Vascular System: Its Role in Cancer and Regeneration*, DOI 10.1007/978-1-4614-0601-3_23,

stromal stem cells from bone marrow, umbilical cord blood stem cells, embryonic stem cells, adipose-derived adult stem cells, etc. [2]. Among these, adult stem cells have received much scientific attention and have avoided medical safety and efficacy issues because they eliminate the ethical, legal, and political concerns associated with embryonic stem cells. Recent studies indicate that nascent stem cells exist within other adult tissues, including the brain [3], dermis [4], periosteum [5], skeletal muscle [6], synovium [7, 8], trabecular bone [5], and vasculature [9] tissues, but the most abundant and accessible source of adult stem cells is adipose tissues [1]. So far, the indices and marks common to distinguish nascent stem cells from progenitor cells, as well as the indices and specific characteristics of oneself, have not been examined closely because close study of stem cells is difficult.

In the early 1960s, after its first observation by Kim, the primo vascular system (PVS, also named the Bonghan system) [10], was thought to be the anatomical basis for traditional acupuncture systems (termed kyungrak theory or meridians) [11, 12]. Kim also claimed that about 0.5–1.0-μm-sized DNA-containing granules (named primo microcells or sanals, Bonghan granules) flowed through the PVS consisting of primo vessels and nodes to internal organs to proliferate and differentiate into specific cell in the organs [13] in a manner functionally similar to that of adult stem cells.

Recently, a series of interesting studies of primo microcells were advanced. Baik et al. [14] reported that 2-μm-sized DNA-containing granules observed in primo nodes might be a sort of cell-free DNA. Ogay et al. [15] studied that fragmented 1.7–2.5-μm-sized DNA-containing granules had cytoplasm and that a tri-laminar plasma membrane might be the origin of adult stem cells. Specifically, Kim et al. [16] suggested that Bonghan corpuscles (primo nodes) might be stem cell niches, hinting that the DNA-containing granules might be a new type of stem cell because of integrin ß1, collagen type 1, fibronectin, and Thy1 expressions. Baik et al. [17] studied the morphological features of primo microcells by using scanning electron microscopy (SEM) and atomic force microscopy (AFM) and suggested that primo microcells were a kind of stem cell because of their being joined by two nuclei to each other. In a previous study, we studied the morphology of organ-surface primo vessels and nodes in rabbits [18] and rats [19] to confirm the existence of primo microcells in primo nodes. In this study, we report primo microcell culture results that support Kim's claim on Sanal theory.

2 Materials and Methods

2.1 Animals

The Sprague Dawley rats of about 250–300 g used in this study were obtained from Samtaco Laboratory Animal Company (Osan, Korea). The animals were housed in a temperature-controlled environment (23°C) with 60% relative humidity, a 12-h light/dark cycle, and ad-libitum access to food and water. All the procedures involving the animals and their care conformed to institutional guidelines, which were in full compliance with current international laws and policies (Guide for the Care and Use of Laboratory Animals, National Academy Press, 1996). This study was approved by the Institute of Laboratory Animal Resources at Wonkwang University.

2.2 *Surgical Procedures*

The rats were injected with urethane (1.5 g/kg) i.m. before the surgical procedure. All surgical procedures were performed under general anesthesia. We incised the skin along the medial alba of the abdomen very carefully and opened the abdomen. The suprahepatic vena cava was clipped with forceps before the hepatectomy. The liver was isolated from the abdominal cavity and was rapidly moved to a phosphate buffer saline (PBS, pH 7.2) pool and washed 3–5 times very carefully. We carefully searched for primo nodes on the liver-lobe surfaces by using small surgical instruments, such as iris scissors, microforceps, and needles, for manipulation. The search for the nodes and the vessels was carried out under a stereoscopic microscope (SZX10, Olympus, Japan).

2.3 *Cell Culture*

The primo nodes isolated on a clean bench were moved to a cell culture medium. After transfer, the samples were cut 3–4 times and spun down at 2,000 rpm for about 5 min to separate the red blood cells (RBCs). The supernatant was cultured in a CO_2 incubator. The results were pictured with a microscope (Olympus, Japan) by using NIS-Elements software (Nikon, Japan).

3 Results and Discussion

3.1 *Organ Surface Primo Vessels and Nodes in Rat Liver*

The PVS suggested with kyungrak by Kim [10, 11] is well known to be different from the vascular system, the nerve system, and the lymph system [20]. Until now, the observation methods for the PVS have varied. Among them, staining methods were generally used, and the staining dyes that were mainly used were trypan blue [21, 22], acridine orange [23], janus green B [20], and alician blue [24–27]. However, until now no specific antibody has been observed on primo nodes and primo vessels, the appearances of the observed primo nodes have varied, and the locations where they have been observed have not been the same.

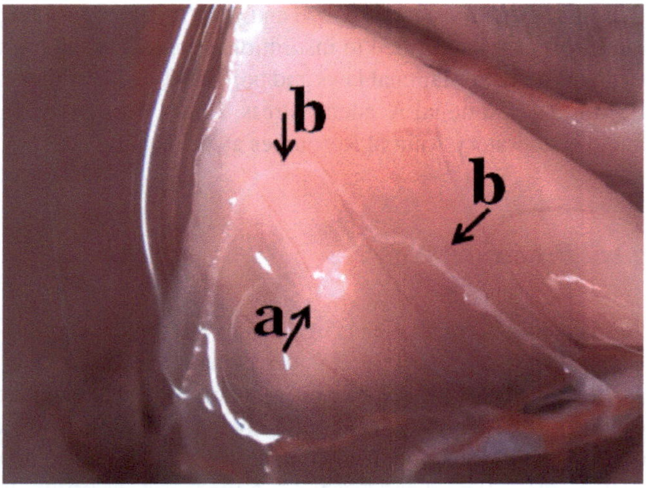

Fig. 23.1 The prominent type of primo node and vessels on the liver surface in rats: (*a*) primo node and (*b*) primo vessel

We formed two hypotheses. First, the PVS existed uniformly but observation methods were not sufficient to observe primo nodes and primo vessels regularly. Second, we did not know what conditions caused primo nodes and primo vessels to be in an activation state or in an inactivation state. Furthermore, we did not know if the PVS to be observed was in an activation state or in an inactivation state. For example, in the living body, the fatty tissue volume and the adipocyte number become larger in nutritive sufficient conditions, but in nutritive insufficient conditions, the volume and the number become smaller.

The primo node and primo vessels in Fig. 23.1 were observed on the surface of a rat's liver by using a visual observation method with stereomicroscopy. The observed primo node and primo vessels were milky in color and semi-transparent, with no blood vessels for nutrition. This is a unique characteristic that distinguishes the PVS from the lymph and the nerve systems.

3.2 *Primo Microcell Growth to a Cell-Like Body*

The sanal theory of Kim BongHan's hypothesis was that the sanal (a living seed in Korean, renamed primo microcell in 2009), which flows through the primo vessels, differentiated into a specific cell for proliferation and reproduction [12, 13]. However, sanal theory as claimed by Kim BongHan was denied by Keller [28], who reported that the PVS really consisted of lymph vessels that had been misidentified.

In our results, the primo microcells gained from primo nodes grew into cells or cell-like bodies after 8–9 days of culture, similar to Kim's claim (Fig. 23.2) [13]. From this interesting result, a few doubts occurred. In Kim's observations, a sanal,

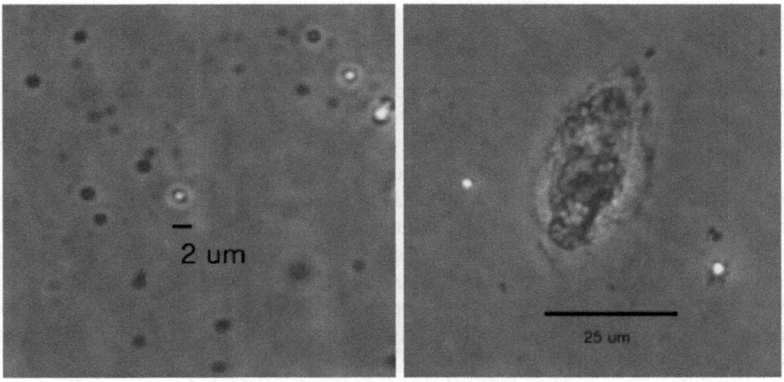

Fig. 23.2 The Kim BongHan's first step result of Sanal: a primo microcell size is about 1.5 μm in diameter (*left*). Eight days later, the primo microcell had grown to be cellbody like, about 30 μm in length and 15 μm in width (*right*). This result is very similar to Kim's results

now renamed primo microcell, had one chromosome, indicating that the DNA in a primo microcell was not perfect. In our observation, the nucleus of a primo microcell was not stained with the DAPI (4′,6-diamidino-2-phenylindole) stain method (data not shown), but the primo microcell grew into a cell-like body without a perfect nucleus. This is similar to an RBC, although an RBC is differentiated without a nucleus.

On the other hand, in Kim's theory [13], a primo microcell produces a daughter cell at the beginning of cell differentiation, and this is repeated so that a primo microcell grows into a cell or tissue. If this hypothesis is correct, how does a primo microcell gain complete DNA. And in our studies, the individual size of some cell-like bodies that have grown from primo microcells is outside the range for normal cell size: ~100 μm in diameter, so we are not convinced whether the cell-like bodies that grow from primo microcell are normal cells or not. Despite these doubts, our conclusion is that the primo microcell grows into cell-like body, and this experimental result is very similar to Kim's observation in the 1960s.

The most critical limit of the current work is that the DNA content in our primo microcells and cell-like body was not tested. In future work, the experiments will be repeated with DNA staining dyes.

4 Conclusion

The primo microcells we observed grew into cell-like bodies. This experiment result is very similar to Kim BongHan's observation in the 1960s and suggests that the primo microcell might be the origin of stem cells or a new type of stem cell. The present report is only preliminary, and further experiments addressing DNA content are needed.

References

1. Gimble J, Guilak F (2003) Adipose-derived adult stem cells: isolation, characterization, and differentiation potential. Cytotherapy 5(5):362–369
2. Weissman IL, Anderson DJ, Gage F (2001) Stem and progenitor cells: origins, phenotypes, lineage commitments, and transdifferentiations. Ann Rev Cell Dev Biol 17:387–403
3. Klein C, Butt SJB, Machold RP, Johnson JE, Fishell G (2005) Cerebellum- and forebrain-derived stem cells possess an intrinsic regional character. Development 132(20):4497–4508
4. Li L, Fukunaga-Kalabis M, Yu H, Xu X, Kong J, Lee JT (2009) Human dermal stem cells differentiate into functional epidermal melanocytes. J Cell Sci 123:853–860
5. Zheng Y-X, Ringe J, Liang Z, Loch A, Chen L, Sittinger M (2006) Osteogenic potential of human periosteum-derived progenitor cells in a PLGA scaffold using allogeneic serum. J Zhejiang Univ Sci B 7(10):817–824
6. Bellayr IH, Gharaibeh B, Huard J, Li Y (2010) Skeletal muscle-derived stem cells differentiate into hepatocyte-like cells and aid in liver regeneration. Int J Clin Exp Pathol 3(7):681–690
7. Fan J, Varshney RR, Ren L, Cai D, Wang DA (2009) Synovium-derived mesenchymal stem cells: a new cell source for musculoskeletal regeneration. Tissue Eng Part B Rev 15(1):75–86
8. Hui J, Ouyang H, Hutmacher D, Goh J, Lee E (2005) Mesenchymal stem cells in musculoskeletal tissue engineering: a review of recent advances at the National University of Singapore. Ann Acad Med Singapore 34:206–212
9. Winter EM, Gittenberger-de Groot AC (2007) Epicardium-derived cells in cardiogenesis and cardiac regeneration. Cell Mol Life Sci 64:692–703
10. Kim B (1961) The Kyungrak system. J Acad Med Sci DPR Korea 108:1–38
11. Kim B (1963) On the Kyungrak system. J Acad Med Sci DPR Korea 90:1–35
12. Kim B (1965) Theory of Kyungrak. J Acad Med Sci DPR Korea 108:1–38
13. Kim B (1965) Sanal theory. J Acad Med Sci DPR Korea 6(108):39–62
14. Baik KY, Sung B-K, Lee B-C et al (2004) Bonghan ducts and corpuscles with DNA-containing granules on the internal organ-surfaces of rabbits. J Int Soc Life Inf Sci 22(2):598–601
15. Ogay V, Baik KY, Lee BC, Soh KS (2006) Characterization of DNA-containing granules flowing through the meridian-like system on the internal organs of rabbits. Acupunct Electrother Res 31(1–2):13–31
16. Kim MS, Hong JY, Hong S et al (2008) Bong-Han corpuscles as possible stem cell niches on organ surfaces. J Pharmacopunct 11(1):5–11
17. Baik KY, Ogay V, Jeoung SC, Soh K-S (2009) Visualization of Bonghan microcells by electron and atomic force microscopy. J Acupunct Meridian Stud 2(2):124–129
18. Ahn SH, Kim MS, Lee SH et al (2009) The morphology study of organ surface BongHan ducts and corpsucle. J Meridian Acupoint 26(1):79–84
19. Lee SH, Ryu Y, Yun Y et al (2010) Anatomical discrimination of the differences between torn mesentary tissue and internal organ-surface primo-vessels. J Acupunct Meridian Stud 3(1):10–15
20. Lee BC, Yoo JS, Baik KY, Kim KW, Soh KS (2005) Novel threadlike structures (Bonghan ducts) inside lymphatic vessels of rabbits visualized with a Janus Green B staining method. Anat Rec B New Anat 286(1):1–7
21. Lee B-C, Kim KW, Soh K-S (2009) Visualizing the network of Bonghan ducts in the omentum and peritoneum by using Trypan blue. J Acupunct Meridian Stud 2(1):66–70
22. Lee B-C, Soh K-S (2010) Visualization of acupuncture meridians in the hypodermis of rat using Trypan blue. J Acupunct Meridian Stud 3(1):49–52
23. Lee B-C, Baik KY, Johng H-M et al (2004) Acridine orange staining method to reveal the characteristic features of an intravascular threadlike structure. Anat Rec B New Anat 278B:27–30
24. Lee C, Lee B-C, Soh K-S (2006) Alcian blue staining method to visualize the lymph intravascular Bonghan duct in the rabbit lymph vessel. J Korean Soc Jungshin Sci 10(1):70–76

25. Lee C, Seol SK, Lee BC, Hong YK, Je JH, Soh KS (2006) Alcian blue staining method to visualize bonghan threads inside large caliber lymphatic vessels and X-ray microtomography to reveal their microchannels. Lymphat Res Biol 4(4):181–190
26. Bk S, Kim MS, Ogay V, Kang DI, Soh KS (2008) Intradermal Alcian-blue injection method to trace acupuncture meridians. J Pharmacopunct 11(2):5–12
27. Yoo JS, Kim MS, Ogay V, Soh KS (2008) In vivo visualization of bonghan ducts inside blood vessels of mice by using an Alcian blue staining method. Indian J Exp Biol 46(5):336–339
28. Kellner G (1966) Bau und Funktion der Haut. Deutche Zeitschrift fur Akupunktur 15:1–31

Chapter 24
Budding Primo Microcells (Sanals) in a Culture Medium with Fertilized Egg Albumen and RPMI Medium

Byung-Cheon Lee, Dae-In Kang, and Kwang-Sup Soh

Abstract We developed a novel culture medium for primo microcells (PM) (sanals) by using a combination of fertilized egg albumen and RPMI-1640 medium (Gibco-Invitrogen, Carlsbad, CA, USA) (1:1). We observed the budding of a PM in real time at 21 min after incubation. At 52 min after incubation the bud of the PM grew to another PM-like structure connected with the first PM. Furthermore, we observed that the bud contained DNA by using the fluorescence imaging with acridine orange, which suggested elongation or growth of DNA from the first PM, named "a mother-PM," in the budding process. However, we were able to observe only one case, so further study to cultivate PM with the same media or modified ones is needed.

1 Introduction

The primo vascular system (PVS), a novel circulatory system corresponding to acupuncture meridians, was first proposed by Kim in the early 1960s [1], and has been extensively reinvestigated to confirm its presence throughout an animal's body since the year 2002 by a Seoul National University group [2]. A primo fluid, which contains hyaluronic acid, hormones, such as adrenaline and noradrenaline, and primo microcells (PM) or sanals among others, flows through the PVS. Kim claimed that the PM played a stem cell-like role in wound healing, regenerating damaged

B.-C. Lee (✉)
Division of Electrical Engineering, Ki Primo Research Laboratory,
KAIST Institute for Information Technology Convergence,
Korea Advanced Institute of Science and Technology,
Daejeon, South Korea

Pharmacopuncture Medical Research Institute,
Korean Pharmacoacupuncture Institute,
Seoul, South Korea
e-mail: donboscolee@gmail.com

K.-S. Soh et al. (eds.), *The Primo Vascular System: Its Role in Cancer and Regeneration*,
DOI 10.1007/978-1-4614-0601-3_24, © Springer Science+Business Media, LLC 2012

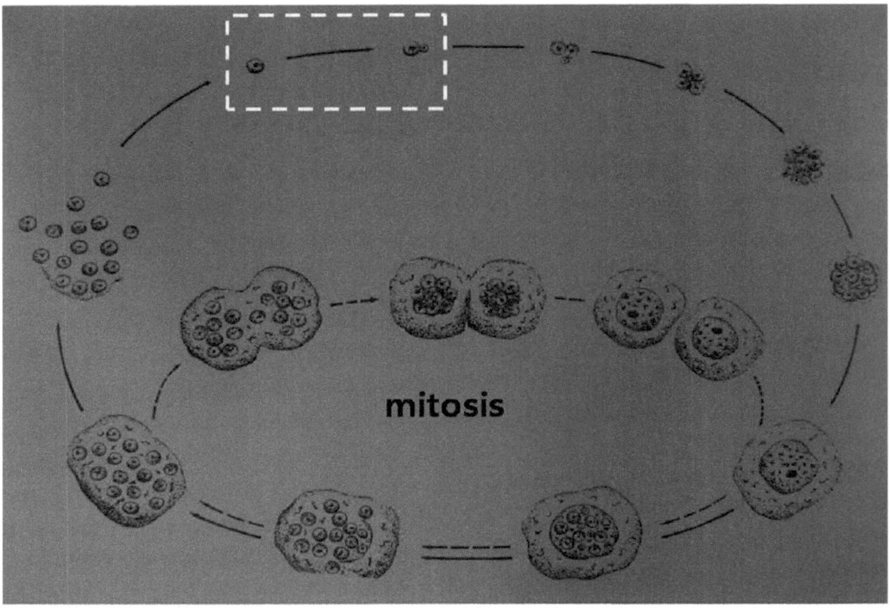

Fig. 24.1 An illustration of mitosis and an alternative cycle from a primo microcell (PM) to a cell via DNA elongation which was first proposed and illustrated by Bonghan Kim in 1960s [3, 4]. In this illustration, the critical step from a PM to be a budding PM is indicated by *dotted rectangle*

liver cells [3], and hematopoiesis [4]. The motion analysis of the PM showed a random motion [5] which was affected by UV-A [6]. The DNA in the PM [7], the physical properties of the membranes of the PM [8, 9], and their presence in nerve primo vessels suspended in brain ventricles and the central canal of spinal cord [10] were investigated. These studies were limited to morphological characteristics, and more important investigations, such as investigation of the inner structure, DNA sequencing, proliferation, physiological function, tagging, and tracing, have not been performed. PMs must be cultivated to obtain in large numbers. In this chapter, we report on a method to culture PMs. We developed a culture medium with fertilized egg albumen and RPMI-1640 medium. We were able to identify the most important initial proliferation process by observing the budding of a PM and the growing bud. The bud contained DNAs as demonstrated by acridine orange staining, which is a critical step for PM proliferation, as shown in Fig. 24.1 suggested in Kim's Sanal papers [3, 4].

2 Materials and Methods

All experiments were performed in accordance with the Principles of Laboratory Animal Care prepared by the National Society of Medical Research, and the Guide for the Care and Use of Laboratory Animals prepared by the Institute of Laboratory Animal Resources and published by the National Institute of Health (NIH Publ.

85–23, Rev. 1985). The study was approved by the Ethical Committee of the Seoul National University. A rat (Wistar, JungAng, Seoul, Korea) was anesthetized by intramuscular injection of urethane (1.5 g/kg) into the right leg. Under deep anesthesia, we dissected the rat's abdomen and exposed the abdominal cavity. Under a stereomicroscope (SZX 12, Olympus) after taking the primo vessels and the primo nodes from the rat's abdominal cavity, we softly homogenated them on 70~100 μm-sized mesh, then about 100 μL of 0.01% acridine orange in phosphate buffered saline, pH 7.4 (PBS), was added at room temperature. The homogenate was incubated in the albumen of a fertilized egg (Pulmuone, Korea) and RPMI media (Gibco-Invitrogen, Carlsbad, CA, USA) (1:1) in the chamber of an inverted microscope, Image Restoration Microscopy (Delta Vision RT, Applied Precision, USA), at 37~38°C. During observation for about 1 h, we took pictures of PM in the homogenate.

3 Results

The budding of a PM was observed for 1 h, as shown in Fig. 24.1. The upper three panels are original fluorescence images of a PM at 0, 21, and 52 min after incubation. The bright white spot depicts a DNA vital stained with acridine orange. The lower panels are contrast-enhanced images, obtained using the software MS office, picture manager, of the original images. The magnified PMs in this panel are circled. At 0 min after incubation, the PM of about 1.5 μm in diameter was round shaped with a bright white DNA signal. At 21 min after incubation, the PM gave birth to the bud, and at 52 min after incubation, the bud of the PM had grown a little bigger than the initial stage of the bud (Fig. 24.2).

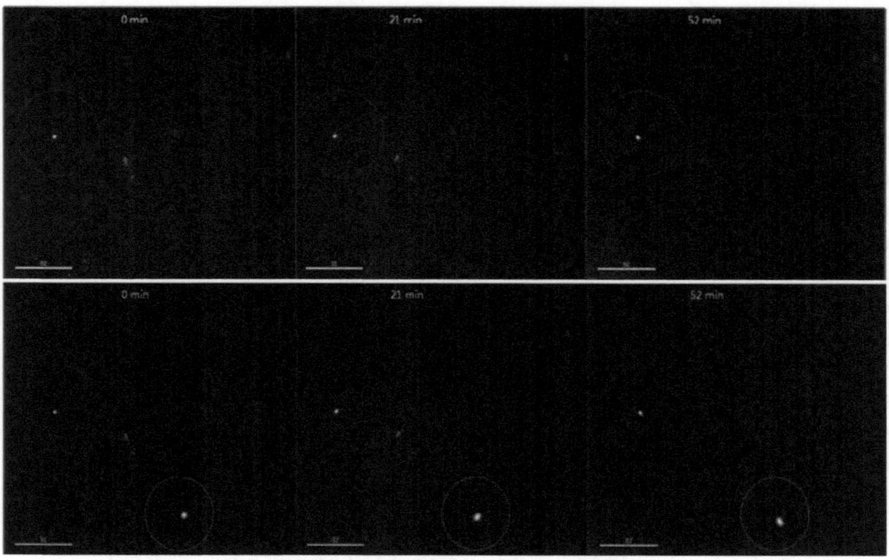

Fig. 24.2 Observation of the budding of a PM

The upper three panels are original fluorescence images of a PM during a 1-h incubation (0, 21, and 52-min incubation). The bright white indicates a DNA signal stained by the DNA-staining dye acridine orange. The scale bar is 32 μm. The lower panels are contrast-enhanced images of the upper original images. The contrast-enhanced PMs in the lower panels are magnified from original PM image and circled. The original PM (about 1.5 μm in diameter) at the first step of incubation is round shaped with bright white DNA signals. At 21 min of incubation, the round PM makes a bud with a DNA signal. At 52 min of incubation, the bud of the PM grew a bit longer. The scale bar is 32 μm.

4 Discussion

Development of a method to culture PMs is a necessary step to pave the way for uncovering the gene-level characteristics and the physiological functions of a PM. In this chapter, we reported on our recipe for a PM-culture medium, which was created for the first time since the unknown one of Kim [3]. With this medium, we were able to observe the budding of a PM within 1 h, which is in good agreement with Kim's claim that a budding occurred in 30–40 min. Our new finding is that the bud from a PM has DNA in it. Even though a budding form can be observed from asexual reproduction in yeast, a PM is generally 1.2~1.5 μm in diameter [3], but the yeast's size is typically 3–4 μm in diameter, and some yeasts can ever reach sizes over 40 μm [11]. About the reason to use egg albumen as a PM-culture medium, we note that the primo fluid which would be an ideal medium contains a high concentration of hyaluronic acids [1]. In addition, hyaluronic acid is a good matrix medium for regeneration of an ischemic heart and a damaged sciatic nerve [12, 13]. With this background knowledge, we developed the PM-culture medium with fertilized albumen that contained a high concentration of hyaluronic acid. Amazingly, the budding of the PM occurred in the cocktail medium of albumen and RPMI. Further variations of the cocktailing with egg yolk and other general media are worth pursuing. The generation of a daughter PM via budding suggests that the proliferation process of the PM is a spore proliferation as in fungi. A similar process of microcells was observed in tumor tissues. Buikis et al. [14] observed that microcells were metabolically active, and intensively accumulated nucleoproteins into their nuclei and cytoplasm. The microcells grew rapidly and transformed into young undifferentiated cells, which implies that stem cells can arise from microcells. An important kind of stem cell whose size is similar to that of a PM is a very small embryonic-like (VSEL) stem cell that was found by Zuba-Surma et al. [15]. A VSEL stem cell is an adult stem cell, but has the characteristics of an embryonic stem cell. The relation between a PM and a VSEL stem cell is an imminent question. In order to answer this critical question, PMs must be cultured, and our report is a first step toward that goal. Due to the limitation of our optical instruments, we were not able to trace the PM beyond 1 h. Thus tracing the PM beyond 1 h in our culture medium for the observation of the generation of a daughter PM from the bud would be worthwhile.

Acknowledgments This work was supported by a National Research Foundation of Korea (NRF) grant funded by the Korea government, Ministry of Education, Science, and Technology (MEST) in 2010 (No. 2010–0025289) and by a Korean Pharmacopuncture Foundation grant funded by the Korean Pharmacopuncture Institute (KPI-2010-003).

References

1. Kim BH (1965) The Kyungrak system. J Acad Med Sci DPR Korea 6:1–38 [in Korean]
2. Soh KS (2009) Bonghan circulatory system as an extension of acupuncture meridians. J Acupunct Meridian Stud 2:93–106
3. Kim BH (1965) Sanal theory. J Acad Med Sci DPR Korea 6:39–62 [in Korean]
4. Kim BH (1965) Sanal and hematopoiesis. J Jo Sun Med 108:1–6 [In Korean]
5. Sung B, Kim MS, Corrigan A et al (2009) In situ microextraction method to determine the viscosity of biofluid in threadlike structures on the surfaces of mammalian organs. Phys Rev E 79(1–3):022901
6. Sung B, Ogay V, Yoo JS et al (2005) UV-A-induced activation of Bonghan granules in motion. J Int Soc Life Inf Sci 23:297–301
7. Ogay V, Baik KY, Lee BC et al (2006) Characterization of DNA-containing granules flowing through the meridian-like system on the internal organs of rabbits. Acupunct Electrother Res Int J 31:13–31
8. Kwon JH, Baik KY, Lee BC et al (2007) Scanning probe microscopy study of microcells from the organ surface Bonghan corpuscle. Appl Phys Lett 90(173903):1–3
9. Baik KY, Ogay V, Jeoung SC et al (2009) Hypothesis on the origin of adult stem cells and Bonghan microcells observed with an electron microscope and an atomic force microscope. J Acupunct Meridian Stud 2:124–129
10. Lee BC, Kim KW, Soh KS (2010) Characteristic features of a nerve primo-vessel suspended in rabbit brain ventricle and central canal. J Acupunct Meridian Stud 3:75–80
11. Walker K, Skelton H, Smith K (2002) Cutaneous lesions showing giant yeast forms of Blastomyces dermatitidis. J Cutan Pathol 29(10):616–618
12. Yoon SJ, Fang YH, Lim CH et al (2009) Regeneration of ischemic heart using hyaluronic acid-based injectable hydrogel. J Biomed Mater Res B Appl Biomater 91:163–171
13. Wang KK, Nemeth IR, Seckel BR et al (1998) Hyaluronic acid enhances peripheral nerve regeneration in vivo. Microsurgery 18:270–275
14. Buikis I, Harju L, Freivalds T (1999) Origin of microcells in the human sarcoma cell line HT-1080. Anal Cell Pathol 18:73–85
15. Zuba-Surma EK, Kucia M, Ratajczak J, Ratajczak MZ (2009) "Small stem cells" in adult tissues: very small embryonic-like stem cells stand up! Cytometry A 75(1):4–13

Part IV
Cancer

Chapter 25
Identification of Primo Vascular System in Murine Tumors and Viscera

Walter Akers, Yang Liu, Gail Sudlow, Joon Lee, Jung Sun Yoo, Byung-Cheon Lee, Kwang-Sup Soh, and Samuel Achilefu

Abstract In the past decade, researchers at Seoul National University of South Korea have validated the existence of the primo-vascular system using animal models, which was originally reported by Bonghan Kim in the 1960s. After exploring a variety of dye stains, Kwang-Sup Soh, et al. found that trypan blue is an effective dye for staining structures of the primo vessels and nodes. Most of the studies reported so far have used relatively large animal models such as rabbits. In this study, we explored the potential of visualizing primo-vascular system in murine models. As compared to conventional models used for primo-vascular studies, murine models are more versatile and affordable. In mice, we found primo vessels extended to and from abdominal viscera, often disappearing in layers of adipose tissue before resurfacing in a more distal region. In addition, we found potential primo-vascular structures on the tumor surface. These structures were loosely attached to the surface of the tumor and some segments appear to be within the clear serosal tissue or invading into the tumor. The experience of identifying primo-vascular system in rodents empowers us to investigate the detailed relationship of primo-vessels and cancer in future studies. A breakthrough in elucidating the complete anatomical and physiological description of the primo-vascular system holds great promise of unraveling the molecular basis of various human diseases.

S. Achilefu (✉)
Department of Biomedical Engineering, Washington University, St. Louis, MO, USA

Department of Biochemistry and Molecular Biophysics, Washington University School of Medicine, St. Louis, MO, USA
e-mail: achilefus@mir.wustl.edu

K.-S. Soh et al. (eds.), *The Primo Vascular System: Its Role in Cancer and Regeneration*, DOI 10.1007/978-1-4614-0601-3_25,

1 Introduction

Primo vessels, also known as Bonghan ducts, are minute anatomical structures worth investigating to improve current understanding of biology, physiology, and medicine. These structures were first identified in reports by Bonghan Kim almost 50 years ago [1]. The perseverance and dedication of Prof. Soh and his team at Seoul National University revived this field of study. Today, primo vessels and primo nodes have been identified on the surface of many internal organs and within the skin. Numerous studies continue to unravel the nature and functions of these vessels.

The true nature of primo vessels remains controversial in some research circles. However, recent studies have identified primo vessels and nodes as unique structures representing a completely separate system [2, 3]. Some reports infer that the primo vessel system may be the anatomical source of the meridian system in oriental medicine [4–7]. These studies demonstrate both circulation and electrophysiologic characteristics of primo-vascular system and its occurrence throughout the body [4–6]. Therefore, understanding the roles of this system in normal and pathophysiology could unravel the basis of many human diseases.

The excitement about the potential involvement of the primo-vascular system in oncology compelled us to engage in this line of research. The structure of primo vessels indicates that growth factors and cytokines could be transported throughout the body, possibly affecting metastasis. Other signaling mechanisms may also be connected to the primo vessel system that will change our understanding of cancer development and progression. Moreover, understanding the role of primo-vascular system in normal physiology may uncover new factors associated with the immune and circulatory systems, which may aid the development of future oncologic therapeutics.

2 Methods

Female Swiss Webster mice, 6–8 weeks old, were used for identification of primo vessels in mice. These outbred mice are immune-competent and grow larger than other strains. For tumor studies, approximately 500,000 4T1 mouse mammary carcinoma cells suspended in PBS were injected subcutaneously in the right flanks. Mice used for these experiments were 30–40 g. For all procedures requiring immobilization and invasive procedures, mice were anesthetized with ketamine and xylazine cocktail (80–100 and 10–15 mg/kg, respectively) by intraperitoneal injection. A surgical plane of anesthesia was achieved as determined by loss of deep pain reflex. Hair was clipped from the abdomen and skin cleaned prior to incision. A ventral midline incision was made from the xyphoid to the pubis and the skin reflected on both sides. Identification of primo vessels associated with the skin and subcutaneous tissues, including tumors, was carried out by topical application of 0.2% trypan blue. Tissues were moistened with 37°C PBS regularly to prevent desiccation.

An Olympus SZX16 stereo microscope with DP71 camera was used for visualization and acquiring images. Animals were humanely euthanized at the end of each experiment prior to waking.

3 Results

Primo vessel structures were difficult to find due to their minute size and variability in location. Staining of the structures frequently required multiple applications of the dye. Mesh-like networks of blue-stained anatomical structures were found, which corresponds to literature findings in other animal models (Fig. 25.1). In broad areas, linear or web-like patterns that indicated the primo vessels were observed. However, it was difficult to positively identify all primo vessels because the non-specificity of trypan blue resulted in the staining of dry or dying tissues.

Using the same method, we identified primo-structures on the subcutaneous surface of the skin in mice (Fig. 25.2a). In addition, primo-structures were also observed on the surface of the liver, stomach, and intestines. These primo vessels often extended to and from abdominal viscera, and disappeared in layers of adipose tissue before resurfacing in a more distal region (Fig. 25.2b). Some of the components of the primo-system identified were several millimeters long and elastic in a network of nodes and vessels, while others were detected as single strand structures that appeared to dive into the organ surface. Moreover, we were also able to isolate primo-vascular structures from its surrounding tissues (Fig. 25.2c).

As we are particularly interested in oncology-related primo-structures, we explored the presence of these structures in a mouse model of cancer. Flank tumors grown from mouse mammary carcinoma cells implanted subcutaneously are a rudimentary but useful model of cancer growth and development. We used the earlier mentioned strategy of trypan blue staining to locate primo vessels on the surface of tumors. Based on trypan blue staining, we were able to find several primo-structures on the tumor surface. Figure 25.3 shows blue-stained vessels and nodes on the

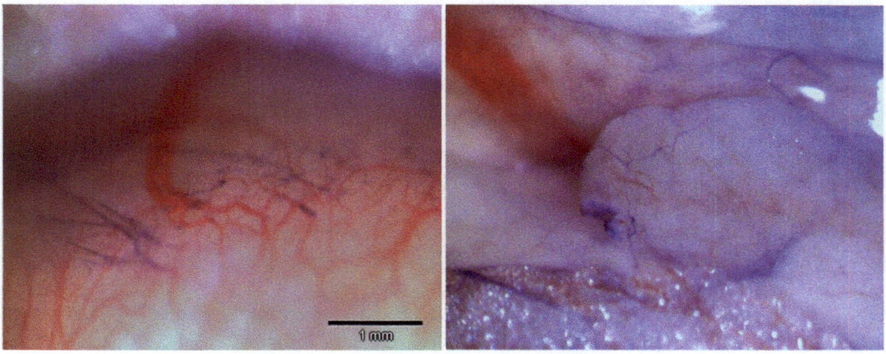

Fig. 25.1 Mesh-like networks of primo-vascular system stained by trypan blue

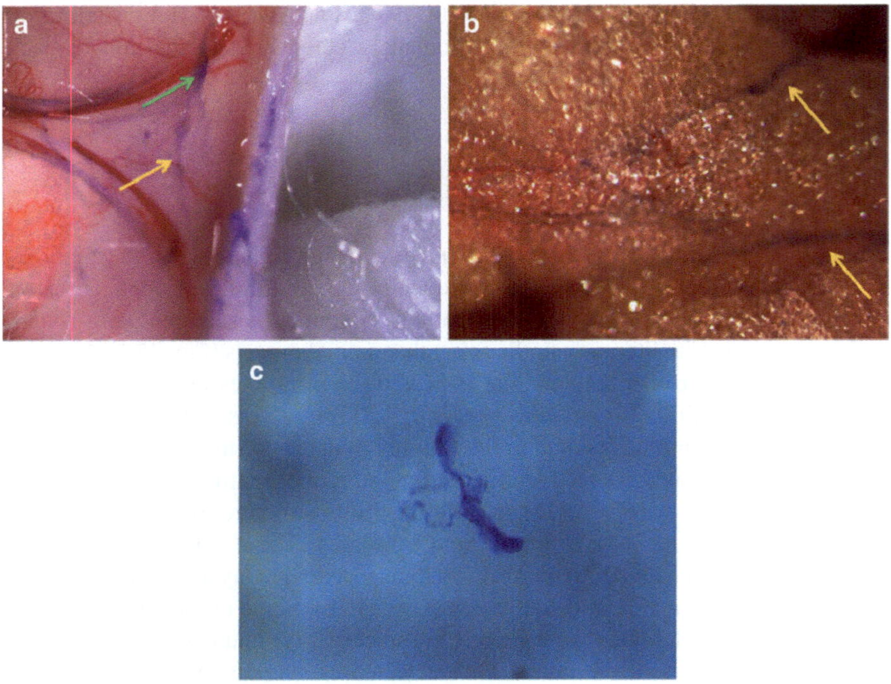

Fig. 25.2 Primo-structures in the mice. (**a**) Primo-vascular system in subcutaneous surface of the skin. *Yellow arrow* indicates vessels, *green arrow* indicates nodes. (**b**) Primo-vascular system in adipose tissues. *Yellow arrow* indicates vessels. (**c**) Primo-structure isolated from surrounding tissues

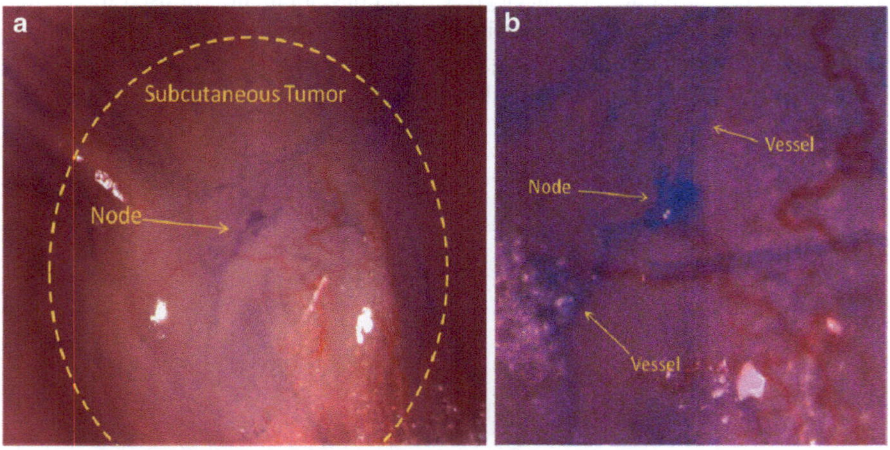

Fig. 25.3 Primo-vascular structures on tumor surface. (**a**) Lower magnification of and (**b**) higher magnification picture of primo node and vessel are shown. Prim- structures can be easily distinguished from blood vessels by blue staining after trypan blue application. These structures were loosely attached to the surface of the tumor

tumor surface that are distinct in color, location, and structure from neighboring blood vessels. Unlike the primo vessels associated with the surface of abdominal organs, these primo vessels were more within the clear serosal tissues. They appeared to be attached to the tumors, invading into them. It is difficult to observe this 3D behavior in the 2D images shown in Fig. 25.3.

4 Conclusions

We have reported our initial investigations into the primo-vascular system. This unique system of vessels and nodes holds great potential for the discovery of new aspects of many diseases. We are primarily interested in the relationship of primo vessels and cancer to discover how they may provide new information for diagnosis and treatment. We have begun our investigations in animal models as these provide the best chance of identifying biological factors relevant to the roles of primo-structures in oncology. We hope to further identify the relationship between the primo-vascular system and cancer in animal models to discover how this relationship can be useful for human patients.

Acknowledgment This work was made possible by financial support from the Mallinckrodt Institute of Radiology at Washington University in St. Louis.

References

1. Kim BH (1962) Study on the reality of acupuncture meridians. J Jo Sun Med 9:5–13 [Korean]
2. Lee BC, Kim KW, Soh KS (2009) Visualizing the network of Bonghan ducts in the omentum and peritoneum by using trypan blue. J Acupunct Meridian Stud 2:66–70
3. Yoo JS, Ayati MH, Kim HB et al (2010) Characterization of the primo-vascular system in the abdominal cavity of lung cancer mouse model and its differences from the lymphatic system. PLoS One 5(4):e9940
4. Lee BC, Soh KS (2010) Visualization of acupuncture meridians in the hypodermis of rat using trypan blue. J Acupunct Meridian Stud 3(1):49–52
5. Jia ZF et al (2010) Fluorescent nanoparticles for observing primo vascular system along sciatic nerve. J Acupunct Meridian Stud 3(3):150–155
6. Han HJ et al (2010) Primo-vessels as new flow paths for intratesticular injected dye in rats. J Acupunct Meridian Stud 3(2):81–88
7. Soh KS (2009) Bonghan circulatory system as an extension of acupuncture meridians. J Acupunct Meridian Stud 2(2):93–106

Chapter 26
Molecular Compositional Differences of the Primo and the Lymphatic Vascular Systems in Murine Melanoma Models

Jung Sun Yoo, Baatartsogt Oyungerel, Il Young Han, Ji Young Kim, Choong Hwan Lee, Kang Duk Choi, Kwang-Sup Soh, and Tae Young Han

Abstract The primo vascular system (PVS) has been found in various cancer models, such as the lung cancer, ovarian cancer, and melanoma models. However, there have been no investigations to discriminate PVS from the similar-looking lymphatic vascular system (LVS) in melanoma models. In the present study, the compositional differences between the PVS and the LVS were investigated in murine melanoma models by using immunofluorescence staining and ultra-performance liquid-chromatography quadrupole time-of-flight mass spectrometry (UPLC-Q-TOF MS). Primo vessels (PVs) and primo nodes (PNs) were found on the surface of a cutaneous melanoma by using Trypan blue staining. An immunofluorescent study revealed that the PVs did not contain LYVE-1, a specific endothelial antigen of lymphatics. UPLC-Q-TOF MS chromatograms clearly showed different ion mass peaks between the two vascular systems. Taken together, the presented data on the compositional differences suggest that the PVS is an entirely different circulation system from the LVS.

1 Introduction

There have been several reports on the surprising findings that, besides the blood and the lymphatic vascular systems (LVSs), there is a new kind of fluid-conducting system, namely the primo vascular system (PVS), inside the abdominal cavity and on the fascia of subcutaneous tumors in murine cancer models such as lung cancer, ovarian cancer, and melanoma models [1–3]. Moreover, in vivo imaging data showed the independent function of the primo system in disseminating cancer cells to distant sites [4]. Here, we show the compositional differences of the PVS and the LVS for B16 F10 murine melanoma models by using immunofluorescence

T.Y. Han (✉)
Banryong Herb Clinic, Nunhyong-dong, Kannam-gu, Seoul, Korea
e-mail: ham1sh@paran.com

K.-S. Soh et al. (eds.), *The Primo Vascular System: Its Role in Cancer and Regeneration*, DOI 10.1007/978-1-4614-0601-3_26, © Springer Science+Business Media, LLC 2012

and ultra-performance liquid-chromatography quadrupole time-of-flight mass spectrometry (UPLC-Q-TOF MS). Immunohistologic examination with LYVE-1 antibody, which is a typical marker for identifying the LVS, was tried with the primo sample on the surface of melanoma tissue from a mouse. Mass spectrometry, in combination with liquid chromatography using a time-of-flight method, provides a powerful tool to analyze small amount of complex biological samples [5]. In this study, an ion mass peak analysis based on UPLC-Q-TOF MS data was applied to discover metabolic differences between the two vascular systems.

2 Methods

2.1 Cell Culture and Melanoma Cancer Model

B16 F10 murine melanoma cells were obtained from the Korean Cell Line Bank (Seoul, Korea). The cells were grown as monolayers in Dulbecco's modified Eagle's medium (DMEM) with 1% penicillin–streptomcyin and 10% fetal bovine serum (FBS) in 95% air and 5% Co_2 at 37°C.

Female nude mice (BALB-c-nu/nu, 5 weeks old, weighing 15–20 g, $n=10$; Japan SLC, Inc., Hamamatsu, Japan) were used in accordance with institutional guidelines under approved protocols. Ten mice were inoculated subcutaneously (s.c.) in the dorsum with 1×10^7 B16 F10 murine melanoma cells (in 1-mL DMEM-1640 medium) for melanoma tumor formation under the skin. All research involving the animals was approved by the Institute of Laboratory Animal Resources of Seoul National University.

2.2 Intraoperative Visualization

To identify PVS on the surface of the melanoma in the skin, four weeks after inoculation with B16 F10 cells, we performed mouse surgery under general anesthesia with intraperitoneal (i.p.) injection of zoletil/rompun. First, the boundary of the skin in the mouse was incised, and the tumor tissue was exposed carefully. Then, a 0.2% Trypan blue solution, previously filtered through 0.22-μm pore-sized filter paper, was applied dropwise on the exposed tumor tissue [6, 7]. After rinsing away the dye with warm saline, primo nodes (PNs) and primo vessels (PVs) were identified by direct visualization of the surgical anatomy with a surgical microscope (SZX12, Olympus, Japan). Finally, the images were captured with a CCD camera (DP70, Olympus, Japan).

2.3 Immunofluorescence Staining

To identify endothelial antigen difference between the PV and the lymphatic vessel (LV) in a melanoma-bearing mouse, we performed immunofluorescence staining with the IgG anti-LYVE-1 polyclonal antibody (pAb), a specific marker of lymphatics. The PV and the LV on the surface of the abdominal organ in a melanoma-bearing mouse were harvested, embedded in OCT, and snap-frozen in liquid nitrogen. Cross sections (10 μm) were fixed in ice-cold acetone for 5 min, washed in PBS, and blocked with CAS-block solution (Invitrogen, USA) for 2 h at 37°C. Slides were incubated with rabbit anti-mouse LYVE-1 (2.5 μg/mL, Abcam, UK) overnight at 4°C, followed by washes and detection with goat-anti-rabbit IgG conjugated to Alexa Fluor 488 (1:500, Molecular Probes, USA). Slides were treated with DAPI for nuclear counterstaining, and coverslipped in antifade reagent (Molecular Probes, USA). Samples were visualized by using fluorescent optical microscopy (Axiophot, ZEISS, Germany).

2.4 *Sample Preparation and Measurement of UPLC-Q-TOF MS*

PVs on the fascia of melanoma tissue inside the skin and the LVs around the inferior vena cava were isolated. The isolated PVs and LVs inside the phosphate buffered saline (PBS) were precipitated in a refrigerated centrifuge (10 min, 17,000 rpm, 4°C). The tissue pellets were resuspended in 100 μL of cold methanol including 0.5 mg/L of lidocaine and camphor sulfonic acid and was again centrifuged for 5 min at 17,000 rpm and 4°C after vortex-mixing for 1 min. Each 50 μL of supernatant with 25 μL of purified water was transferred into a vial for the UPLC-Q-TOF MS analysis.

UPLC-Q-TOF MS experiments were performed using a Waters ACQUITY UPLC™ system (Waters, Milford, MA, USA) and Waters Q-TOF Premier system (Micromass MS Technologies, Manchester, UK). The mass spectrometer was operated in both the positive (2.5 kV) and the negative (3.0 kV) electrospray ionization (ESI) modes. Mobile phase A was 0.1% formic acid in water, and mobile phase B was 0.1% formic acid in acetonitrile. A linear gradient was used: 10% B for 1 min, 40% B for 4 min, 70% B for 6 min, 90% B for 3 min, and 10% B for 2 min. The total run time was 16 min at 0.3 mL/min. The column was 100×2.1 mm with an internal diameter of 1.7 mm (Waters ACQUITY BEH C_{18}), and the column temperature was held at 35°C. Mass spectra were collected from *m/z* 100 to 1,500.

3 Results

3.1 *Visualization of the Primo vascular System Around the Melanoma*

To visualize PVs and PNs in cutaneous melanoma models, we performed careful intraoperative imaging four weeks after the inoculation with B16 F10 cells. Figure 26.1a presents representative examples of the PVS distribution on the surface

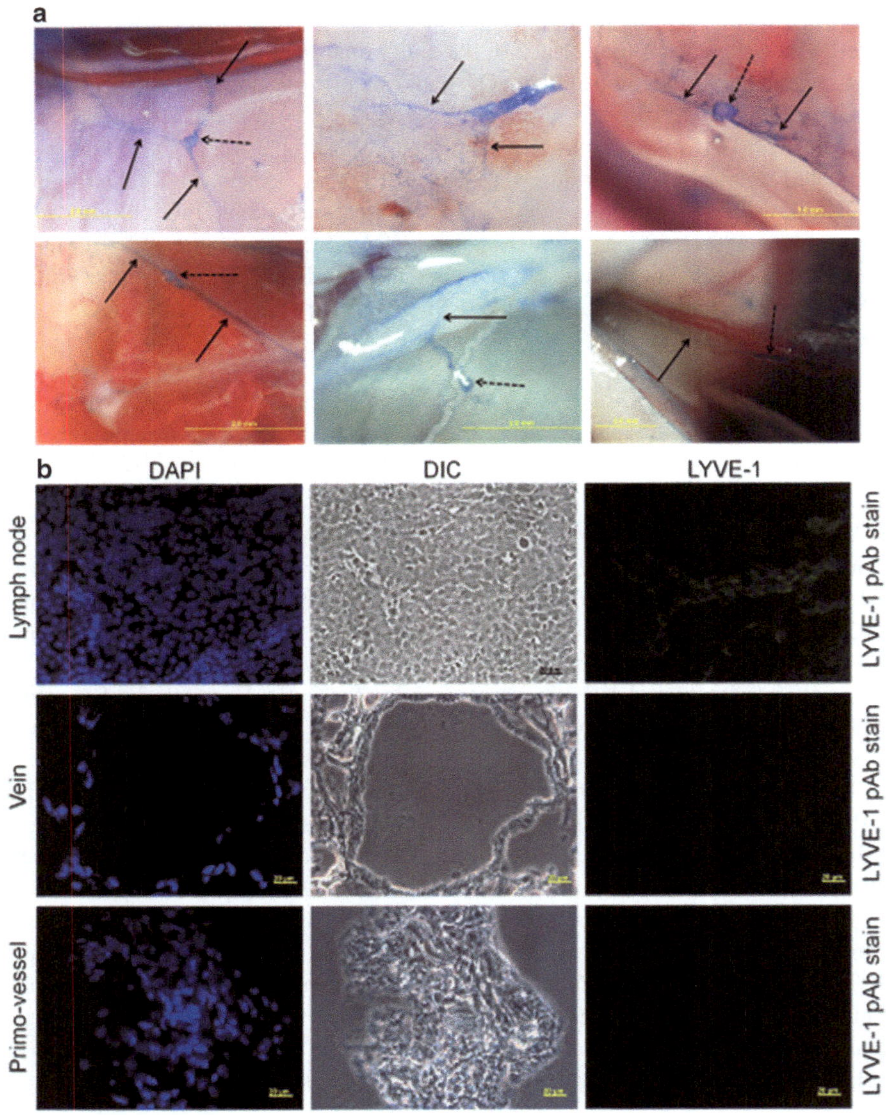

Fig. 26.1 (**a**) Visualization of the primo vessels (*arrows*) and the primo nodes (*dotted arrows*) on the surface of melanoma tissue by using in situ Trypan blue staining. (**b**) Immunofluorescence staining of the lymph node (the *first panel*), the vein (the *middle panel*), and the primo vessel (the *last panel*) with anti-LYVE-1 polyclonal antibody (*green*)

of the melanoma tissue. Trypan blue spreading under in vivo conditions revealed thin lines of PVs (arrows) and round-shaped PNs (dotted arrows) with optimum contrast.

3.2 *Immunofluorescence Staining with Anti-LYVE-1 Antibody*

To test for the existence of LYVE-1, a member of the link protein superfamily, in the PV, we conducted immunofluorescence staining with the anti-LYVE-1 antibody. Figure 26.1b depicts the LYVE-1 immunoreactivities of the lymph node (the first panel), the vein (the middle panel), and the PV (the last panel) with green fluorescence. Fluorescence images of DAPI (blue)-exhibiting nuclei and phase contrast images were also taken. The lymph node was positive for LYVE-1, whereas the vein and the PV showed no presence of LYVE-1.

3.3 *Ion Mass Analysis Based on UPLC-Q-TOF MS*

From the extracts of PVs and LVs in melanoma-bearing mice, multiple ion mass peaks were analyzed from the UPLC-Q-TOF MS data set. As shown in Fig. 26.2, more than four ion mass peaks of the sample which was extracted from PVs showed significantly higher intensity than the ones from LVs. Especially the ion mass peak of PVs at *m/z* 415 in the positive ESI mode and *m/z* 459 in the negative ESI mode revealed a major metabolite of PVs, whereas LVs do not have. Furthermore, the ion mass peaks showed broad differences during retention time from 13 to 15 min by different molecular identity of metabolites although they were unidentified because of low resolution of the peaks. Overall the ion mass peak comparison using UPLC-Q-TOF MS data effectively demonstrated the compositional difference between the PVs and the LVs, which indicated different metabolite compositions.

4 Conclusions

In this work, we revealed distinct compositional differences between the PVS and the LVS in murine melanoma models by using immunofluorescence staining and a UPLC-Q-TOF MS analysis. A specific marker of lymphatics, LYVE-1, was not expressed at all in PVs, and the ion mass peak pattern in the chromatograms of UPLS-Q-TOF MS was very different between the two vascular systems, showing a clear distinction between PVS and the LVS. These findings will introduce the PVS as a new anatomical and functional structure related to melanoma formation and metastasis and simultaneously provide new insight into future therapy for melanomas. This observation on the molecular compositional difference between primo

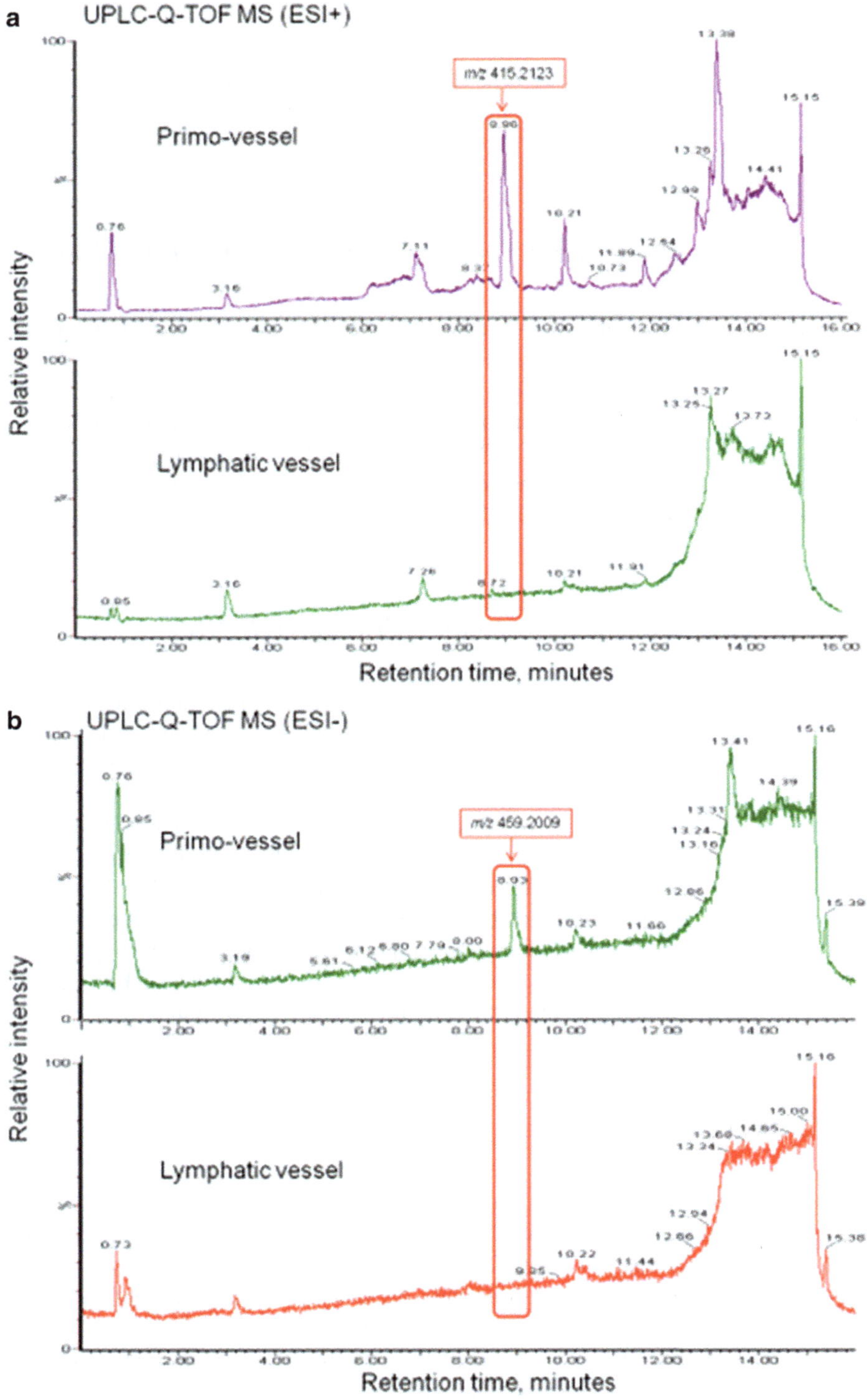

Fig. 26.2 UPLC-Q-TOF MS chromatograms for extracts from primo vessels and lymphatic vessels. (**a**) Specific ion mass peak difference (*red box*) between primo vessels and lymphatic vessels in the positive ion mode. (**b**) Analogous data in the negative mode

vessels and lymph vessels is significant because it is the first molecular study of the PVS [7] after the proteomic analysis of the rabbit's PVS [8].

Acknowledgements We thank Prof. Mison Chun and Prof. Hee Jae Joo at the Ajou University School of Medicine in Korea for their useful discussions. This work was supported in part by the Korea Institute of Oriental Medicine. Prof. Soh thanks the Association of Korean Oriental Medicine for support.

References

1. Yoo JS, Kim HB, Ogay V et al (2009) Bonghan ducts as possible pathways for cancer metastasis. J Acupunct Meridian Stud 2(2):118–123
2. Yoo JS, Ayati H, Kim HB et al (2010) Characterization of the primo-vascular system in the abdominal cavity of lung cancer mouse model and its differences from the lymphatic system. PLoS One 5(4):e9940
3. Heo C, Hong M, Jo A et al. The study of primo-vascular system (PVS) by utilizing melanoma tumor model with green fluorescence protein (GFP) expressing mouse, J Acupunct Meridian Stud, submitted
4. Yoo JS, Kim HB, Won N et al (2010) Evidence for an additional metastatic route: in vivo imaging of cancer cells inside the primo-vascular system around tumors and organs. Mol Imaging Biol 13(3):471–80. doi:10.1007/s11307-010-0366-1
5. Yang S, Sadilek M, Synovec RE et al (2009) Liquid chromatography-tandem quadrupole mass spectrometry and comprehensive two-dimensional gas chromatography-time-of-flight mass spectrometry measurement of targeted metabolites of *Methylobacterium extorquens* AM1 grown on two different carbon sources. J Chromatography A 1216:3280–3289
6. Lee BC, Kim KW, Soh KS (2009) Visualizing the network of bonghan ducts in the omentum and peritoneum by using Trypan blue. J Acupunct Meridian Stud 2:66–70
7. Soh KS (2009) Bonghan circulatory system as an extension of acupuncture meridians. J Acupunct Meridian Stud 2:93–106
8. Lee ST, Lee BC, Nam CH et al (2008) Proteomic analysis for tissues and liquid from Bonghan ducts on rabbit intestinal surfaces. J Acupunct Meridian Stud 1:1–13

Chapter 27
Using Human Observations to Gain Biologic Insights and New Treatments; Discovery of a Quadruplex-Forming DNA Aptamer as an Anticancer Agent

Donald M. Miller, Shelia D. Thomas, Kara Sedoris, Ashraful Islam, David Muench, Cortney Clarkson, and Charles A. Koller

Abstract The discovery and characterization of the primo vascular system represents a major new development in our understanding of human biology. The current understanding of this novel system is based on careful observations over an extended period of time. We have, in an analogous fashion, used clinical observations in patients with malignant diseases to develop new insights which have led to novel types of therapies. Beginning with the observation, in 1979, of in vivo differentiation of chronic myelogenous leukemia cells in a patient treated with the RNA synthesis inhibitor, mithramycin, we have characterized the ability of DNA binding compounds to inhibit gene expression. As a result of this work, we have discovered a quadruplex-forming, DNA aptamer, AS1411, which has significant anticancer activity with very little toxicity. AS1411 is currently in Phase IIb studies. We have recently shown that genomic quadruplex-forming sequences are selectively toxic to transformed cells and have therapeutic potential.

1 Background

During the past 30 years, the treatment of cancer and hematologic malignancies has changed dramatically. It is easy to overlook the fact that our current detailed molecular understanding of the basis of cancer is quite recent. In fact, it was not until 1906 (104 years ago) that Peyton Rous discovered the first cancer causing virus (gene), the Rous sarcoma virus. This work represented the first suggestion that cancer might have a genetic etiology and eventually led to a Nobel Prize for Dr. Rous. Although, the structure of DNA was not known to be a double helix until 1956, the entire human genome has now been sequenced (2003). More recent milestones include the

D.M. Miller (✉)
Department of Medicine, James Graham Brown Cancer Center,
University of Louisville School of Medicine, Louisville, KY 40202, USA
e-mail: donaldmi@ulh.org

K.-S. Soh et al. (eds.), *The Primo Vascular System: Its Role in Cancer and Regeneration*,
DOI 10.1007/978-1-4614-0601-3_27,

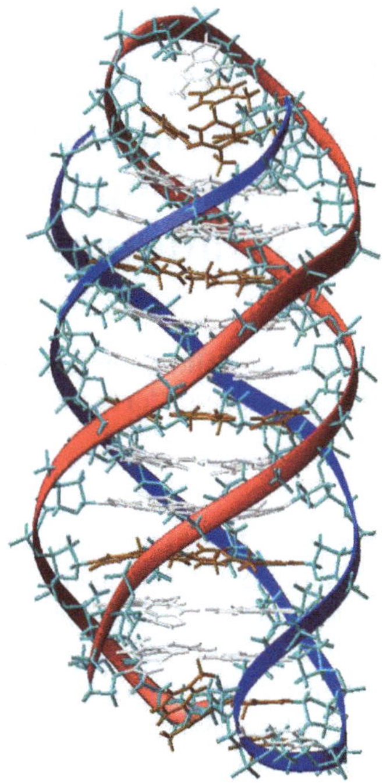

Fig. 27.1 Structure of AS1411. AS1411 is a 26 bp G-rich oligonucleotide which forms a stable quadruplex dimer

first synthesis of cDNA (39 years ago), the cloning of the first human gene (34 years ago), and the discovery of the first human transforming oncogene (H-ras) (29 years ago). Remarkably, it was announced earlier in 2010 that almost complete sequence data had been obtained on 100 cancer genomes (Fig. 27.1).

The development of modern molecular biology has provided great hope for the creation of targeted therapies which take advantage of the current understanding of cancer biology. This area has developed very rapidly. Gleevec, which targets the Bcr/Abl translocation in chronic myelogenous leukemia (CML), was approved for human use in 2002, since that time, several second generation tyrosine kinase inhibitors, including Dasatanib and Nilotinib, have been licensed. These drugs were developed very quickly in response to reports of Gleevec resistance. Currently, more than 200 targeted therapies are in various stages of clinical and preclinical development. Our group has been very fortunate to be part of a 30-year research project which was based on a clinical observation in a single patient which has now led to a new class of compounds which are being tested in humans.

2 Mithramycin Induces In Vivo Differentiation in CML

In 1979, a 43-year-old female who had been diagnosed with blast phase CML presented to the Hematology Oncology service at the University of Michigan. Later that year, she was hospitalized with complications of blast phase CML.

On admission, she had leukemia-related hypercalcemia, for which she received mithramycin, an RNA synthesis inhibitor. Within 2 days after receiving mithramycin, her Ca^{++} had returned to near normal values. Surprisingly, her WBC went from nearly 200,000 to normal (5,000) and her percentage of blast cells from 95 to 0% during the first week of treatment [1]. She had a dramatic improvement in her symptoms, returning to an essentially normal status. Over the next few weeks she experienced a gradual recurrence of her leukemia, but subsequently had a second response to mithramycin treatment. She was maintained in mithramycin response for more than a year with three times weekly treatment. This unusual clinical observation prompted us to investigate the molecular mechanism by which mithramycin induces leukemic cell death.

At that time, the overall survival of patients with CML was roughly 3 years, while the survival of patients with blast phase CML was similar to that of patients with other forms of acute leukemia, averaging less than 3 months. The response rate of patients with blast phase CML to chemotherapy was <25% and most clinical responses were short-lived. It is important to remember that, in 1979, effective treatments for blast phase CML including Interferon and targeted therapies had not yet become available. Based on this clinical observation, a pilot study of the efficacy of mithramycin in blast phase CML was performed at the University of Michigan. This study demonstrated that 10 of 13 patients had clinical evidence of response to treatment [2]. Although a subsequent multi-institutional study by the Southeast Oncology Group showed that mithramycin was not overwhelmingly effective in blast phase CML (26% response rate), treatment of CML cells with mithramycin has provided an important biologic model to study the mechanism by which DNA binding compounds induce transcriptional inhibition in leukemic cells.

3 Molecular Mechanisms of Mithramycin Action

The observation of clinical responses to mithramycin in patients with blast phase CML prompted experiments which demonstrated that mithramycin induces in vitro and in vivo differentiation of leukemic cells resulting in leukemic cell death [3]. We showed that this differentiation is accompanied by selective inhibition of transcription of genes with G–C rich promoters (including c-myc) [4]. This effect reflects decreased transcription of these genes mediated by inhibition of regulatory protein binding to G–C rich regions of DNA [3]. As a result of its selective binding of guanine and cytosine nucleotides, mithramycin selectively inhibits expression of the c-Myc proto-oncogene, thus inhibiting cellular proliferation [5, 6]. Mithramycin was one of the first drugs shown to inhibit myointimal proliferation of the carotid artery after injury [7]. We also demonstrated that mithramycin inhibits the expression of other genes with G–C rich promoters, including dihydrofolate reductase [8]. However, the gene specificity of mithramycin transcriptional inhibition was limited by the fact that it binds to any combination of adjacent cytosine and guanine nucleotides. We now know that mithramycin inhibits the expression of hundreds, if not thousands, of genes in eukaryotic cells. Although our efforts to make a gene-specific mithramycin derivative proved unsuccessful, it has represented a prototype drug which has allowed studies of transcriptional inhibition.

4 Triple Helical DNA Provides Sequence-Specific Interactions with Regulatory Sequences

Because of the lack of sequence- and gene-specificity of contemporary DNA binding drugs, we initiated an effort to design sequence-specific DNA binding compounds, focusing specifically on the formation of triplex helical DNA sequences. It has been known since 1957 [9] that polypurine-polypyrimidine DNA sequences form three stranded DNA structures stabilized by Hoogstein base pairing. The binding of the third strand is "sequence-specific" in triple helical DNA, as reflected by the fact that replacement of even one base in a 25-base oligonucleotide reduces binding affinity [10]. This provides the opportunity to use "gene-targeted" triplex-forming oligonucleotides to inhibit expression of individual genes.

The promoters of the human c-myc, EFG-R, dihydrofolate reductase, K-ras and IL-2 receptor genes contain triplex DNA forming sequences. We were able to show that triplex formation in target sequences can effectively inhibit protein binding and gene expression [11–13]. Using in vitro transcription systems, we demonstrated that triplex-forming oligonucleotides targeted to cancer-related genes including c-myc, c-myb, k-ras, and DHFR could bind specifically to important promoter sequences and prevent regulatory protein binding to these important sequences [11]. We could also demonstrate that triplex formation also inhibits transcription initiation in in vitro systems [13, 14]. However, efforts to demonstrate gene-specific inhibition of transcription in cell systems were, generally, unsuccessful. We now believe that intracellular triplex DNA is repaired by cellular defense mechanisms and does not inhibit gene expression reliably.

5 G-Quadruplex Oligonucleotide Aptamers Inhibit Cell Growth

In 1997, our laboratory noticed that G-rich oligonucleotides, which were being used as controls for experiments with TFO's (triplex-forming oligonucleotides), had striking growth inhibitory activity for cancer cells, but did not affect the growth of nontransformed cells [15]. We have characterized the ability of these oligonucleotides to induce S-phase arrest and apoptosis and have begun to understand their mechanism of action. The G-quadruplex oligonucleotide aptamer, AS1411, inhibits the growth of a wide variety of tumor cell lines. AS1411 forms a highly stable four stranded dimer, stabilized by G-quartet formation, which is resistant to exonuclease and endonuclease digestion (Fig. 27.2).

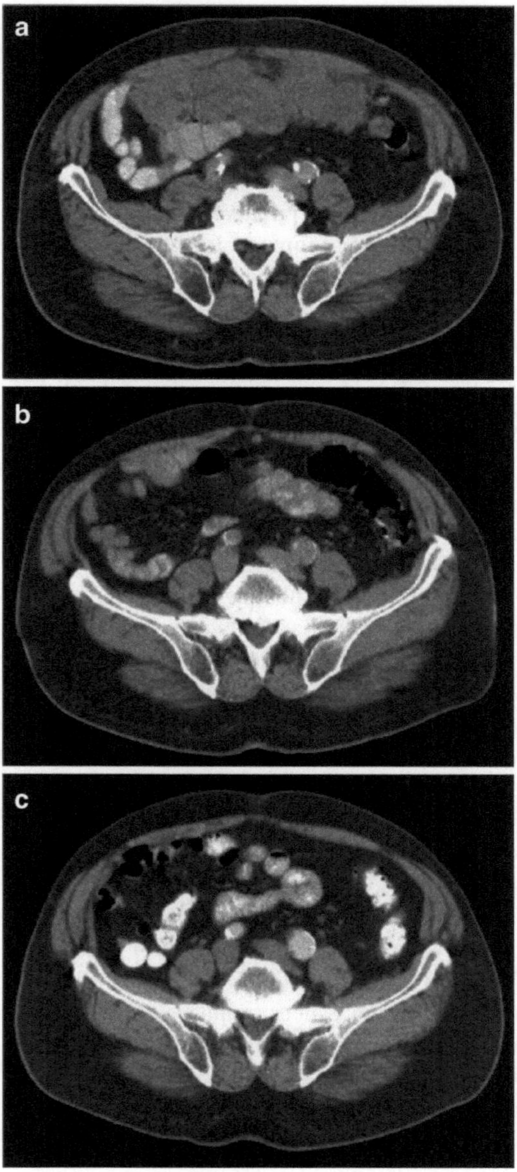

Fig. 27.2 Serial CT scans of Patient 7 with renal cell carcinoma treated with AS1411. (**a**). Pretreatment scan showing large peritoneal mass. (**b**) CT scan at 15 weeks after a single treatment. (**c**) CT scan at 44 weeks after treatment. Scan at 48 months after treatment is normal

Quadruplex DNA is a four stranded structure which is stabilized by G-quartets, four guanines which interact via Hoogstein bonding to form a planor ring [16, 17]. These structures may contain one or more individual nucleic acid strands in parallel or antiparallel configuration. Quadruplex DNA is very stable once formed, deriving its chemical stability from hydrogen bonding within each quartet, stacking of the hydrophobic quartets upon one another and coordination of a non-covalent cation in the central channel[18]. We have shown that divalent cations play a critical role in stabilizing quadruplex DNA under physiologic conditions [19]. Because of its stable tertiary structure, quadruplex DNA is resistant to attack by the enzymes which target single-stranded or duplex DNA. Quadruplex DNA is remarkably stable in serum containing medium and within tumor cells.

The existence of putative quadruplex-forming structures in the genome underlines the potential importance of four stranded DNA structures in growth regulation [20–22]. These sequences are disproportionately represented in the promoters of eukaryotic genes, particularly growth regulatory genes. There is strong evidence that quadruplex-forming sequences in the c-myc [23], c-Myb [24], K-Ras [25], Bcl-2 [26, 27], retinoblastoma [28], and HIF1α [29] gene promoters play important roles in the regulation of these important growth regulatory genes. These sequences have been shown to be protein binding sequences and to play important roles in regulation of gene expression [30]. Quadruplex-forming sequences also occur in telomeric DNA [31, 32], immunoglobulin switch region sequences [33], human immunodeficiency type I RNA [34], and the fragile X repeat sequences [35].

Although the exact mechanism by which AS1411 inhibits cell growth and induce apoptosis is unknown, it has been shown to inhibit the growth of a wide variety of tumor cell lines. Its growth inhibitory activity correlates with its ability to form G-quartets, as well as to bind to the ubiquitous cellular protein, nucleolin [15, 36]. AS1411 induces apoptosis following S-phase arrest, but does not inhibit protein or DNA synthesis [36]. It is a member of a group of oligonucleotides which have non-sequence-specific effects which are mediated by mechanisms other than the classic antisense mechanism [15, 36–41]. It has been shown that quadruplex-forming oligonucleotides inhibit activation of NF-kappaB (NFκB) [42]. AS1411, which functions as an aptamer, demonstrates nearly complete growth inhibition of breast, cervical, and prostate cancer cell lines [15, 36]. AS1411 has also shown antitumor activity in vivo, in xenograft bearing mice. This activity includes activity against a broad spectrum of tumors, including non-small cell lung cancer, renal cell carcinoma, small cell lung cancer, and breast cancer. In xenograft models, AS1411 was able to completely inhibit the growth of the xenografts at a dose of 1 or 5 mg/animal.

6 AS1411 Has Anticancer Activity with Minimal Toxicity in a Human Phase I Clinical Trial

Because of the striking growth inhibitory activity of the quadruplex-forming oligonucleotides and their activity in animal models with no apparent toxicity, we felt that it was important to characterize their anticancer activity and toxicity in humans. The preclinical data suggested activity in a wide variety of solid tumors and leukemias with no apparent toxicity. Therefore, an open label Phase I clinical trial was initiated in 2003. In this study, 31 patients with advanced solid tumors refractory to conventional/standard treatment were treated with a four or seven day continuous infusion of AS1411 at escalating dosages [43]. The goal of this study was to establish the maximum tolerated dose (MTD) and dose-limiting toxicity (DLT), determine pharmacokinetic distribution and profile, and obtain preliminary evidence of biological/clinical response.

Thirty patients were treated, 29 of whom were evaluable for efficacy. The MTD could not be determined because no DLTs occurred before or at the maximum dosage tested. No treatment-related serious adverse events occurred and only one adverse event was treatment related (grade 2 hypomagnesemia). One patient with renal cell carcinoma (RCC) had a complete response which has lasted for >60 months (shown in Fig. 27.2) and one had a partial response which has lasted 32 months. The other ten patients with RCC and 58.6% of all patients had stable disease for >4 weeks. We concluded that in patients with metastatic/unresectable cancer, AS1411 at a dosage of 1–40 mg/kg/day is well tolerated with no serious toxicity. AS1411 is currently being tested in a Phase IIb clinical trial in acute myelogenous leukemia.

7 The c-myc Promoter Quadruplex-Forming Sequence (Pu27) Inhibits Cellular Proliferation and Induces Apoptosis

As described earlier, we have extensively characterized the antiproliferative activity of AS1411, a quadruplex-forming oligonucleotide which we discovered serendipitously when it was used as a control for an experiment with triplex-forming oligonucleotides. Since the exact role of quadruplex-forming sequences in the promoters of eukaryotic genes remains unclear, we decided to characterize the cellular effect of the c-myc promoter quadruplex-forming sequence. Initial experiments compared the effect of these sequences (in the single-stranded form) on cellular proliferation and cell cycle-arrest to those of AS1411. Our initial experiments demonstrating that the c-myc promoter quadruplex-forming sequence binds α-enolase, suggested that this sequence may be competing for some of the same proteins which are bound by AS1411. Therefore, we characterized the effect of an unmodified phosphodiester oligonucleotide encoding Pu27 on proliferation of U937 leukemia cells. We have shown that the c-myc quadruplex-forming sequence (Pu27) inhibits the growth of U937 cells in a concentration dependent manner. There appear to be distinct differences in tumor specificity of each of the genomic sequences that we have tested, although each exhibits growth inhibitory activity. In each case, mutations which abrogate quadruplex formation (as confirmed by circular dichroism analysis) have little or no growth inhibitory effect. Interestingly, very preliminary experiments have shown that these naturally occurring sequences have little effect on the growth of nontransformed cells.

8 Lessons for the Primo-Vascular Field

There are several important lessons from our experience which are relevant to the primo vascular field. In order for the relevance of the PVS to human disease to be fully appreciated, it will be necessary to document the presence and function of this system in human tissue. To that end, we have performed preliminary experiments in a mouse xenograft model of human tumors. We have used Trypan blue staining to demonstrate the existence of primo vascular vessels in this tissue. Current experiments are characterizing the cellular nature of these vessels and characterizing the PV nodes which we have isolated.

References

1. Koller CA et al (1985) In vivo differentiation of blast-phase chronic granulocytic leukemia. Expression of c-myc and c-abl protooncogenes. J Clin Invest 76(1):365–369
2. Koller CA, Miller DM (1986) Preliminary observations on the therapy of the myeloid blast phase of chronic granulocytic leukemia with plicamycin and hydroxyurea. N Engl J Med 315(23):1433–1438
3. Ray R et al (1989) Mithramycin blocks protein binding and function of the SV40 early promoter. J Clin Invest 83(6):2003–2007
4. Miller DM et al (1987) Mithramycin selectively inhibits transcription of G-C containing DNA. Am J Med Sci 294(5):388–394
5. Snyder RC et al (1991) Mithramycin blocks transcriptional initiation of the c-myc P1 and P2 promoters. Biochemistry 30(17):4290–4297
6. Snyder RC, Miller DM (1992) Regulation of c-myc transcription initiation and elongation. Crit Rev Oncog 3(3):283–291
7. Chen SJ et al (1994) Mithramycin inhibits myointimal proliferation after balloon injury of the rat carotid artery in vivo. Circulation 90(5):2468–2473
8. Blume SW et al (1991) Mithramycin inhibits SP1 binding and selectively inhibits transcriptional activity of the dihydrofolate reductase gene in vitro and in vivo. J Clin Invest 88(5):1613–1621
9. Felsenfeld G, Rich A (1957) Studies on the formation of two- and three-stranded polyribonucleotides. Biochim Biophys Acta 26(3):457–468
10. Jain A, Wang G, Vasquez KM (2008) DNA triple helices: biological consequences and therapeutic potential. Biochimie 90(8):1117–1130
11. Ebbinghaus SW et al (1993) Triplex formation inhibits HER-2/neu transcription in vitro. J Clin Invest 92(5):2433–2439
12. Gee JE et al (1992) Triplex formation prevents Sp1 binding to the dihydrofolate reductase promoter. J Biol Chem 267(16):11163–11167
13. Helm CW et al (1993) A unique c-myc-targeted triplex-forming oligonucleotide inhibits the growth of ovarian and cervical carcinomas in vitro. Gynecol Oncol 49(3):339–343
14. Mayfield C et al (1994) Triplex formation by the human Ha-ras promoter inhibits Sp1 binding and in vitro transcription. J Biol Chem 269(27):18232–18238
15. Bates PJ et al (1999) Antiproliferative activity of G-rich oligonucleotides correlates with protein binding. J Biol Chem 274(37):26369–26377
16. Gellert M, Lipsett MN, Davies DR (1962) Helix formation by guanylic acid. Proc Natl Acad Sci U S A 48:2013–2018

17. Burge S et al (2006) Quadruplex DNA: sequence, topology and structure. Nucleic Acids Res 34(19):5402–5415
18. Gilbert DE, Feigon J (1999) Multistranded DNA structures. Curr Opin Struct Biol 9(3):305–314
19. Blume SW et al (1997) Divalent transition metal cations counteract potassium-induced quadruplex assembly of oligo(dG) sequences. Nucleic Acids Res 25(3):617–625
20. Du Z et al (2007) Enrichment of G4 DNA motif in transcriptional regulatory region of chicken genome. Biochem Biophys Res Commun 354(4):1067–1070
21. Eddy J, Maizels N (2006) Gene function correlates with potential for G4 DNA formation in the human genome. Nucleic Acids Res 34(14):3887–3896
22. Huppert JL, Balasubramanian S (2007) G-quadruplexes in promoters throughout the human genome. Nucleic Acids Res 35(2):406–413
23. Ambrus A et al (2005) Solution structure of the biologically relevant G-quadruplex element in the human c-MYC promoter. Implications for G-quadruplex stabilization Biochemistry 44(6):2048–2058
24. Palumbo SL et al (2008) A novel G-quadruplex-forming GGA repeat region in the c-myb promoter is a critical regulator of promoter activity. Nucleic Acids Res 36(6):1755–1769
25. Cogoi S, Xodo LE (2006) G-quadruplex formation within the promoter of the KRAS protooncogene and its effect on transcription. Nucleic Acids Res 34(9):2536–2549
26. Dai J et al (2006) NMR solution structure of the major G-quadruplex structure formed in the human BCL2 promoter region. Nucleic Acids Res 34(18):5133–5144
27. Dai J et al (2006) An intramolecular G-quadruplex structure with mixed parallel/antiparallel G-strands formed in the human BCL-2 promoter region in solution. J Am Chem Soc 128(4):1096–1098
28. Murchie AI, Lilley DM (1992) Retinoblastoma susceptibility genes contain 5′ sequences with a high propensity to form guanine-tetrad structures. Nucleic Acids Res 20(1):49–53
29. De Armond R et al (2005) Evidence for the presence of a guanine quadruplex forming region within a polypurine tract of the hypoxia inducible factor 1alpha promoter. Biochemistry 44(49):16341–16350
30. Fry M (2007) Tetraplex DNA and its interacting proteins. Front Biosci 12:4336–4351
31. Kan ZY et al (2007) G-quadruplex formation in human telomeric (TTAGGG)4 sequence with complementary strand in close vicinity under molecularly crowded condition. Nucleic Acids Res 35(11):3646–3653
32. Sundquist WI, Klug A (1989) Telomeric DNA dimerizes by formation of guanine tetrads between hairpin loops. Nature 342(6251):825–829
33. Sen D, Gilbert W (1988) Formation of parallel four-stranded complexes by guanine-rich motifs in DNA and its implications for meiosis. Nature 334(6180):364–366
34. Sundquist WI, Heaphy S (1993) Evidence for interstrand quadruplex formation in the dimerization of human immunodeficiency virus 1 genomic RNA. Proc Natl Acad Sci U S A 90(8):3393–3397
35. Fry M, Loeb LA (1994) The fragile X syndrome d(CGG)n nucleotide repeats form a stable tetrahelical structure. Proc Natl Acad Sci USA 91(11):4950–4954
36. Xu X et al (2001) Inhibition of DNA replication and induction of S phase cell cycle arrest by G-rich oligonucleotides. J Biol Chem 276(46):43221–43230
37. Anselmet A et al (2002) Non-antisense cellular responses to oligonucleotides. FEBS Lett 510(3):175–180
38. Buckheit RW Jr et al (1994) Potent and specific inhibition of HIV envelope-mediated cell fusion and virus binding by G quartet-forming oligonucleotide (ISIS 5320). AIDS Res Hum Retroviruses 10(11):1497–1506
39. Burgess TL et al (1995) The antiproliferative activity of c-myb and c-myc antisense oligonucleotides in smooth muscle cells is caused by a nonantisense mechanism. Proc Natl Acad Sci USA 92(9):4051–4055

40. Mata JE et al (1997) A hexameric phosphorothioate oligonucleotide telomerase inhibitor arrests growth of Burkitt's lymphoma cells in vitro and in vivo. Toxicol Appl Pharmacol 144(1):189–197
41. Dapic V et al (2002) Antiproliferative activity of G-quartet-forming oligonucleotides with backbone and sugar modifications. Biochemistry 41(11):3676–3685
42. Girvan AC et al (2006) AGRO100 inhibits activation of nuclear factor-kappaB (NF-kappaB) by forming a complex with NF-kappaB essential modulator (NEMO) and nucleolin. Mol Cancer Ther 5(7):1790–1799
43. Laber DA, Taft BS, Kloecker GH, Bates PJ, Trent JO, Miller DM (2006) Extended phase I study of AS1411 in renal and non-small cell lung cancers. J Clin Oncol, ASCO Annual Meeting Proceedings. 24(No 18S):13098

Chapter 28
Translational Development of Therapeutic Vaccines for Lymphoma

Larry W. Kwak

Abstract Follicular lymphoma expresses a unique immunoglobulin molecule termed idiotype that has been used as a tumor-specific antigen for vaccine development. Early stage clinical studies revealed that vaccination consisting of keyhole limpet hemocyanin-conjugated lymphoma idiotype protein in combination with granulocyte-macrophage colony-stimulating factor induced tumor-specific immune responses and molecular remission in patients with follicular lymphoma. Three double-blind, randomized, phase III trials were conducted to further determine the clinical benefits of this vaccine therapy. Compared to the placebo, prolonged disease-free survival in vaccinated patients was concluded only in one study where all the patients enrolled in the trial already had complete remission from induction chemotherapy. Next generation idiotype vaccines are being developed with the focus on simplifying vaccine formulation and potentiating tumor-specific immunity. This category includes genetically modified idiotype single chain DNA liposome-encapsulated idiotype vaccine and dendritic cell vaccine. Although preclinical data supported the immunogenicity and therapeutic advantage of these new vaccines, their clinical benefits remain to be tested. Optimizing next generation idiotype vaccines may require combination with immune adjuvants that potentiate vaccine-induced antitumor immunity, have direct effects against tumor or block immune regulatory checkpoints. Moreover, identification of a universal follicular lymphoma antigen is important for future development of vaccine therapy against this disease.

L.W. Kwak (✉)
University of Texas M. D. Anderson Cancer Center, Houston, TX, USA
e-mail: lkwak@mdanderson.org

K.-S. Soh et al. (eds.), *The Primo Vascular System: Its Role in Cancer and Regeneration*, DOI 10.1007/978-1-4614-0601-3_28,

1 Discovery Research

Immunoglobulin (Ig) molecules are composed of heavy and light chains, which possess highly specific variable regions at their amino termini. These variable regions contain determinants (molecule shapes) that can themselves be recognized as antigens or idiotypes. Thus the idiotypic determinants of the surface Ig of a B-cell lymphoma or myeloma can serve as a tumor-specific marker for the malignant clone. The preclinical rationale for idiotype vaccination was first demonstrated by Lynch and Eisen in 1970s and later confirmed by others in different tumor models. Kwak, working in the laboratory of Ronald Levy at Stanford University, pioneered the first study translating therapeutic Id vaccines to human patients [1]. This small pilot study demonstrated that specific antibody response could be elicited in patients with low-grade lymphoma. Subsequently, the Kwak laboratory at the National Cancer Institute (NCI) discovered that the addition of granulocyte macrophage colony-stimulating factor (GM-CSF) substantially improved vaccine potency and successfully induced T-cell responses in murine lymphoma models.

2 NCI Investigational New Drug Development

With dedicated process development support by the Biological Resources Branch at NCI, the first investigational new drug (IND)-supported study of the prototype Id vaccine product (patient-specific, hybridoma-generated, Id protein chemically linked with keyhole limpet hemocyanin (KLH)) was a phase II clinical trial for previously untreated patients with follicular lymphoma by the Kwak laboratory [2]. This trial used soluble GM-CSF as an integral part of the vaccine. All patients on this study received treatment with a uniform chemotherapy regimen, which produced a homogeneous group of 20 patients who were all in first complete remission (minimal residual disease state). After a 6-month break following induction therapy (to allow for immune reconstitution after the immunosuppressive effects of chemotherapy), soluble recombinant GM-CSF was mixed with Id-KLH and the complete vaccine was injected subcutaneously in 5 monthly doses. This study was the first to show convincing lymphoma-specific CD8+ T-cell responses in the vast majority of patients (85%). In addition, evidence for molecular remissions was obtained in most of the subset of patients whose lymphomas could be detected by this technique. Particularly compelling were the clinical outcomes of this phase II study. After a median follow-up of 9.2 years, the median disease-free survival (DFS) was 8 years and overall survival was 95.2%. The DFS was superior to that of a historic, ProMACE chemotherapy-treated control group (median DFS, about 2.2 years). These results were subsequently confirmed in other, non-IND studies of various vaccine formulations using hybridoma-generated Id proteins.

3 Pivotal Phase III Trial

Three randomized, double-blind, placebo-controlled, multicenter clinical trials were designed to formally determine the clinical efficacy of therapeutic Id vaccination in follicular lymphoma (FL). The first phase III trial was initiated by the NCI, and subsequently sponsored by Biovest International Inc. Patients with previously untreated advanced-stage FL were treated with cisplatin, adriamycin, and cyclophosphamide (PACE) chemotherapy regimen until clinical remission. Patients achieving complete remission were randomized at a ratio of 2:1 to receive Id-KLH (BiovaxID) plus GM-CSF or KLH plus GM-CSF. The primary endpoint for this trial was DFS. Out of 234 enrolled patients, 177 achieved complete response and were subsequently randomized to receive either active or control vaccine. Of these, 117 maintained remission for the 6-month rest period and received at least one dose of vaccine. Seventy-six patients received Id-KLH plus GM-CSF and 41 patients received KLH plus GM-CSF. Both arms were balanced for the International Prognostic Index and other relevant clinical factors. After a median follow-up of 56.6 months (range 12.6–89.3 months), median time to relapse after randomization for the Id-KLH/GM-CSF arm was 44.2 months, compared to 30.6 months for the control arm ($p=0.045$; hazard ratio = 1.6) [3].

The other two phase III trials reported negative results but differed in terms of induction therapy, eligibility of the patients, and method of Id production (Myvax, Genitope; Mitumprotimub-t, Favrille).

4 Conclusions and Future Directions

Several differences in the above phase III trials may account for their disparate results. First, only patients who achieved complete remission were randomized in the BiovaxID clinical trial whereas MyVax trial included patients in partial response and the Mitumprotimub-t trial included patients in partial and stable disease. Moreover, the choice of a doxorubicin-containing chemotherapy regimen likely resulted in a higher proportion of patients achieving complete remission in the BiovaxID trial. Results from these trials suggest that a minimal residual disease state may be required for the vaccine-induced immune responses to be effective. Second, BiovaxID was prepared by heterohybridoma method whereas MyVax and Mitumprotimub-t are recombinant proteins, suggesting that the immunogenicity of the Id vaccines may be different between the different formulations. Finally, it is possible that B-cell depletion, resulting from the use of rituximab in the Mitumprotimub-t trial, may have had a deleterious effect on subsequent vaccine efficacy. Taken together, we speculate that minimal residual disease status, as exemplified by complete clinical remission, appears to be necessary for the clinical antitumor effect of idiotype vaccines, as demonstrated by the success of BiovaxID in a pivotal phase III trial.

An effective vaccine requires three factors: tumor antigen, delivery platform, and vaccine adjuvant. Additional vaccine therapies are being developed, including those based on DNA, liposome, and dendritic cells as delivery systems, and CpG and checkpoint inhibitors. It is hoped that immunotherapeutic agents may in the future allow effective treatment of lymphoid malignancies in combination with conventional cytotoxic therapies. Various strategies are being developed to optimize these vaccines. Lymphoma vaccines will have to be optimally integrated with standard therapies, in particular with rituximab which is now extensively used in follicular lymphoma. It is notable that T cell responses in absence of B cells were demonstrated in patients with mantle cell lymphoma, the majority of which developed lymphoma-specific Th1 cells following treatment with rituximab [4]. Finally, definitive characterization of anti-idiotype cellular immune responses have defined the unique immunodominant peptide epitopes within the hypervariable complementarity-determining regions, but not framework regions of immunoglobulin heavy chain, recognized by lymphoma-specific T cells [5]. Taken together, these results justify the use of Id vaccines after rituximab-based induction regimens.

Finally, other questions remain such as which patients are most likely to benefit from vaccination. Additional immune response analysis is in progress using blood and serum samples from vaccinated patients enrolled on this phase III trial. In addition, banked residual tumor samples are available from each patient from excisional lymph node harvests originally required as starting material for vaccine production. These residual tumor samples will be used for genomic profiling and biomarker discovery, based on correlation with clinical outcome data.

References

1. Kwak LW, Campbell MJ, Czerwinski DK et al (1992) Induction of immune responses in patients with B-cell lymphoma against the surface-immunoglobulin idiotype expressed by their tumors. N Engl J Med 327:1209–1215
2. Bendandi M, Gocke CD, Kobrin CB et al (1999) Complete molecular remissions induced by patient-specific vaccination plus granulocyte-monocyte colony-stimulating factor against lymphoma. Nat Med 5:1171–1177
3. Schuster SJ, Neelapu SS, Gause BL et al (2011). Vaccination with patient-specific tumor-derived antigen in first remission improves disease-free survival in follicular lymphoma. J Clin Oncol. 29(20):2787–2794
4. Neelapu SS, Kwak LW, Kobrin CB et al (2005) Vaccine-induced tumor-specific immunity despite severe B-cell depletion in mantle cell lymphoma. Nat Med 11:986–991
5. Baskar S, Kobrin CB, Kwak LW (2004) Autologous lymphoma vaccines induce human T cell responses against multiple, unique epitopes. J Clin Invest 113:1498–1510

Chapter 29
Oxygen Transport to Tumors: Pathophysiology and Clinical Implications

Peter Vaupel

Abstract Tumor oxygenation is heterogeneous and severely compromised as compared to normal tissue. Hypoxia, i.e., oxygen depletion as a consequence of an imbalance between oxygen supply and cellular oxygen consumption, mostly results from inadequate diffusion and perfusion within tumors and from reduced O_2 transport capacity in anemic patients. Detection of tumor hypoxia has been performed using minimally invasive approaches, imaging PET and MRI techniques, and immunohistochemistry on tissue biopsies. The development of tumor hypoxia is independent of a series of relevant tumor characteristics (e.g., clinical size, stage, histology, and grade) and various patient demographics. Overall median pO_2 in cancers of the uterine cervix, head and neck, and breast is about 10 mmHg with the overall hypoxic fraction ($pO_2 \leq 2.5$ mmHg) being approximately 25%. Metastatic lesions do not substantially deviate from the oxygenation status of (their) primary tumors. Identification of tumor hypoxia may allow an assessment of a tumor's potential to develop an aggressive phenotype or acquired treatment resistance, both of which lead to poor prognosis. Detection of hypoxia in the clinical setting may therefore be helpful in selecting high-risk patients for individual and/or more intensive treatment schedules.

1 Introduction

The physiology of tumors is uniquely different from that of normal tissues. It is characterized inter alia by oxygen (O_2) depletion (hypoxia or anoxia), extracellular acidosis, high lactate and adenosine levels, glucose and energy deprivation, significant interstitial fluid flow, and interstitial hypertension [1, 2]. This hostile microenvironment is largely determined by an abnormal tumor microcirculation. The tumor is

P. Vaupel (✉)
Department of Radiotherapy and Radiooncology, Klinikum rechts der Isar, Technical University of Munich, Ismaninger Strasse 22, 81675 Munich, Germany
e-mail: peter.vaupel@lrz.tum.de

K.-S. Soh et al. (eds.), *The Primo Vascular System: Its Role in Cancer and Regeneration*, DOI 10.1007/978-1-4614-0601-3_29, © Springer Science+Business Media, LLC 2012

characterized by vigorous proliferation leading to immature, structurally defective and, in terms of perfusion, ineffective microvessels which lack the signals to mature. Tumor vasculature is often described as an "aberrant monster" [3]. Consequently, tumor blood flow is chaotic, inadequate, and heterogeneous [1, 2, 4, 5].

2 Causes of Tumor Hypoxia

The pathogenesis of tumor hypoxia is multifactorial. Tumor hypoxia, i.e., critical O_2 levels below which clinical, biological, and/or molecular effects are progressively observed (e.g., slowing of proliferation rate, changes of the transcriptome, proteome and genome, metabolic hypoxic stress response, and development of an aggressive phenotype, ATP depletion, binding of hypoxic markers, acquired treatment resistance) frequently results from the inadequate perfusion. *Perfusion-limited O_2 delivery* leads to *ischemic hypoxia* which is often transient. For this reason, this type of hypoxia is often called "acute" hypoxia, a term that does not take into account the mechanisms underlying this condition [6]. "Acute" hypoxia also can result as a consequence of transient flow with plasma only due to the very low O_2 content in plasma or strong variations in RBC fluxes.

Hypoxia in tumors can also be caused by an increase in diffusion distances, so that cells far away (>70 μm) from the nutritive blood vessels receive less oxygen than required. This condition is termed *diffusion-limited hypoxia*, also known as "chronic" hypoxia. In addition to enlarged diffusion distances, an adverse diffusion geometry (e.g., concurrent vs. countercurrent tumor microvessels) can also cause hypoxia [7].

Tumor-associated or therapy-induced anemia can lead to a reduced O_2 transport capacity of the blood, a major (systemic) factor contributing to the development of hypoxia (*anemic hypoxia*). This type of hypoxia is especially pronounced in tumors or tumor areas exhibiting low perfusion rates. A similar condition can be caused by carboxyhemoglobin (HbCO) formation in heavy smokers, which leads to a *functional anemia*, since hemoglobin blocked by carbon monoxide (CO) is no longer capable of transporting oxygen.

Hypoxia can rapidly develop in (primary or metastatic) liver tumors that are preferentially supplied by branches of the portal vein (*hypoxemic hypoxia*).

There is abundant evidence for the existence of steep, fluctuating and unstable oxygen gradients leading to a substantial spatial and temporal heterogeneity (*4D-heterogeneity*) in the development and extent of tumor hypoxia due to pronounced intra-tumor (and inter-tumor) variabilities in vascularity and perfusion rates (for reviews see refs. [1, 7, 8]).

3 Characterization of Tumor Oxygenation

Current knowledge concerning the oxygenation status of tumors in terms of O_2 partial pressure distributions is predominantly derived from clinical studies on the pretreatment O_2 status of solid tumors using minimally invasive oxygen microsensors [9].

Table 29.1 Pretherapeutic oxygenation status of locally advanced human tumors (selection of data [8, 9])

Tumor type	Median pO_2 (mmHg)	HF 2.5 (%)	HF 5 (%)	HF 10 (%)
Soft tissue sarcomas	14	13	21	
Gliomas	13	26	30	49
Vulvar cancers	11	25	40	
Breast cancers	10	30	47	50
Head and neck cancers	10	21	32	50
Renal cell cancers	10		10	50
Cervix cancers	9	28	44	59
Prostate cancers	7	26	46	
Biliary tract cancers	4–8		95	100
Pancreas cancers	2	59		98

HF 2.5 = fraction of pO_2 values ≤2.5 mmHg; HF 5 = fraction of pO_2 values ≤5 mmHg; HF 10 = fraction of pO_2 ≤10 mmHg

Table 29.2 Oxygenation status of normal tissues [8, 9]

Tumor type	Median pO_2 (mmHg)	HF 2.5 (%)	HF 5 (%)	HF 10 (%)
Pancreas	57	2		
Breast	52	0	0	0
Subcutis	51	0	0	4
Cervix	42	8	13	20
Kidney	31	0	3	8
Liver	30	0	5	13
Skeletal muscle	25–30	0	4	12
Brain	24–27	0	3–8	13

This technique allows the direct and reliable measurement of O_2 partial pressures (pO_2 values) in accessible tissues. It provides quantitative measures.

Investigations carried out in the clinical setting over the last two decades demonstrated that the presence of hypoxic tissue areas with pO_2 values ≤2.5 mmHg is a characteristic pathophysiological property of locally advanced solid tumors, and such areas have been found in a wide range of human malignancies. Evidence has accumulated showing that at least 50–60% of locally advanced solid tumors may exhibit hypoxic and/or anoxic tissue areas that are heterogeneously distributed within the tumor mass [9] (Anoxia describes the pathophysiological state where no O_2 is detectable (measurable) in the tissues ($pO_2 = 0$ mmHg)).

Relevant parameters of the pretherapeutic oxygenation status of locally advanced human tumors are presented in Table 29.1. There is clear evidence that the oxygenation of solid tumors is distinctly poorer than that of the respective normal tissues (see Table 29.2).

When the available data on pretreatment tumor oxygenation of 150 locally advanced cancers of the uterine cervix is summarized [8, 9], there is evidence that:

1. Oxygenation in tumors is heterogeneous and compromised as compared to normal tissues.
2. Tumor oxygenation is not regulated according to the metabolic demands as is the case in normal tissues.

3. On average, the median pO_2 values in primary cancers of the uterine cervix are lower than those in the normal cervix.
4. Many cervical cancers contain hypoxic tissue areas (at least 60%).
5. There is no characteristic topological distribution of O_2 tensions within cervix cancers (periphery vs. center).
6. Tumor-to-tumor variability in oxygenation is greater than intra-tumor variability.
7. Tumor oxygenation is independent of various patient demographics (e.g., age, menopausal status, parity).
8. Anemia (found in approximately 30% of patients at diagnosis) considerably contributes to the development of hypoxia, especially in low-flow tumor areas.
9. In cervix cancers of moderately/severely anemic patients, hypoxic areas are more frequently found than in nonanemic patients.
10. Tumor oxygenation and the extent of hypoxia are independent of clinical size, FIGO stage, histological type (squamous cell carcinomas vs. adenocarcinomas), grade, and lymph node status.
11. Tumor oxygenation is weakly dependent on the pathological tumor stage (pT stage).
12. Local recurrences of cervix cancers have a higher hypoxic fraction than the primary tumors.
13. Hypoxia in cervical cancers has been found to be of prognostic significance in many investigations.

4 Clinical Importance of Tumor Hypoxia

Cells exposed to chronic hypoxia respond by reducing their overall protein synthesis, which in turn leads to restrained proliferation and a change in the cell cycle distribution. Under anoxia, most cells undergo immediate arrest in whichever cell cycle phase they are currently in. Additionally, hypoxia can induce apoptosis and – below a critical energy state – hypoxia may result in necrotic cell death.

In contrast, hypoxia has been recognized as an important driving force in *malignant progression* [8, 10, 11]. Hypoxia – as an inherent consequence of unregulated growth – can promote local invasion, intravasation of cancer cells and finally metastatic spread to distant sites whereby this occurs in a cooperative manner:

- On the *transcriptome* level leading to hypoxia-induced *changes in gene expression* coordinated by a special set of transcription factors, such as HIFs, NF-κB, AP-1, Sp1, indicating redundancy in biological mechanisms in malignant tumors below 1% O_2 [8]. Besides these mechanisms, two additional oxygen-sensitive pathways have been identified, namely signaling through inhibition of the mammalian target of rapamycin (mTOR) kinase and signaling through activation of the unfolded protein response (UPR). Although these different signaling pathways are activated independently, there is evidence that HIF-, mTOR- and UPR-dependent responses to hypoxia act in an integrated way.

- On the *proteome* and *metabolome* level via adaptive gene expression, posttranscriptional and posttranslational modifications [12].
- On the *genome/epigenome* level by increasing genomic and epigenomic instability (below 0.1% O_2).
- On the level of cell populations by *clonal selection* and *clonal expansion* according to phenotype fitness [11].

Tumor cell variants with adaptations favorable to survival under hypoxic stress may have growth advantages over nonadapted cells in the hypoxic microenvironment and may subsequently develop a more aggressive phenotype which in turn is responsible for malignant progression. Cyclic hypoxia and pronounced spatio-temporal heterogeneities (4D-heterogeneities) in hypoxia may be the most powerful factors promoting an aggressive tumor phenotype.

Hypoxia is known to restrict the success of tumor therapy. Hypoxia-driven resistance to therapy – like that triggered by other microenvironmental factors – adds to the "classical" treatment resistance based on the molecular biology of tumors. Tumor hypoxia plays a pivotal role in *acquired treatment resistance*, since O_2 depletion in solid tumors is classically associated with *resistance to radiotherapy*, but has also been shown to reduce the efficacy of certain forms of *chemotherapy*, *photodynamic therapy*, *immunotherapy*, and *hormonal therapies* [13]. Hypoxia directly and/or indirectly confers resistance to therapy. Direct effects are mediated through reduced generation of free radicals (some chemotherapy, photodynamic therapy) or lacking fixation of DNA damage (X- and γ-rays). Indirect hypoxia-driven effects are mostly based on cell cycle effects and on slowing of cellular proliferation kinetics, on changes in the transcriptome, on differential regulations of gene expression, on alterations of the proteome, on enhanced mutagenesis, genomic instability and clonal selection, i.e., on hypoxia-associated tumor progression [10].

There is an increasing evidence that tumor hypoxia – through favoring malignant progression and reducing the therapeutic response – may act as an *adverse prognostic factor* [10, 13]. Independent of standard prognosticators, such as tumor stage and nodal status, hypoxia has been suggested as a negative prognostic factor for patient outcome. Studies of tumor hypoxia involving the direct assessment of the oxygenation status have suggested worse disease-free survival for patients with hypoxic cancers of the uterine cervix or soft tissue sarcomas. In head and neck cancers, available studies suggest that hypoxia is prognostic for survival and local control. In many – albeit not in all – studies endogenous markers (HIF-1α, GLUT-1, CA IX) also showed prognostic significance for patient outcome. Noninvasive assessment of hypoxia using imaging techniques (e.g., PET, MRI) with regard to patient prognosis is so far limited. In the clinical studies performed up until now, the lack of homogeneous patient cohorts and of standardized treatment protocols, inconsistencies of the endpoints characterizing the oxygenation status and methodological differences may compromise the power of the prognostic parameter used.

Acknowledgments This work has been supported by a grant from the Deutsche Krebshilfe (106758).

References

1. Vaupel P, Kallinowski F, Okunieff P (1989) Blood flow, oxygen and nutrient supply, and metabolic microenvironment of human tumors: a review. Cancer Res 49:6449–6465
2. Vaupel P (2004) Tumor microenvironmental physiology and its implications for radiation oncology. Semin Radiat Oncol 14:198–206
3. Shchors K, Evan G (2007) Tumor angiogenesis: cause or consequence of cancer? Cancer Res 65:7059–7061
4. Vaupel P (2006) Abnormal vasculature and defective microcirculatory function in solid tumors. In: Siemann DW (ed) Vascular-targeted therapies in oncology. Wiley, Chichester, pp 9–29
5. Molls M, Vaupel P (eds) (2000) Blood perfusion and microenvironment of human tumors. Implications for clinical radiooncology. Springer, Berlin
6. Vaupel P, Mayer A, Höckel M (2004) Tumor hypoxia and malignant progression. Methods Enzymol 381:335–354
7. Vaupel P, Harrison L (2004) Tumor hypoxia: causative factors, compensatory mechanisms, and cellular response. Oncologist 9(suppl 5):4–9
8. Vaupel P (2009) Pathophysiology of solid tumors. In: Molls M, Vaupel P, Nieder C, Anscher MS (eds) The impact of tumor biology on cancer treatment and multidisciplinary strategies. Springer, Berlin, pp 51–92
9. Vaupel P, Höckel M, Mayer A (2007) Detection and characterization of tumor hypoxia using pO_2 histography. Antioxid Redox Signal 9:1221–1235
10. Vaupel P (2004) The role of hypoxia-induced factors in tumor progression. Oncologist 9(suppl 5):10–17
11. Giaccia AJ (1996) Hypoxic stress proteins: survival of the fittest. Semin Radiat Oncol 6:46–58
12. Höckel M, Vaupel P (2001) Tumor hypoxia: definitions and current clinical, biologic and molecular aspects. J Natl Cancer Inst 93:266–276
13. Vaupel P (2009) Physiological mechanisms of treatment resistance. In: Molls M, Vaupel P, Nieder C, Anscher MS (eds) The impact of tumor biology on cancer treatment and multidisciplinary strategies. Springer, Berlin, pp 273–290

Chapter 30
Stress Responses of Pancreatic Cancer Cells and Their Significance in Invasion and Metastasis

Yoshiharu Motoo, Qi-Sheng Xia, Naoki Nakaya, Takeo Shimasaki, Hideo Nakajima, and Yasuhito Ishigaki

Abstract Stress responses of pancreatic cancer cells are detected in case of chemotherapy. We have reported the expression of various stress proteins in pancreatic cancer cells with special reference to tumor aggressiveness. Here, we elucidated a previously unknown function of SMG-1, one of PI3 kinases, in the signal transduction of stress responses. Gemcitabine induces epithelial–mesenchymal transition (EMT), producing EMT-inducing factors. Regulation of stress responses might lead to the control of metastasis and the overcome of drug resistance in pancreatic cancer.

1 Introduction

Pancreatic cancer cells express stress proteins when they are exposed to various stresses including anticancer agents. Such stress responses might be related to invasion and metastasis of cancer. We have reported the expression of pancreatic stress proteins such as p8 [1, 2], PAP (pancreatitis-associated protein) [3], and TP53INP1 (tumor protein 53-induced nuclear protein1) [4] in human pancreatic cancer cells. PAP and p8 were overexpressed in pancreatic cancer cells, especially in cases with invasion and metastasis, whereas the expression of TP53INP1 was reduced in cancer cells. PAP over-expression is significantly correlated with survival. TP53INP1 expression is lost in pancreatic cancer and precursor lesions. TP53INP1 was induced by an anticancer agent, gemcitabine. We reported a newly found heat shock-inducible tumor small protein (HITS) [5]. The expression of HITS in tumor cells was much lower than in normal epithelial cells.

Y. Motoo (✉)
Department of Medical Oncology, Kanazawa Medical University,
Uchinada, Ishikawa 920-0293, Japan
e-mail: motoo@kanazawa-med.ac.jp

K.-S. Soh et al. (eds.), *The Primo Vascular System: Its Role in Cancer and Regeneration*,
DOI 10.1007/978-1-4614-0601-3_30,

Human primary-microRNA-155 (pri-miR-155)/B-cell integration cluster (BIC) transcript is the precursor of miR-155. BIC and miR-155 are often over-expressed in the progression of various types of tumors.

Here, we present our recent research on the effect of anticancer agents on the expression of stress proteins and BIC, as well as relevant signal pathways. We studied on the involvement of a phosphatidylinositol 3 kinase (PI3K)-related kinase, suppressor of morphogenesis in genitalia-1 (SMG-1).

2 Methods

2.1 Cell Culture and Treatments

Human pancreatic cancer PANC-1 cells were cultured in Dulbecco's modified Eagle's medium (DMEM) with fetal bovine serum (FBS) and antibiotics. Gemcitabine was purchased from Eli Lilly. To inhibit PI3K-related kinases, cells were treated with wortmannin for 1 h prior to gemcitabine treatment. Other inhibitors were as follows: U0126 (mitogen-activated protein kinase/ERK kinase (MEK1/2) inhibitor), C-Jun N-terminal kinase (JNK) inhibitor II (JNK inhibitor, SP600125), SB 218078 (Chk1 inhibitor), and bisindolylmaleimide I (protein kinase C [PKC] inhibitor). All inhibitors were from Calbiochem.

2.2 RNA Purification and Quantitative Real-Time Reverse Transcriptase-Polymerase Chain Reaction

Total RNA was extracted using the miRNeasy Kit (Qiagen). Expression levels of BIC and beta-actin were quantified using Taqman FAM/MGB probes on an ABI 7900HT Fast Real-Time PCR System (Applied Biosystems). Reverse transcription of RNA to cDNA was carried out using the Taqman Reverse Transcription Reagents Kit (Applied Biosystems). BIC PCR products were detected using the Taqman gene expression assays.

2.3 Small Interfering RNA Transfection

Transfection was performed using Lipofectamine RNAiMAX (Invitrogen). The human ataxia-telangiectasia mutated (ATM), ataxia telangiectasia and Rad3-related (ATR), and SMG-1 small interfering RNA (siRNA) duplexes were purchased from Dharmacon (Thermo Fisher Scientific Inc.). A Stealth RNAi™ Negative Control Med GC siRNA (Invitrogen) and a luciferase siRNA (CGTACGCGGAATACTTCGA, B-Bridge) were used as negative controls.

2.4 Western Blot Analysis

Whole cell lysates were prepared in RIPA buffer with protease-inhibitor cocktail (Roche). Cellular protein samples were diluted in loading buffer (Bio-Rad). The separated proteins were transferred to Immobilon-P polyvinylidene fluoride (PVDF) membranes (Millipore). After blocking with skim milk solution, the membranes were treated with anti-ATM (clone 2C1, Abcam) or ATR (Santa Cruz Biotechnology). Anti-SMG-1 rabbit antiserum was produced by immunization with a synthesized human SMG-1 C-terminal polypeptide. The membranes were probed with the secondary antibodies conjugated with horseradish peroxidase (Amersham Biosciences). The bound antibodies were detected with a Supersignal West Femto Maximum sensitivity substrate (Pierce) and then visualized with a Fuji Xerox LAS4000 image analyzer (Fujifilm).

3 Results

BIC expression was induced by GEM in pancreatic cancer PANC-1 cells. The MEK1/2 and JNK inhibitors, but not the Chk1 and PKC inhibitors, suppressed the upregulation of BIC, which was also significantly inhibited by the PI3-kinase inhibitor, wortmannin. RNAi studies showed that wortmannin-sensitive SMG-1, not ATM or ATR, was involved in the upregulation of BIC. BIC RNA was induced by gemcitabine in PANC-1 cells. Mature miR-155 is also induced by gemcitabine in PANC-1 cells, but the maximal effect was seen after 96 h. BIC was also induced by 5-FU, hydrogen peroxide, or bleomycin treatment.

Gemcitabine-induced BIC upregulation is inhibited by MEK1/2 and JNK inhibitors, but not by Chk1 or PKC inhibitor. Gemcitabine-induced BIC up-regulation is also inhibited by Wortmannin or SMG1 siRNA, but not by ATM or ATR siRNA.

4 Discussions

We presented our recent research results on the molecular mechanisms of stress responses of pancreatic cancer cells when they are exposed to anticancer agents, especially gemcitabine. BIC and miR-155 expressions were induced by gemcitabine in pancreatic cancer PANC-1 cells. The MEK1/2, JNK inhibitors, and PI3-kinase inhibitor, wortmannin, suppressed the upregulation of BIC. We elucidated the involvement of SMG-1, PI3K-related kinase, in the stress responses of pancreatic cancer cells for the first time (Fig. 30.1).

Gemcitabine induces epithelial–mesenchymal transition (EMT) [6], producing EMT-inducing factors. It is suggested that pancreatic cancer cells release stress proteins and EMT-inducing factors into the PVS, and that these factors are involved in distant metastasis.

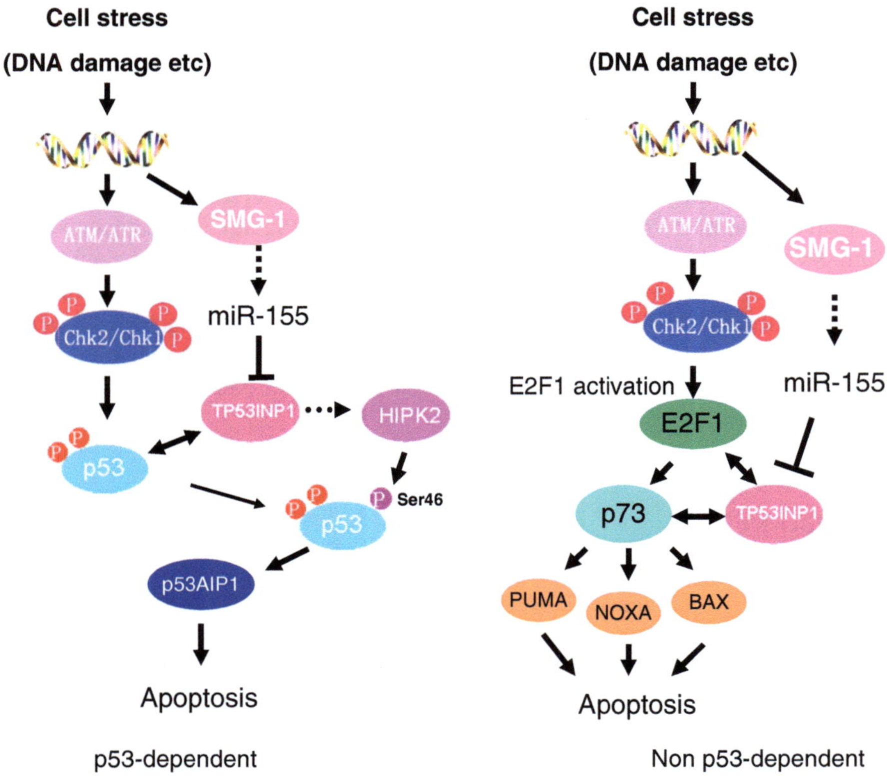

Fig. 30.1 Pathways to apoptosis in response to cellular stress in human pancreatic cancer cells. (**a**) Pathways in a p53-dependent manner; (**b**) pathways in a p53-independent manner

5 Conclusions

In this study, we elucidated a previously-unknown function of SMG-1, one of PI3K-related kinases, in the signal transduction of stress responses [7]. Regulation of stress responses might lead to the control of metastasis and the overcome of drug resistance in pancreatic cancer.

References

1. Su SB, Motoo Y, Iovanna JL et al (2001) Expression of p8 in human pancreatic cancer. Clin Cancer Res 7:309–313
2. Su SB, Motoo Y, Iovanna JL et al (2001) Overexpression of p8 is inversely correlated with apoptosis in pancreatic cancer. Clin Cancer Res 7:1320–1324
3. Xie MJ, Motoo Y, Iovanna JL et al (2003) Overexpression of pancreatitis-associated protein (PAP) in human pancreatic ductal adenocarcinoma. Dig Dis Sci 48:459–464

4. Gironella M, Seux M, Xie MJ et al (2007) Tumor protein 53-induced nuclear protein 1 expression is repressed by miR-155, and its restoration inhibits pancreatic tumor development. Proc Natl Acad Sci U S A 104:16170–5
5. Nakajima H, Ishigaki Y, Xia QS et al (2010) Induction of HITS, a newly identified family with sequence similarity 107 protein (FAM107B), in cancer cells by heat shock stimulation. Int J Oncol 37:583–593
6. Wang Z, Li Y, Kong D, Banerjee S et al (2009) Acquisition of epithelial-mesenchymal transition phenotype of gemcitabine-resistant pancreatic cancer cells is linked with activation of the notch signaling pathway. Cancer Res 69:2400–2407
7. Xia QS, Ishigaki Y, Zhao X et al (2011) SMG-1 is involved in gemcitabine-induced pri-miR-155/BIC upregulation in human pancreatic cancer PANC-1 cells. Pancreas 40:55–60

Chapter 31
Human Urine Extract (CDA-2) Eliminates Cancer Stem-Like Cells and Inhibits Metastasis: Its Potential Role on the Microenvironment of Primo Vascular System

Chih-Jung Yao, Ping-Hsiao Shih, Chi-Tai Yeh, and Gi-Ming Lai

Abstract The concept of "cancer stem cell" may renew the notion of designing cancer therapy. The progression and relapse of tumor may be initiated by the remaining cancer stem cell that escapes from conventional cancer treatments such as chemotherapy and radiation. The research on therapeutics targeting this small population of cancer stem cells is badly needed. By UV laser-equipped flow cytometer and cell-permeable DNA binding dye Hoechst 33342, a distinct side population (SP) cells expressing high-level of ATP-binding cassette transporter protein ABCG2/Bcrp1, could be identified and sorted. These SP cells possess characteristics of stem cells such as self-renewal and expression of stemness genes. Using this stem-like SP cells as a model, we had found that a human urine extract CDA-2 could eliminate SP cells in Huh7 hepatoma cells and downregulate the expression of stemness genes such as ABCG2, Gli, and β-catenin. The mRNA levels of DNA methyltransferase 1, 3A, and 3B were obviously higher in SP cells and could be markedly downregulated by CDA-2 in a dose-dependent manner, suggesting the effects of CDA-2 on epigenetic modification. In addition, CDA-2 and its active component (P23.2) were also found to inhibit the EGF-induced epithelial–mesenchymal transition (EMT) in A549 lung cancer cells. It is proposed that the microenvironment may drive the cancer stem cell to undergo EMT and then migrate into blood stream as circulating tumor cells. Recently, a striking finding described that the Bonghan system (primo-vascular system) is a possible stem cell niche and pathway for cancer metastasis. Based on our results, it is interesting to study the potential role of CDA-2 on the inhibitory regulation of microenvironment existed in the Bonghan system. Further investigation on the manipulation of the Bonghan system by CDA-2 for developing a novel therapeutics is warranted.

G.-M. Lai (✉)
Associate Investigator, National Institute of Cancer Research,
National Health Research Institutes, Associate Professor in Medicine,
Taipei Medical University, Deputy Director, Cancer Center, Wan Fang Hospital,
Taipei Medical University, Taipei 116, Taiwan
e-mail: gminlai@nhri.org.tw

K.-S. Soh et al. (eds.), *The Primo Vascular System: Its Role in Cancer and Regeneration*,
DOI 10.1007/978-1-4614-0601-3_31,

1 Introduction

In recent years, there has been growing evidence that a small subset of cells termed cancer stem cells are wholly responsible for the sustained growth and propagation of the tumor [1]. However, research has been hampered by the lack of distinct molecular markers on cancer stem cells. Fortunately, analysis of the hematopoietic system has shown that bone marrow stem cells contain a subpopulation that effluxes the DNA binding dye, Hoechst 33342, out of the cell membrane [2]. These cells are called side population (SP) cells and are shown to have stem cell characteristics and enrich the stem cell population [3]. The SP analysis has been recently used in the attempt to isolate cancer stem cells from several types of cancers, and cancer stem cells seem to be enriched in the SP population [4].

In addition to the medicinal use of urine in ancient time, urine extract had been studied prospectively for anticancer effects by Burzynski [5] and scientists elsewhere in the world [6]. In 1994, a human urinary preparation, CDA-2 was developed by Dr. M.C. Liau in China and showed significant effects in improving the chemotherapy responses in non small-cell lung cancer, and patients with breast cancer. More recently, human urine extract CDA-2 has been found to induce differentiation in pheochromocytoma and glioma cells [7, 8]. Since differentiation induction is one of the best ways to eliminate cancer stem cells, in this study, we examined the effects of CDA-2 and its active component P23.2 on the cancer stem-like SP cells and investigate the underlying mechanisms. The profound inhibitory effects of CDA-2 on SP cells implied its potential on the microenvironment existed in the Bonghan system for anticancer modulation.

2 Materials and Methods

CDA-2 was provided by EverLife Pharmaceutical Co. Ltd. Hefei, Anhui, China. The active component P23.2 was obtained by *n*-hexane extraction followed by silica gel chromatography and HPLC separation. Human hepatoma cell line Huh7 and lung cancer cell line A549 were chosen for this study. The cell viability was determined by Sulforhodamine B staining. UV laser-equipped flow cytometer and cell-permeable DNA-binding dye Hoechst 33342 were employed to detect the cancer stem-like SP cells in cancer cell lines. RT-PCR was used to analyze the expression of stemness and EMT-related genes.

3 Results and Discussion

After sorting, both the SP (R1) and non-SP (R2) cells from Huh7 hepatocellular carcinoma cells were collected for the subsequent experiments (Fig. 31.1a). We used RT-PCR to analyze the differentially expressed genes in these SP and non-SP

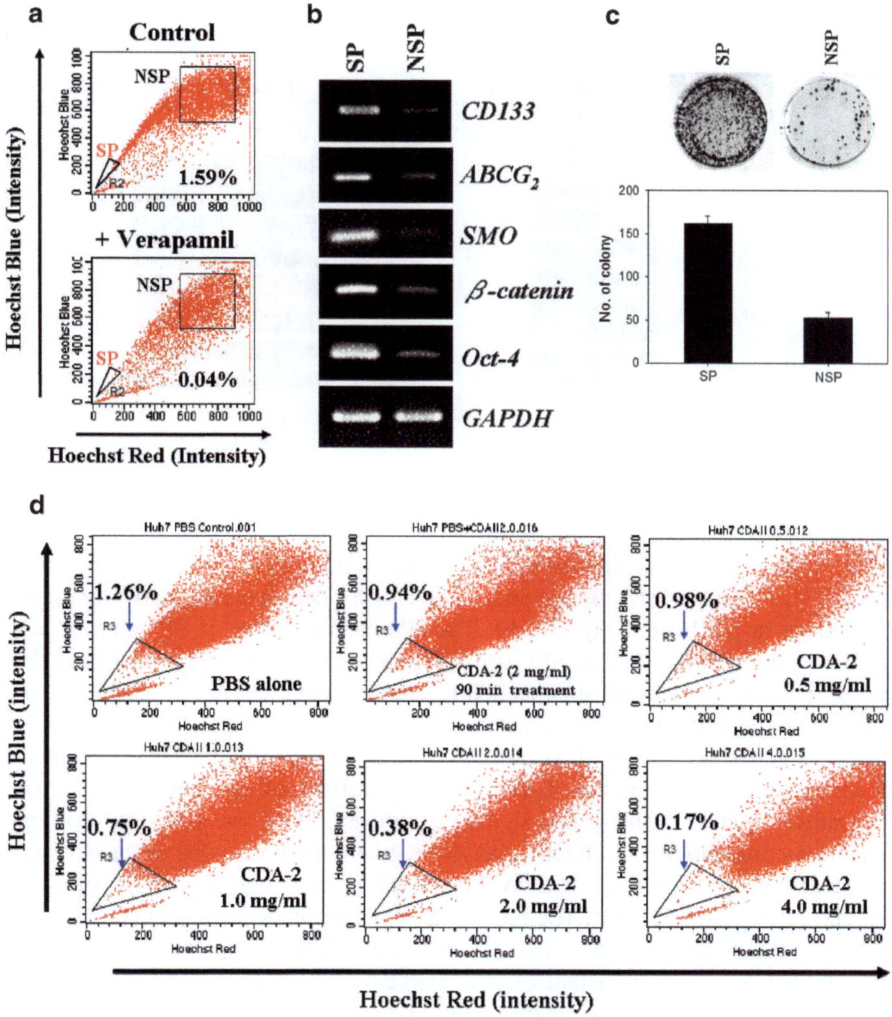

Fig. 31.1 Huh7 side population (SP) cells possess higher clonogenicity and stem cell markers than non-SP (NSP) cells. (**a**) Sorting of cancer cells using Hoechst 33342. The R1 gate represents the SP cells (1.59 % of total cells) and the R2 gate represents the NSP cells. Both of them were collected for the subsequent research. (**b**) RT-PCR assay showed that the SP cells possessed higher stem cell markers. (**c**) Clonogenic assay showing the clonogenicity of SP cells is markedly higher than that of NSP cells. (**d**) CDA-2 dose-dependently decreased the percentage of SP cells (the *triangle* gate area) in Huh7 hepatoma cells. Hepatocellular carcinoma cells were incubated with 5 μg/mL Hoechst 33342 in the presence of varying concentrations of CDA-2 for 2 days. Verapamil served as the positive control for ABCG2 inhibition. Representative data are shown. (**e**) Effects of CDA-2 on the mRNA expression of ABCG2, β-catenin and Gli in Huh7 SP cells. SP cells were treated with varying concentration of CDA-2 for 48 h. mRNA levels were analyzed by RT-PCR analysis. The reference GAPDH was used as the loading control. (**f**) RT-PCR analysis of DNMT1, DNMT3a, and DNMT3b mRNA expression in Huh7 SP and NSP cells. Significantly higher DNMT1, DNMT3a, and DNMT3b mRNA expression was observed in SP cells when compared with NSP cells. The reference GAPDH was used as the loading control. (**g**) Alteration of mRNA expression of DNMT1, DNMT3a, and DNMT3b in Huh7 SP cells treated with 0.05, 0.25, and 0.5 mg/mL of CDA2 for 48 h. These DNMTs mRNA levels were significantly suppressed by relatively lower concentration of CDA-2. The reference GAPDH was used as the loading control

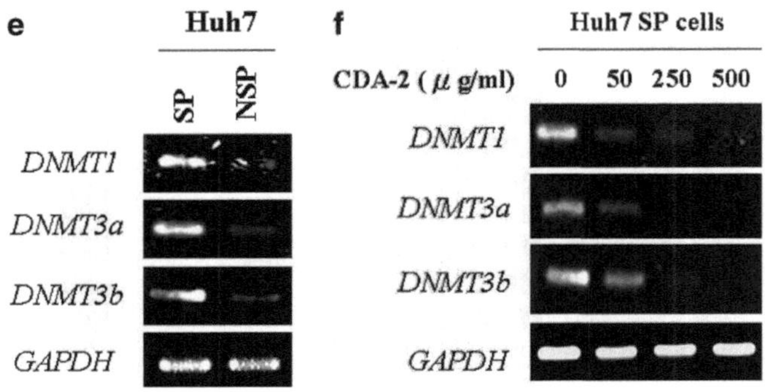

Fig. 31.1 (continued)

(NSP) cells. The results showed that the expression levels of stemness genes such as CD133, ABCG2, SMO, β-catenin, and Oct4 in SP cells were all markedly higher than those in NSP cells (Fig. 31.1b). The clonogenic ability of SP cell was also markedly higher than that of NSP cells (Fig. 31.1c). Interestingly, we found that these SP cells could be dose-dependently eliminated by CDA-2 (Fig. 31.1d). To investigate the underlying mechanisms, we examined the effects of CDA-2 on the stemness genes expression of SP cells. As shown in Fig. 31.1e, CDA-2 obviously downregulated the expression of ABCG2 (drug resistance), β-catenin (Wnt signaling pathway), and Gli (hedgehog signaling pathway). As it was proposed that epigenetic modification is an effective way to regulate the phenotype of cancer stem cell, we explored the effects of CDA-2 on the DNA methyltansferase 1 (DNMT1), 3a, and 3b in Huh7 SP cells. The results showed that the SP cells expressed much higher levels of DNMT1, 3a, and 3b as compared with NSP cells (Fig. 31.1f). All of these three DNA methyltransferases were markedly suppressed by CDA-2 treatment (Fig. 31.1g), suggesting the involvement of epigenetic modification in CDA-2-induced elimination of cancer stem-like SP cells. It is interesting to find any active component from CDA-2 that plays the key role. After solvent extraction and HPLC separation, a peak in the HPLC profile named as P23.2 was found to possess the activities of SP cell elimination and EMT inhibition. After 48 h incubation, P23.2 reduced the Huh7 SP cells from 1.69 to 0.18% at dose of 25 μg/mL and similar result was also observed in A549 SP cells (Fig. 31.2a). Moreover, the P23.2 also exerted inhibitory effects on the EGF-induced epithelial–mesenchymal transition (EMT) in A549 lung cancer cells as that did by CDA-2. The elevated mesenchymal marker *Snail* (Fig. 31.2b) and the decreased epithelial marker *E-cadherin* (Fig. 31.2c) by EGF treatment were almost completely reversed by 50 μg/mL of P23.2. (Fig. 31.2). Recently, pluripotent stem cells have been found in the Bonghan corpuscle [9]. In addition to the possible stem cell niches [10], the Bonghan corpuscle had also been proposed as pathways for cancer metastasis [11]. Regarding the profound effects of CDA-2 on SP cell elimination and EGF-induced EMT, it might be able to modulate the microenvironment existed in Bonghan corpuscle for antitumor therapy.

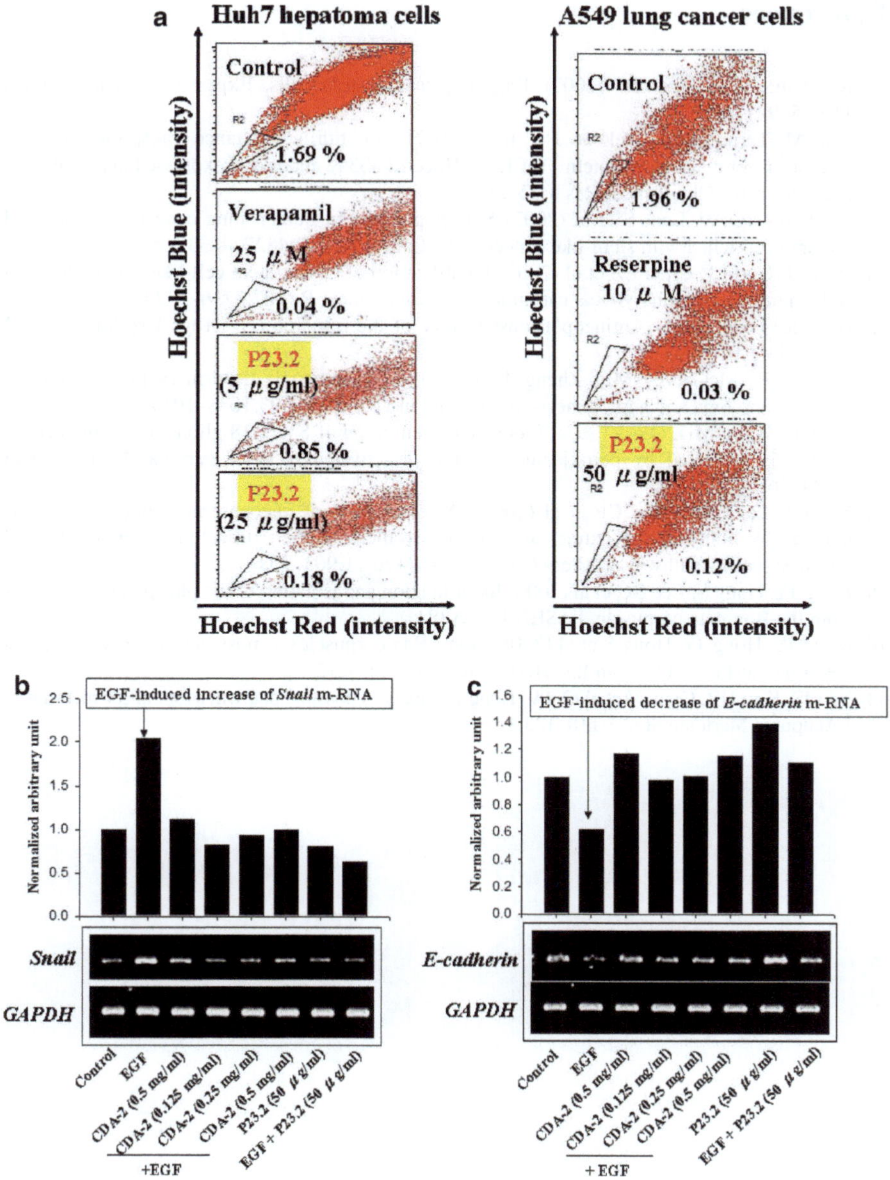

Fig. 31.2 (**a**) P23.2 decreased the SP percentage in Huh7 hepatoma and A549 lung cancer cells. Cells were incubated with P23.2 for 48 h and then the percentages of SP cells were analyzed by UV laser-equipped flow cytometer and DNA-binding dye Hoechst 33342. Verapamil served as the positive control for ABCG2 inhibition. Representative data are shown. CDA-2 and its active component P23.2 inhibited EGF (100 ng/mL)-induced increase of Snail (**b**) and decrease of E-cadherin m-RNAs (**c**) in A549 lung cancer cells. Cells were treated with the indicated agents for 72 h and then the expression of Snail and E-cadherin m-RNAs was analyzed by RT-PCR

References

1. Kvinlaug BT, Huntly BJ (2007) Targeting cancer stem cells. Expert Opin Ther Targets 11:915–927
2. Kim M, Turnquist H, Jackson J et al (2002) The multidrug resistance transporter ABCG2 (breast cancer resistance protein 1) effluxes Hoechst 33342 and is overexpressed in hematopoietic stem cells. Clin Cancer Res 8:22–28
3. Ho MM, Ng AV, Lam S et al (2007) Side population in human lung cancer cell lines and tumors is enriched with stem-like cancer cells. Cancer Res 67:4827–4833
4. Wang J, Guo LP, Chen LZ et al (2007) Identification of cancer stem cell-like side population cells in human nasopharyngeal carcinoma cell line. Cancer Res 67:3716–3724
5. Burzynski SR (1986) Antineoplastons: history of the research (I). Drugs Exp Clin Res 12 (Suppl 1):1–9
6. Masood R, McGarvey ME, Zheng T et al (1999) Antineoplastic urinary protein inhibits Kaposi's sarcoma and angiogenesis in vitro and in vivo. Blood 93:1038–1044
7. Lin CL, Wang MH, Qin YF et al (2009) Differentiation of SWO-38 glioma cells induced by CDA-2 is mediated by peroxisome proliferator-activated receptor gamma. J Neurooncol 95:29–36
8. Yao CJ, Lai GM, Chan CF et al (2005) Differentiation of pheochromocytoma PC12 cells induced by human urine extract and the involvement of the extracellular signal-regulated kinase signaling pathway. J Altern Complement Med 11:903–908
9. Hong JY, Hong S, Lee BC et al (2006) Identification and isolation of the pluripotent stem cells from the Bonghan corpuscle. FASEB J 20:A884
10. Kim MS, Hong JY, Hong S et al (2008) Bong-Han corpuscles as possible stem cell niches on the organ-surfaces. J Korean Inst Herbal Acupunct 11:5–12
11. Yoo JS, Kim HB, Ogay V et al (2009) Bonghan ducts as possible pathways for cancer metastasis. J Acupunct Meridian Stud 2:118–123

Part V
Imaging, Oxygen, Physiology and Others

Chapter 32
Mapping PVS by Molecular Imaging with Contrast Agents

Kyung A. Kang

Abstract The primo vascular system (PVS) is now verified for its existence in the animal body. Some of its anatomical information and its role for a few diseases were also revealed. One of the important future tasks for better understanding the PVS may be its distribution (map) inside the body. Most studies already performed on the PVS visualization are limited to the vessels on the surface of various organs, and inside the lymph/blood vessels. Thorough mapping of PVS will be valuable because it may reveal the mode of the communication among the organs connected via this system. In addition, the changes in map of the PVS system in the course of disease progression may provide us with important information that can be utilized for better health management.

Because the diameters of small PVS vessels are only in the range of tens of micrometers, for existing biomedical imaging modalities to be effective for imaging the system, external agents generating very high contrast combined with highly PVS-specific targeting agent will be required. In this chapter, a futuristic design of a single contrast agent guided by highly PVS-specific targeting molecule for MRI, X-ray/CT, optical, and TEM imaging is discussed.

1 Introduction

For the past 8 years, the study progress on the primo vascular system (PVS) study by Kwang-Sup Soh has been extraordinary. His accomplishments have been especially meaningful because he went through enormous difficulties for the PVS research status established up to the current level. Because Bong-Han Kim [1], the pioneer on the PVS research never revealed the dye used for observing PVS, Soh and his colleagues had to restart the study from the beginning. Even after Soh group

K.A. Kang (✉)
Department of Chemical Engineering, University of Louisville, Louisville, KY, USA
kyung.kang@louisville.edu

K.-S. Soh et al. (eds.), *The Primo Vascular System: Its Role in Cancer and Regeneration*, DOI 10.1007/978-1-4614-0601-3_32, © Springer Science+Business Media, LLC 2012

was successful in rediscovering the system, it has been difficult to convince other scientists on the existence of the system, because accepting that there is another new vascular system not uncovered until now was too unbelievable to them. Research results by Soh group are now fully recognized as the evidence of the existence of the system that Kim had previously observed, e.g., the vessels and corpuscles in and on various organs [2–4], its morphological characterization of the vessels [5], and the new type of cells transported inside the vessels [6–8]. In addition, Soh group also revealed densely formed PVS vessels on adipose tissues [7] and the formation of the vessels around the tumors and cancer cell transportation via the vessels [9–11].

One of the many remaining important studies to be performed for better understanding this still new system may be fully mapping it in the animal body. This task is important because it may be able to tell what and how organs communicate via the system, how the communication system changes during the pathological condition and most importantly how we utilize the system to manage human health. Until now, most of the studies on visualizing PVS system are performed by optical microscopy, for the vessels exposed on the surface of various organs [3] or inside blood or lymph vessels [12, 13]. Since the diameter of the primo vessels is in the range of tens of micrometer in the animals tested, it was possible for Soh group to observe them by optical microscope with [12] and without applying dyes [13], or by electron microscopy [6]. Observing surface-bound PVS by optical microscopy alone, however, cannot show its distribution under/in the tissue or organs. To efficiently map the system, imaging modalities capable of penetrating tissue/organs may be required.

Currently available biomedical imaging modalities, with deep penetration capability, are optical, MRI, X-ray/CT, single photon emission computed tomography (SPECT), positron emission tomography (PET), and ultrasound. The small diameter of the PVS (as small as ~10 μm) [12, 13] may be at or less than the lower limit of these imaging techniques and therefore, appropriate contrast agent will definitely be required for adequate PVS mapping. For the contrast agent to be efficaciously functional, it needs to be delivered specifically to the PVS, requiring targeting molecules only for PVS.

2 Discussions on Methods and Materials

2.1 Mapping of PVS by Various Imaging Modalities

The two important aspects in biomedical imaging are the sensitivity and the resolution. SPECT and PET may have the highest sensitivity but they require administering radioactive materials and it may be difficult to use them for an animal very frequently. Ultrasound may not be the best choice due to the poor resolution. Optical imaging has a higher sensitivity than MRI (10^{-9} vs. 10^{-5} M) [14, 15] although its resolution is not the best. MRI has an advantage of providing a high spatial resolution and the structural and functional information in three-dimensions with excellent contrast between the different soft tissues of the body. X-ray is useful in the detection of pathology of the skeletal system, and imaging of soft tissues is better with computed axial tomography (CT scanning) [14]. Although not often for biomedical use, electron microscopy, such as transmission or scanning electron microscopy (TEM or SEM) can also be used for bio-sample analysis and it can provide a resolution level of subnanometer.

For mapping the PVS, MRI and X-ray/CT fit better for the imaging of the entire body because of their good penetration depth and resolution. Fluorophore-mediated optical imaging and TEM imaging may be good for smaller samples, such as tissues or organs, through the three-dimensional image reconstruction technique of their two dimensional images.

2.2 *Targeting Molecules for PVS*

The imaging contrast agents can be effective when they are applied only to the entity of the interest (in our case, PVS). The most frequently used targeting molecules for this type of study are either monoclonal antibodies or aptamers specifically against the molecule of interest. For known biomarkers, it is not very difficult to generate these targeting molecules. The biomarkers specific only for PVS system, however, are not currently known. Therefore, at this time, the endothelial cells of PVS vessels may have to be used for these molecule generations, with the endothelial cells from lymph and blood vessels for cross-reactivity tests, assuming that these two have the most similar molecular constituents and that there are few other tissues or organs with similar molecular properties to PVS. Until more thorough molecular analysis of the PVS endothelial cells are completed, this may be the best approach.

2.3 *Nanoparticle Contrast Agent for Multi-Imaging Modalities*

Studies on the diameter and length of the PVS vessels of the animals (mice, rats, and rabbits) were already performed by Soh group. The vessel diameter inside the rabbit lymph vessel are in the range of 18–44 μm [12, 13]. The resolution of MRI, which is the best among the three suggested imaging modalities can go down to 50 μm using hardware for the small animal MRI [16]. Therefore, in order for these modalities to be used in PVS mapping in the whole body of an animal, highly effective external contrast agents are certainly required.

Metal nanoparticles have unique properties that are not in their bulk form and when these properties are smartly used they can be converted to image contrast agents for various imaging modalities. The size "nano" is small enough to incorporate to bio- or physiological systems in the cellular or molecular level. It is large enough to increase the circulation time in blood and the uptake rate by cells, increasing efficiency for the delivery [17, 18]. The most frequently used metal nanoparticles for

a

Spacers	Estimated Length (nm)
HS-$(CH_2)_{12}$-Poly$(EG)_4$-ONH_2 [C12PEG4]	3.2
HS-$(CH_2)_{12}$-Poly$(EG)_6$-ONH_2 [C12PEG6]	3.9
HS-$(CH_2)_{12}$-Poly$(EG)_8$-ONH_2 [C12PEG8]	4.6

b

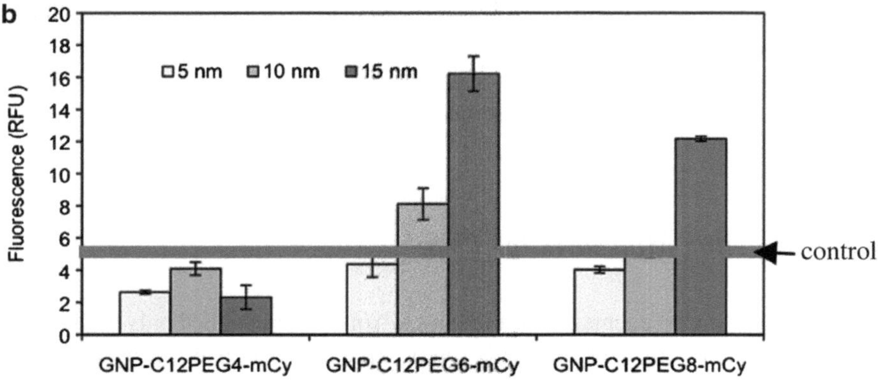

Fig. 32.1 (**a**) Spacers at various lengths, used to conjugate Cypate to GNPs and (**b**) fluorescence level when Cypate is conjugated to a GNP via the spacers

biomedical purpose are gold nanoparticles (GNPs) and iron oxide nanoparticles (IONP) because of their non-toxic nature. Their surface properties are good for conjugating targeting molecules, other biologicals, or flurophores.

Optical contrast agent. Most optical imaging is done with fluorescent contrast agents and the two mostly used for this purpose are organic fluorophores and quantum dots. If the imaging is to be done for human, FDA approved organic fluorophores in the near infrared (NIR) range are appropriate because NIR is a range outside the wavelengths of most intrinsic chromophores or fluorophores in the tissue and also because it penetrates deeper into the tissue. NIR contrast agents that are currently used for human is indocyanine green (ICG). The quantum yield of ICG is only 0.0028 and 0.012, in water and plasma [19], respectively, and therefore, artificially enhancing its quantum yield is highly desired. Since fluorescence of a fluorophore is initiated by optically exciting its electrons, one way of changing fluorescence level is by placing it near an electric field.

GNPs generate strong electric (plasmon) field on and around them upon receiving light [20]. Therefore, they can be good candidates for influencing electron status of a fluorophore [20, 21]. The strength of the plasmon field generated by a GNP is the highest on its surface and decays rapidly with the distance from the GNP surface [22, 23]. If the field strength that increases the fluorescence of a fluorophore is known, then one can achieve a higher fluorescence by placing the fluorophore at the distance from a GNP with this particular field strength [20, 21]. Figure 32.1 shows the fluorescence of Cypate [24], an ICG-based NIR fluorophore, when it was

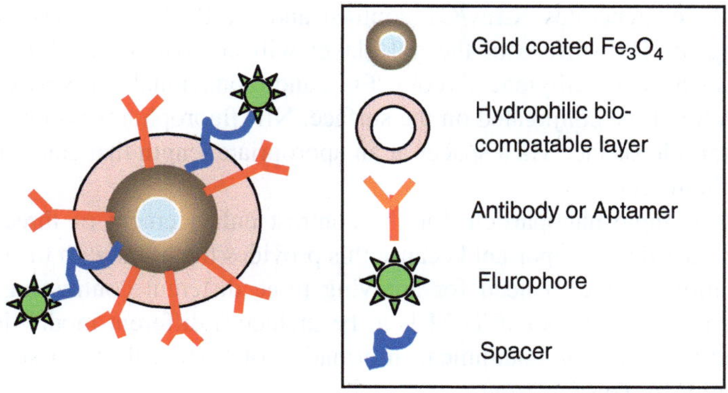

Fig. 32.2 A single nano-contrast agent for multi-image modalities. Gold layer and IONP core enhance X-ray/CT and optical (fluorescence) contrasts, and MRI contrasts, respectively. Antibody or aptamer against PVS will guide the agent specifically to PVS. IONP core may also be used for hyperthermic treatment

conjugated on the surface of GNPs at various sizes. As can be seen in the figure, Cypate fluorescence can be highly enhanced by GNP extensively by selecting the GNPs size and spacers between the GNP and Cypate.

Nano-contrast agent for X-ray/CT. GNPs have shown to enhance the contrast of X-ray/CT [25, 26]. Recently, there has been numerous attempts of using GNPs to enhance contrast of X-ray/CT and the results are highly promising [27, 28].

Nano-contrast agent for MRI. IONPs are effective MRI contrast agent [29, 30]. There are already clinically available IONPs as MRI contrast agent (e.g., Fridex I.V. by AMAG Pharmaceutials). Using IONPs, Flint et al. [31] was able to successfully obtain MRI neuroimages at a resolution up to 4–8 μm.

TEM contrast agent. Electron microscopy is used for observing the response of a system of interest to the electrons as the disturbance. Metals respond to the electrons well and therefore, gold has been used as electron microscopy contrast agent for a long time. Both NGP and IONPs will be appropriate for TEM contrast enhancement for PVS with the targeting agents.

The contrast agents will then be conjugated to the PVS targeting molecules and the resulting PVS targeting agents will be applied to the body before imaging.

2.4 *Nanoparticle Contrast Agent for Multi-Imaging Modalities*

The properties of the contrast agents for the imaging modalities discussed above can be placed in one, single nano-entity. It is an NIR fluorophore linked, PVS targeting molecule conjugated, gold coated IONP (Fig. 32.2): The thin gold layer can generate high plasmon field and, therefore, it can be used for NIR fluorescence enhancement.

The gold layer generates X-ray/CT contrast and the IONP core acts as an MRI contrast agent. The surface of the gold layer will be coated with biocompatible polymer, such as polyethylene glycol (PEG), and monoclonal antibody or aptamer against PVS will be conjugated on the surface. NIR fluorophores will be linked on the gold particle surface via a spacer at an appropriate length that can generate the maximal fluorescence.

Having a single nanoparticle for the contrast enhancement of three different imaging modalities is important because this provides the users with the sensitivity and resolution without a need for applying three different contrast agents. The images from three (or four if TEM is to be included) different modalities can be overlapped to obtain the anatomical information of PVS with the best sensitivity and resolution at the same time.

Also, in the future, if there is a need for destroying PVS on certain areas, as a treatment, the IONP core can be used to guide to heat the system utilizing an alternating electromagnetic (AEM) field, noninvasively, since IONPs have been used for cancer hyperthermia for a long time [32].

2.5 Strategy of Mapping PVS with Targeting Contrast Agent

Since we have not yet started the imaging with the proposed contrast agent, we do not know exactly which imaging modalities will provided the PVS mapping capabilities yet. It is desired to use MRI and CT for the entire body PVS mapping. The small animal imaging studies will be first performed. The contrast agent will be applied to the animal, the targeting molecules will react with the PVS and the nanoparticles will be placed in the PVS. Then the sensitivity and the resolution of these two modalities will be tested. If these two do not provide enough sensitivity, then we may have to use optical and TEM imaging techniques. In this case, body parts or organs will be frozen first. Then the imaging will be taken in two dimensions, slice-by-slice. The obtained images will be used for the three dimensional image reconstruction to obtain the PVS distribution in the animal body.

3 Conclusions

The mapping for PVS in an animal body is highly important to understand the role of the system in animal body and also designing strategy of disease management in the body using this novel system. This information can also verify whether PVS is related to the meridian path, which has been the theory that Bong-Han Kim insisted 50 years ago.

References

1. Kim BH (1963) On the Kyungrak system. J Acad Med Sci 10:1–41
2. Shin H, Johng H-M, Lee B-C, Cho S-I, Soh K-S, Baik K-Y, J-S YOO, Soh K-S (2005) Feulgen reaction study of novel threadlike structures (Bonghan ducts) on the surfaces of mammalian organs. Anat Rec B New Anat 284:35–40
3. Sung B, Kim MS, Lee B-C, Yoo JS, Lee S-H, Kim Y-J, Kim K-W, Soh K-S (2007) Measurement of flow speed in the channels of novel threadlike structures on the surfaces of mammalian organs. Naturwissenschaften 95(2):117–124
4. Lee B-C, Kim S, Soh K-S (2008) Novel anatomic structures in the brain and spinal cord of rabbit that may belong to the Bonghan system of potential acupuncture meridians. J Acupunct Meridian Stud 1(1):29–35
5. Ogay V, Bae KH, Kim KW, Soh K-S (2009) Comparison of the characteristic features of Bonghan ducts, blood and lymphatic capillaries. J Acupunct Meridian Stud 2(2):107–117
6. Kwon J, Baik KY, Lee B-C, Soh K-S, Lee NJ, Kang CJ (2007) Scanning probe microscopy study of microcells from the organ surface Bonghan corpuscle. Appl Phys Lett 90:173903. doi:10.1063/1.2732183
7. Lee B-C, Bae K-H, Gil-Ja Jhon G-J, Soh K-S (2009) Bonghan system as mesenchymal stem cell niches and pathways of macrophages in adipose tissues. J Acupunct Meridian Stud 2(1):79–82
8. Baik KY, Ogay V, Jeoung SC, Soh K-S (2009) Visualization of Bonghan microcells by electron and atomic force microscopy. J Acupunct Meridian Stud 2(2):124–129
9. Yoo JS, Kim HB, Ogay V, Lee B-C, Ahn S, Soh K-S (2009) Bonghan ducts as possible pathways for cancer metastasis. J Acupunct Meridian Stud 2(2):118–123
10. Yoo JS, Kim HB, Won N, Bang J, Kim S, Ahn S, Lee B-C, Soh K-S (2010) Evidence for an additional metastatic route: in vivo imaging of cancer cells in the primo-vascular system around tumors and organs. Mol Imaging Biol 13(3):471–480. doi:10.1007/s11307-010-0366-1
11. Yoo JS, Ayati MH, Kim HB, Zhang W-B, Soh K-S (2010) Characterization of the primo-vascular system in the abdominal cavity of lung cancer mouse model and its differences from the lymphatic system. PLoS One 5(4):e9940
12. Lee C, Seol S-K, Lee B-C, Hing Y-K, Je J-H, Soh K-S (2006) Alcian blue staining method to visualize Bonghan threads inside large caliber lymphatic vessels and X-ray microtomography to reveal their microchannels. Lymphat Res Biol 4:181–189
13. Lee B-C, Soh K-S (2008) Contrast-enhancing optical method to observe a Bonghan duct floating inside a lymph vessel of a rabbit. Lymphology 41:178–185
14. Cassidy PJ, Radda GK (2005) Molecular imaging perspectives. J R Soc Interface 2:133–144
15. Culver J, Akers W, Achilefu S (2008) Multimodality molecular imaging with combined optical and SPECT/PET modalities. J Nucl Med 49:169–172
16. Vande Velde G, Baekelandt V, Dresselaers T, Himmelreich UQ (2009) Magnetic resonance imaging and spectroscopy methods for molecular imaging. J Nucl Med Mol Imaging 53(6):565–585
17. Davis ME, Chen Z, Shin DM (2008) Nanoparticle therapeutics: an emerging treatment modality for cancer. Nat Rev Drug Discov 7:771–782
18. Jiang W, Kim BYS, Rutka JT, Chan WCW (2008) Nanoparticle-mediated cellular response is size-dependent. Nat Nanotechnol 3:145–150
19. Benson RC, Kues HA (1978) Fluorescence properties of indocyanine green as related to angiography. Phys Med Biol 23:159–163
20. Kang KA, Hong B (2006) Biocompatible nano-metal particle fluorescence enhancers. Crit Rev Eukaryot Gene Expr 16:45–60
21. Wang J, Nantz MH, Achilefu S, Kang KA (2010) FRET-like fluorophore-nanoparticle complex for highly specific cancer localization. Adv Exp Med Biol 662:407–414
22. Neeves AE, Birnboim MH (1989) Composite structures for the enhancement of nonlinear-optical susceptibility. J Opt Soc Am B 6:787–796

23. Bharadwaj P, Anger P, Novotny L (2007) Nanoplasmonic enhancement of single-molecule fluorescence. Nanotechnology 18:044017
24. Achilefu S, Dorshow R, Bugaj J, Rajagopalan R (2000) Novel receptor-targeted fluorescent contrast agents for in vivo tumor imaging. Invest Radiol 35:479–485
25. Hainfeld JF, Slatkin DN, Focella TM, Smilowitz HM (2006) Gold nanoparticles: a new X-ray contrast agent. Br J Radiol 79:248–253
26. Xu C, Tung GA, Sun S (2008) Size and concentration effect of gold nanoparticles on X-ray attenuation as measured on computed tomography. Chem Mater 20(13):4167–4169
27. Popovtzer R, Agrawal A, Kotov NA, Popovtzer A, Balter J, Carey TE, Kopelman R (2008) Targeted gold nanoparticles enable molecular CT imaging of cancer. Nano Lett 8(12):4593–4596
28. Li J, Chaudhary A, Chmura SJ, Pelizzari C, Rajh T, Wietholt C, Kurtoglu M, Aydogan B (2010) A novel functional CT contrast agent for molecular imaging of cancer. Phys Med Biol 55(15):4389–4397
29. Weissleder R, Elizondo G, Wittenberg J, Babito CA, Bengele HH, Josephson L (1990) Ultrasmall superparamagnetic iron oxide: characterization of a new class of contrast agents for MR imaging. Radiology 175:489
30. Hoehn M, Wiedermann D, Justicia C, Ramos-Cabrer P, Kruttwig K, Farr T, Himmelreich U (2007) Cell tracking using magnetic resonance imaging. J Physiol 584(pt 1):25–30
31. Flint JJ, Lee CH, Hansen B, Fey M, Schmidig D, Bui JD, King MA, Vestergaard-Poulsen P, Blackband SJ (2009) Magnetic resonance microscopy of mammalian neurons. Neuroimage 46(4):1037–1040
32. Hilger I, Hergt R, Kaiser WA (2005) Use of magnetic nanoparticle heating in the treatment of breast cancer. IEE Proc Nanobiotechnol 152(1):33–39

Chapter 33
Unusual Optical Properties of Collagen and Implications for the Primo Vascular System

Eduard van Wijk, Margo Groeneveld, Jan van der Greef, and Roeland van Wijk

Abstract The aim of this study was to extend previous investigations on the property of artificially prepared collagen gels to modify ultraweak photon emission emanating from a biological source. The interaction of collagen gels with enzyme-dependent ultra weak photon emission facilitated by the Xanthine oxidase–Xanthine–MCLA (XO–X–MCLA) enzyme system was studied. This enzyme system is frequently associated with reactive oxygen species (ROS) production. Collagen was also tested in combination with a light-emitting diode (LED) with similar spectral properties as the XO–X–MCLA system. The data demonstrate that a collagen gel is capable of increasing photon emission of the enzyme system. In contrast, weak photon emission stimulated by a LED was not increased, but rather decreased. It was concluded from these data that collagen may exert a remarkable influence on distant molecular activation wherein chemical and electrical signaling is impossible. This capacity may also have significant consequences for understanding the proposed properties of primo-vessels in their particular role as optical channels throughout a living being.

1 Introduction

Primo-vessels are a newly discovered type of microconduit (channel) forming a novel circulatory system throughout a living being [1]. These vessels are characterized by their multilumen collagen structure with openings and pores at the outer boundary. Primo-vessels, an ordered network of water molecules interspersed within the protein fibrillar matrix of collagen, possesses special optical properties [2].

E. van Wijk (✉)
Sino Dutch Center for Preventive and Personalized Medicine,
Leiden University, Leiden, The Netherlands

Meluna Research, Amersfoort, The Netherlands
e-mail: eduard.vanwijk@sinodutchcentre.nl

K.-S. Soh et al. (eds.), *The Primo Vascular System: Its Role in Cancer and Regeneration*,
DOI 10.1007/978-1-4614-0601-3_33, © Springer Science+Business Media, LLC 2012

This network is hypothesized to support rapid jump conduction of protons and is possibly responsible for its semiconduction as demonstrated by dielectric measurements.

The conductivity of collagen is a function of the collagen fibrillar structure [3]. Conductivity increases with greater volume of absorbed water [4]. Conduction along a fiber is at least 100 times more than that across a fiber [5]. This orientation of a collagen liquid crystalline disposition helps to explain its electrical intercommunication among its primo-channels. Moreover, its crystalline structure facilitates its sensitivity to weak input of mechanical pressure [6].

The electromagnetic properties of collagen have led to the speculation that collagen has similar collective properties of photon emission dynamics. Evidence came from photo-induced delayed luminescence characteristics estimated from bovine Achilles' tendon of rats [7]. This delayed luminescence cannot be described simply in terms of excitation/deexcitation of electronic levels of single molecules. Instead, it is necessary to consider the existence of collective electronic states [8]. The special photon emission properties are also confirmed by using collagen gels wherein the collagen structure both conduct and modify the photon pulses coming from a biological source [9].

The aim of this preliminary study is to extend previous investigations of the property of artificially prepared collagen gels in order to modify ultraweak photon emission emanating from a biological source. The authors studied the interaction of collagen gels with enzyme-dependent ultraweak photon emission facilitated by the Xanthine–Xanthine oxidase enzyme system. This enzyme system is frequently associated with ROS (O_2^- and H_2O_2) production. These particular radicals do not independently produce measurable photon emission. However, when Methylated Cypridina Luciferina Analog (MCLA) is added, as a catalyst, substrate is oxidized. MCLA is a specific detector of both singlet oxygen ($^1O_2$) and O_2^- [10] and is in itself in a highly excited state. MCLA in combination with the Xanthine oxidase–Xanthine system can be utilized as a biological photon source to study the interaction with collagen gel.

2 Material and Methods

2.1 *Xanthine Oxidase, Xanthine, and MCLA Mixture*

The autooxidative chemiluminescence of 0.5/1 μM MCLA in 2.5 mL phosphate buffer (0.05M; pH 7.4) is dependent on the concentration of MCLA. This emission is not significantly different after adding only Xanthine. However, photon emission is higher when also adding the enzyme Xanthine oxidase. When Xanthine and Xanthine oxidase are both present, superoxide is produced activating MCLA. A 2.5 mL reaction mixture containing 100 μM Xanthine, 10^{-2} units Xanthine oxidase, and 0.5/1 μM MCLA in buffer in plastic petri dishes of 35 mm diameter results in an initial emission of about 20,000 counts per second and subsequently decreases.

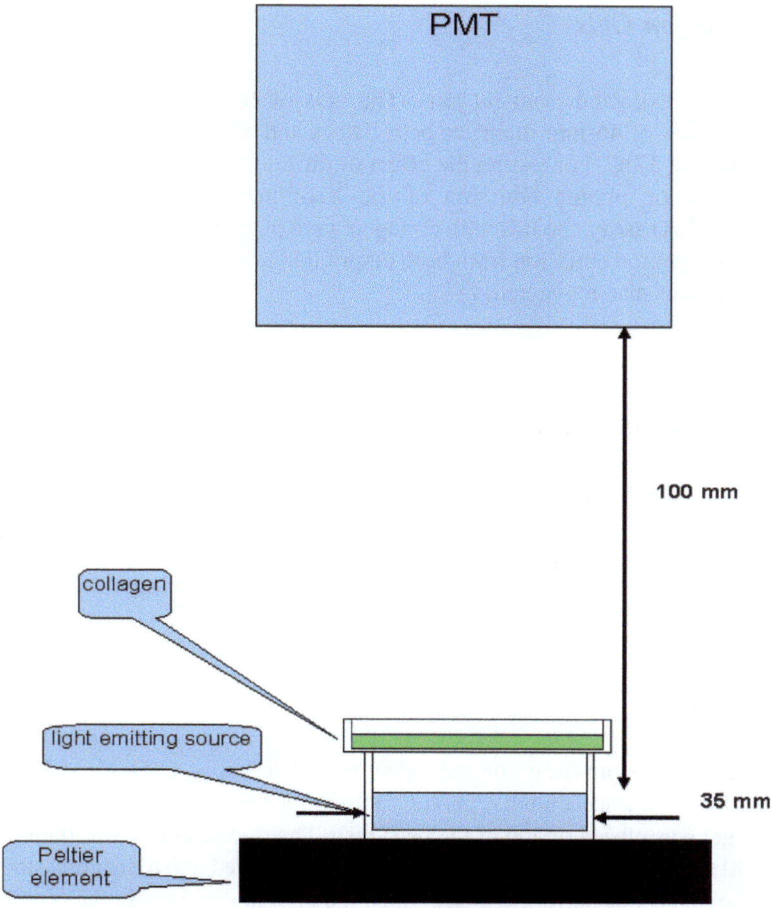

Fig. 33.1 Schematic representation of experimental setup equipment. The light-emitting biochemical source (Xanthine–Xanthine oxidase–MCLA) was placed under the photomultiplier (PMT); the distance between the sample and PMT was 10 cm. The temperature of the dish was controlled by a Peltier-element. A plastic lid with collagen is placed on top of the light source

2.2 *Photomultiplier*

Photon emission of the mixtures was measured below a vertically placed, cooled (–25°C) photomultiplier (H-R550, Hamamatsu Photonics Co. Ltd, Hamamatsu, Japan). The opening window of the photomultiplier was 46 mm. Figure 33.1 includes a schematic representation of the experimental equipment: the sample was placed under the PMT; the distance between the sample and photomultiplier tube was 10 cm. The temperature of the dish was controlled by a Peltier-element.

2.3 Collagen Gels

Collagen was extracted from rat tails. The gels of collagen were poured inside polystyrene lids of 46-mm diameter petri dishes and allowed to polymerize for at least an hour at 37°C. For testing the effect of different, in concentrations varying, collagen gels on photon emission of the Xanthine oxidase–Xanthine–MCLA (XO–X–MCLA) assay, the lids with collagen were placed above the assay mixture. As a control, photon emission from both empty lids as well as lids filled with phosphate buffer was also registered.

2.4 Light-Emitting Diode

To compare the MCLA-dependent photon emission with the photon emission from an artificial light source, a light-emitting diode (LED) (455–495 nm; Roithner Laser Technik, GmbH; Type EJL-465-211) was used.

3 Results

The photon emission from the mix that included buffer, XO–X–MCLA, was recorded for 90 s. Subsequently, photon emission measurement was stopped and a collagen gel was placed on top of the petri dish. The measurement was then continued for 30 s, after which the collagen sample was removed and a final measurement without collagen was performed. Generally, the photon emission in the presence of collagen was higher than expected from the curve without collagen (Fig. 33.2).

The effect of collagen on photon emission was calculated by a two-step procedure. Initially, a theoretical curve for the time between the first and second measurement of the reaction without collagen was estimated utilizing a second-order polynomial. Subsequently, this theoretical curve was compared with the actual values measured. The difference was calculated by subtracting the photon emission calculated by the above best fit from the photon emission measured with collagen. The outcome (Δ) was expressed as a percentage [(measured – fit)/measured] × 100% (Table 33.1). The table demonstrates that photon emission was increased in the presence of collagen gel.

To make a comparison between a biologic and artificial light-emitting system, a LED with emission properties similar to MCLA (peak emission approximately 455–495 nm) was utilized as a photon source. When a collagen gel was placed on top of the LED in a manner similar to the XO–X–MCLA assay, the photon emission from the LED was decreased.

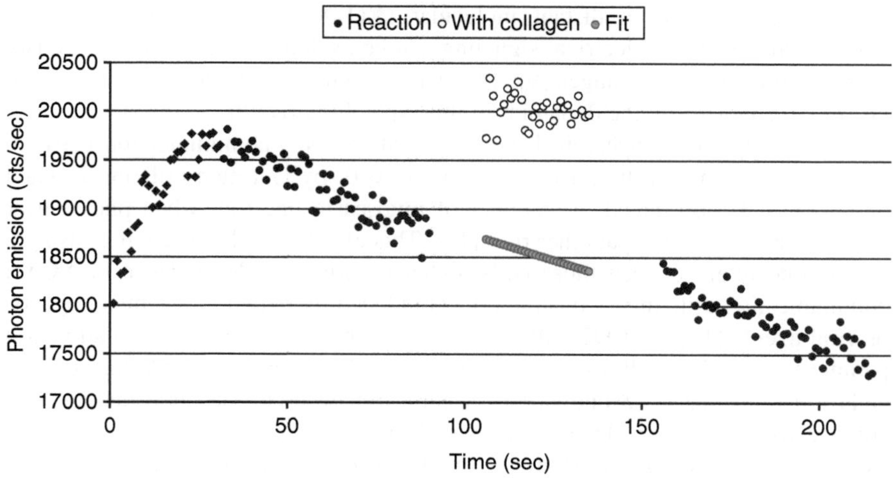

Fig. 33.2 The effect of collagen on photon emission. Photon emission kinetics of a XO–X–MCLA reaction mixture (*filled circle*). At zero time, XO was added to initiate the reaction resulting in a short increase in photon emission. During the subsequent, steady, and slow decrease, the effect of collagen was tested. The *middle part* (90–150 s) demonstrates the photon emission during collagen (*circle*). The theoretical best fit during this period is indicated by the (*filled square*) symbol

Table 33.1 Effect of different collagen concentrations on biochemical (XO–X–MCLA) and artificial (LED) source of photon emission. The outcome (Δ) was expressed as a percentage: [(measured − fit)/measured] × 100%

	Effect on photon emission (Δ%)	
Collagen (mg/mL)	MCLA	LED
1	6.6 ± 1.9 (n = 6)	−18.6 ± 1.7 (n = 3)
0.5	14.7 ± 1.8 (n = 6)	−19.1 ± 1.7 (n = 3)
0.1	6.9 ± 4.1 (n = 6)	−21.9 ± 1.7 (n = 3)
Controls		
Buffer	−1.8 ± 3.2 (n = 6)	−7.5 ± 0.6 (n = 2)
Empty lid	−3.2 ± 3.8 (n = 6)	−11.4 ± 1.4 (n = 3)

The outcome (Δ) was expressed as a percentage: [(measured − fit)/measured] × 100%

4 Discussion and Conclusions

The data demonstrate that a collagen gel placed between the photomultiplier and the XO–X–MCLA system is capable of increasing the photon emission. In contrast, when a collagen gel was placed on top of a LED, the photon emission recorded from the LED has been decreased. In the presence of the plastic layer or with buffer as a control, neither an increase nor a decrease was registered. In contrast, weak photon emission from a LED with similar wavelength as MCLA was not increased, but rather decreased. This decrease is expected from the absorbance when no interaction takes place. It was concluded from these data that collagen may exert a remarkable influence in distant molecular (XO–X–MCLA system) activation wherein chemical and electrical signaling can be excluded. It confirms previously observed properties of collagen [9]. This experimental model may, in the future, be utilized to characterize the dynamics of this type of interaction.

The principle of nonchemical, nonelectrical (NCNE) communication between biological systems can be traced back to the work by Gurvitch utilizing isolated onion roots. He estimated the number of mitoses in one root when placed near actively dividing cells in another root [11]. This effect was abolished after using a UV-opaque filter between those cells. Other reports provide evidence of NCNE communication utilizing cytopathic effects induced by radiation, chemical agents, or viral agents [12]. In 1992, Albrecht-Buehler observed that cell orientation is a parameter for NCNE cell-to-cell communication [13]. Baby hamster kidney (BHK) cells were inoculated on one side of a glass film wherein the opposite side was covered with a confluent layer of BHK cells. Cells on one side aligned their long axes in the direction of the whorls of the opposite confluent cells. The effect was inhibited by a thin metal coating on the glass films

Popp et al. utilized spontaneous emission as a parameter to detect NCNE signals [14]. They placed two cuvettes containing the luminescent maritime bacteria in front of two photomultipliers and recorded the increase in synchronized photon emission between the two vessels. In 1994, a similar experimental design was performed utilizing pig neutrophil granulocytes [15]. Farhadi et al. [16] investigated NCNE signaling between GI inducer cells exposed to oxidants and distant GI cells (separated by air).

Collagen is an abundant molecule with special properties and data indicate that collagen has properties of interference with photon emission emanating from biomolecular sources. In addition, collagen may also play a role in the tuning between different photon-emitting events. Collagen in connective and fascia tissues is present throughout the organism and seems to form one single immense structure. It can, therefore, be speculated that this property of collagen to interfere with photon-emitting processes facilitates the possibility of tuning photon emissions throughout the organism. This is one step closer to the hypothesis that metabolism is regulated by a (bio-) photon field. The revolutionary view that collagen plays a special role in the biophoton field vis-à-vis its NCNE communication capacity may also have significant consequences for understanding the proposed properties of primo-vessels in their particular role as optical channels throughout a living being.

Acknowledgments This work was supported by an independent research grant from the Samueli Institute of Information Biology and the Rockefeller-Samueli Center for Research in Mind-Body Energy. The collagen was kindly provided by Varsha Thakoersing, LACDR, Leiden University. The authors also thank Dr. John Ackerman for his assistance in editing the text.

References

1. Kim BH (1963) On the kyungrak system. J Acad Med Sci 10:1–41
2. Han YH, Yang JM, Yoo JS et al (2007) Measurement of the optical properties of in-vitro organ-surface Bonghan corpuscles of rats. J Kor Phys Soc 49:2239–2246

3. Sasaki N (1984) Dielectric properties of slightly hydrated collagen: time-water content superposition analysis. Biopolymers 23:1724–1734
4. Ho MW, Knight DP (1998) The acupuncture system and the liquid crystalline collagen fibers of the connective tissues. Am J Chin Med 26:251–263
5. Pethig R (1996) Dielectrophoresis. Crit Rev Biotechnol 16:331–348
6. Mikhailov AS, Ertl G (1996) Nonequilibrium structures in condensed systems. Science 272:1596–1597
7. Ho MW, Musumeci F, Scordino A et al (2002) Delayed luminescence from bovine Achilles' tendon and its dependence on collagen structure. J Photochem Photobiol B 66:165–170
8. Brizhik L, Scordino A, Triglia A et al (2001) Delayed luminescence of biological systems arising from correlated many-soliton states. Phys Rev E Stat Nonlin Soft Matter Phys 64:031902
9. Troshina TG, Loochinskaia NN, Van Wijk E et al (2006) Absorption and emission of photons by collagen samples. In: Beloussov LV et al (eds) Biophotonics and coherent systems in biology. Springer, New York
10. Kambayashi Y, Ogino K (2003) Reestimation of the cypridina analogs (MCLA) as a chemiluminescence probe to detect active oxygen species. J Toxicol Sci 28:139–148
11. Gurvitsch AG (1926) Das Problem der Zellteillung Physiologish Betrachtet. Springer, Berlin
12. Trushin MV (2004) Distant non-chemical communication in various biological systems. Riv Biol 97:409–442
13. Albrecht-Buehler G (1992) Rudimentary form of cellular vision. Proc Natl Acad Sci U S A 89:8288–8292
14. Popp FA, Chang JJ, Gu Q et al (1994) Nonsubstantial biocommunication in terms of Dicke's theory. In: Ho MW et al (eds) Bioelectrodynamics and Biocommunication. World Scientific Publishing, Singapore
15. Shen X, Mei W, Xu X (1994) Activation of neutrophils by a chemically separated but optically coupled neutrophil population undergoing respiratory burst. Experientia 50:963–968
16. Farhadi A, Forsyth C, Banan A et al (2007) Evidence for non-chemical, non-electrical intercellular signalling in intestinal epithelial cells. Bioelectrochemistry 71:142–148

Chapter 34
Basic Electrophysiological Properties of Cells in the Organ Surface Primo Vascular Tissues of Rats

Jae-Hong Choi, Tae Hee Han, Chae Jeong Lim, So Yeong Lee, and Pan Dong Ryu

Abstract The primo vascular system (PVS) is a novel circulatory system composed of primo vessels and primo nodes, and its morphological and functional properties are largely unknown. In this study, we characterized basic electrophysiological properties of the cells in the primo vessels and primo nodes on the surface of abdominal organs. The electrophysiological activities of the cells in the primo vessels and nodes were studied by intracellular recording technique with sharp electrodes and by whole-cell slice patch recording techniques, respectively. The cells of the primo vessels and nodes did not exhibit spontaneous activities or action potentials. The resting membrane potentials of the primo vessel and node cells were −21.14 ± 2.24 mV (n = 35) and −36.69 ± 1.38 mV, respectively (n = 67). The current-voltage relations of the primo vessel cells were linear, but those of the primo node cells were outwardly rectifying. Morphologically, most primo vessel cells showed a round shape, but a small portion of cells showed a longitudinal one. In contrast, most cells in the primo node slices were round-shaped, and the longitudinal cells were not observed. Taken together, the results indicate that the cells in the PVS tissues are electrophysiologically and morphologically heterogenous, and that the small round cells that are most abundant in the primo vessels and nodes are nonexcitable.

1 Introduction

The primo vascular system (PVS; Bonghan system) is a novel circulatory system consisting of primo vessels and primo nodes, as well as primo fluid circulating within the system [1]. Kim, who first discovered the PVS, reported on spontaneous

P.D. Ryu (✉)
Laboratory of Veterinary Pharmacology, Research Institute of Veterinary Science, College of Veterinary Medicine, Seoul National University, Seoul, Korea
e-mail: pdryu@snu.ac.kr

K.-S. Soh et al. (eds.), *The Primo Vascular System: Its Role in Cancer and Regeneration*, DOI 10.1007/978-1-4614-0601-3_34,

electrical activity, as well as spontaneous mechanical contractility, in the Bonghan tissues [2, 3]. In his report, the electrical activity of Bonghan tissue was reduced or disappeared at temperatures lower than 27°C and was altered by cholinergic drugs such as pilocarpine [2]. In addition, previous studies have shown the presence of endothelial cells in the PVS [4] and a fluid flow that was demonstrated by injecting fluorescent nanoparticles [5, 6] or Alcian blue into the primo nodes [7]. All these observations support the idea that the PVS can function as a circulatory system.

In the vascular, cardiac, or intestinal smooth muscle cells exhibiting motility, it is well known that mechanical contractility is preceded by electrical activity. Recently, Park et al. [8] and Choi [9] showed that the resting membrane potential (RMP) of the PVS tissue cells was about −40 mV, and that a small portion of recorded cells showed spontaneous electrical activity in the PVS tissue. However, the detailed electrophysiological properties of PVS tissue cells and the mechanism of the excitation-contraction coupling in the PVS are not yet understood. In this study, we examined the passive membrane property of the cells of primo vessels by using the intracellular recording technique and the cells of the primo nodes by using the slice patch clamp technique. For the latter, we developed a slice preparation of the primo node [10].

2 Materials and Methods

2.1 Isolation of Organ Surface PVS Tissues and Preparation of Slices

Male and female Sprague-Dawley rats weighing 200–400 g were purchased from Orient Bio (Kyunggi-do, Korea). The rats were housed in a temperature- (24–26°C) and light-controlled (12:12 h light-dark cycle) room. The animals had ad libitum access to food and water. All animal experiments were carried out in agreement with the guidelines of the Laboratory Animal Care Advisory Committee of Seoul National University. The rats were anesthetized with an anesthetic cocktail (ketamine 75 mg/kg and xylazine 10 mg/kg) administered intramuscularly to their right femoral regions. Under general anesthesia, the PVS tissues were isolated from the rats according to methods reported previously [11, 12]. The harvested PVS tissues were kept in ice cold (4°C) Ca^{2+}-free Krebs solution and continuously oxygenated with gas (95% O_2, 5% CO_2). Then, primo nodes were sliced using a vibrating microtome, as reported previously [10]. A 4% low-melting agarose (Lonza, Rockland, ME, USA) was dissolved at 70°C and poured into a cubic frame (25 × 25 × 25 mm). The isolated PVS tissues including the primo nodes were planted in the agarose solution at 35°C. Then, the agarose solution was rapidly chilled on ice. After the agarose was solidified, the agarose block was fixed onto a slicing chamber with adhesive. The primo node slices (200 μm) were sectioned with a vibrating microtome (Vibratome 1000 plus, Vibratome, St. Louise, MO, USA). The slices were incubated in a chamber containing oxygenated Krebs solution at 31°C [13]. The composition of the Krebs solution was (in mM) NaCl, 120.35; KCl, 5.9; $NaHCO_3$, 15.5; NaH_2PO_4, 1.2; $MgSO_4$, 1.2; $CaCl_2$, 2.5; and glucose, 11.5 [14]. EGTA was substituted for $CaCl_2$ in the Ca^{2+}-free Krebs solution [15].

2.2 Intracellular Recording

The isolated PVS tissues were transferred to a recording chamber (1 mL) and were pinned firmly to the bottom of the Sylgard-lined chamber while being continuously perfused (3 mL/min) with oxygenated (95% O_2, 5% CO_2) Krebs solution at 30–33°C. The PVS tissues were penetrated with pins, and the cells in the primo vessel were penetrated by the microelectrodes for intracellular recording. Microelectrodes were pulled from borosilicate glass capillaries of 1.0 mm outer diameter and 0.78 mm inner diameter, and the open resistance of the electrodes used was 30–50 MΩ when filled with 3 M KCl. Membrane potential was recorded using a DC amplifier (IE-251, Warner Instruments, San Francisco, CA, USA). After low-pass filtering (cutoff frequency, 1 kHz), membrane potential changes were digitized using an analog-digital converter (Digidata 1200, Axon Instruments, Inc., Foster City, CA, USA) and analyzed using pClamp8 software (Axon Instruments). The membrane input resistance was obtained by dividing the potential changes (mV) with applied current pulses (from −0.4 to +0.4 or 0.9 nA in 0.1 nA increments).

2.3 Patch Clamp Recording

To examine the basic electrical property of primo node cells, a primo node slice was transferred to a recording chamber (0.7 mL) and fixed with a grid of nylon stocking threads supported by an O-shaped silver wire weight while being perfused (3 mL/min) with oxygenated (95% O_2, 5% CO_2) Krebs solution at 30–33°C. Pipettes were pulled from borosilicate glass capillaries of 1.7 mm in diameter and 0.5 mm wall thickness. The K^+-gluconate-rich solution contained (in mM): 135 K-gluconate, 5 KCl, 20 HEPES, 0.5 $CaCl_2$, 5 EGTA, and 5 MgATP [16]. Its pH was adjusted to 7.2 with KOH. Individual primo node cells were identified using a light microscope with differential interference contrast (BX50WI, Olympus, Tokyo, Japan) for the whole-cell patch recording. After a target primo node cell was selected for recording, the usual approach of the patch electrode was performed with the aid of a three-dimensional micromanipulator (MP-225, Sutter Instruments, Novato, CA, USA). The open resistance of the pipette ranged from 3 to 6 MΩ, and the seal resistances ranged from 1 to 3 GΩ. Electrical signals were recorded using an Axoclamp 2B amplifier (Axon Instruments, Foster City, CA, USA). The signals were filtered at 1 kHz and digitized at 10 kHz using an analog-digital converter (Digidata 1320A) and the pClamp software (Version 7.0, Axon). RMPs were corrected for the liquid junction potential (~14.3 mV). The membrane input resistance was calculated by the potential changes (mV) after applied hyperpolarizing current pulses (−60 pA). Currents were activated using a series of depolarizing voltage commands (from −80 mV to +80 mV, in 10 mV increments). The pictures of primo node cells in slice were visualized using a microscope (BX50WI) equipped with CCD camera (VPC-170IR, Japan) [17].

3 Results and Discussion

3.1 Basic Electrophysiological Properties of Primo Vessel Cells

In the intracellular recording experiments, the primo vessel cells were targeted without the visual identification of individual cells. Based on a total of 35 primo vessel cells, the RMP, membrane resistance (R_{input}), and membrane capacitance (C_{input}) were −21.14±2.24 mV, 96.30±13.17 MΩ, 9.62±0.45 pF, respectively (mean±S.E.M.). The RMP level was stable, and none of the cells recorded showed spontaneous activity. The current-voltage (I-V) relations obtained by injecting current (−0.4 to 0.9 nA) were linear (Fig. 34.1a). Action potentials were not induced at depolarizing pulses up to +50 mV. Based on the RMP, the primo vessel cells could be classified into two types: Type A with a mean RMP of −13.13±0.66 mV, and Type B with a mean RMP of −38.64±2.96 ($p<0.001$, by *t*-test). Type A and B cells accounted for about 70% ($n=23$) and 30% ($n=12$), respectively, of the total primo vessel cells recorded. Type A cells have smaller membrane capacitance than Type B cells (8.7±0.61 vs. 13.3±1.30 pF, $p<0.001$), but Type A and B cells were not different in membrane resistance ($p>0.05$). The results indicate that Type B cells are larger in membrane potential as well as in its surface area (or size). In addition, the results also show that the primo vessel cells in both types are nonexcitable because they showed neither spontaneous potentials nor action potentials. Our observation on the RMP is largely in agreement with the results from previous intracellular recording studies [8, 9]. For example, Choi [9] reported that mean RMP was about −22 mV. However, they showed spontaneous potentials or action potentials in 2 out of 17 primo vessel cells. They reported the presence of spontaneous potentials (~20 mV) of 4.1 Hz and the reduction of frequency by acetylcholine. The lack of spontaneous activity in the primo vessel cells in the present study might have arisen due to the difference in the type of cells recorded. As shown in Fig. 34.1c, there are numerous small round cells in the primo vessel. Therefore, it is likely that these cells showed the nonexcitable electrophysiological property recorded in the present and previous reports. In addition, there is a small population of fiber-like long structures or cells in Fig. 34.1c (marked by white arrows). These cells might be a type of excitable cells reported in previous studies [8, 9]. Although the proportion of cells showing action potentials was very small, the presence of spontaneous activity itself is especially worthy of notice because such spontaneous activity or excitability is necessary for the mechanical activity of the PVS.

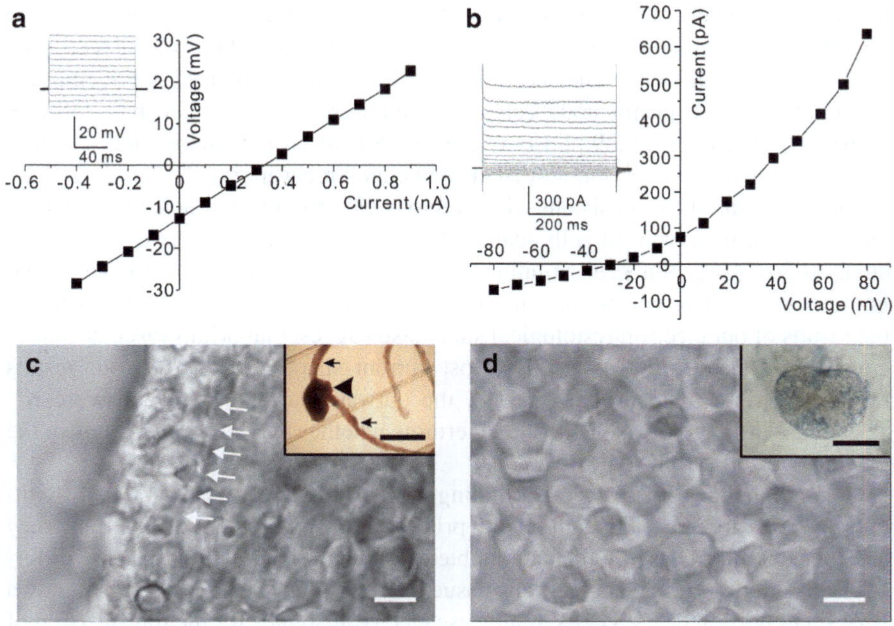

Fig. 34.1 Representative current-voltage relations of primo vascular tissue cells and microscopic images of the primo vessel (PV) and primo node (PN). (**a**) Current-voltage (I-V) relationship of a primo vessel cell. The inset illustrates the voltage responses to step current pulses (100 ms) from −0.4 to 0.9 nA in a PV cell with a sharp electrode. (**b**) I-V relationship recorded from a cell in the PN slice with patch electrode. The I-V curve was acquired by depolarizing step pulses from −80 to 80 mV. Scale bars are 20 ms and 10 mV for (**a**), and 100 ms and 200 pA for (**b**). (**c**) A typical example of PV cells in tissue. Note the round cells and longitudinal cells or structures pointed to by the arrows. The *inset* shows PV tissue at a lower magnification in the recording chamber (*arrowhead*: PN, *arrow*: PV). (**d**) A representative example of a PN slice (200 m). Note the numerous round cells without processes in the PN slice. The *inset* shows a PN slice at a lower magnification in the recording chamber. Scale bars are 10 μm for (**c**, **d**) and 1 mm for (**c**, *inset*) and 400 μm for (**d**, *inset*)

3.2 Basic Electrophysiological Properties of Primo Node Cells

To examine the basic electrophysiological properties of primo node cells, we prepared primo node slices (200 μm) as shown in Fig. 34.1d (Inset). The PN slices, viable for at least 4 h under our experimental conditions, were suitable for studying the electrophysiological properties of individual primo node cells [10]. The most abundant cells in the slices were small round cells (~10 m, Fig. 34.1d). The outward currents at positive (depolarizing) voltages were much larger than the inward currents at negative (hyperpolarizing) voltages in response to voltage commands from −80 to +80 mV for 600 ms in 10 mV increments, showing an outward rectification. There was little time-dependent current activated during the voltage command (Fig. 34.1b). Collectively, the results indicate that the primo node cells are similar to the primo vessel cells in that the majority is nonexcitable, but different in that the current-voltage relations are linear in the primo vessel cells but outwardly rectifying in the primo node cells. The mean RMP, membrane resistance, and membrane capacitance of the primo node cells recorded were -36.69 ± 1.38 mV, 758.14 ± 47.61 MΩ, and 19.60 ± 0.89 pF ($n = 67$, mean ± S.E.M.). Spontaneous action potentials were not observed in whole-cell current clamp recording.

Our observation is consistent with previous electron microscopic study, which has shown various types of cells scattered in the extracellular matrix of primo vascular tissues [18]. Numerous immune cells existed, including macrophages and granulocytes such as eosinophils, mast cells, monocytes, and fibroblasts. Likewise, the results of our electrophysiological study also suggest that primo vascular tissues consist of different kinds of cells. Most current shapes recorded in our results resembled those of macrophages [19], the outward currents of which were K^+-selective. Further study is needed to determine whether the outward currents in the primo node cells are K^+-selective.

Taken together, the most salient finding of the study is that the round cells without processes are most abundant in both primo vessel and primo node cells, and they are nonexcitable. These findings undoubtedly show that the PVS tissues are different from smooth muscle and nervous tissue neurons. Our results suggest that only a very small portion of PVS cells, if any, use electrical signals for the transmission of information as in excitable cells in smooth muscle or nervous tissue. Therefore, it is likely that there is a way other than electrical activity for information transmission or the flow of the substance (primo fluid) through the PVS. For example, primo vessels can be influenced by the movement of other organs, including blood vessels, lymphatic vessels, intestines, and so on.

Acknowledgments This work was supported by a grant (No. 2008–0059382) in the Mid-career Researcher Program of the NRF funded by the Korean Government (MEST).

References

1. Soh KS (2009) Bonghan circulatory system as an extension of acupuncture meridians. J Acupunct Meridian Stud 2:93–106
2. Kim BH (1963) On the kyungrak system. J Acad Med Sci DPR Korea 90:1–41
3. Kim BH (1965) Kyungrak system. J Acad Med Sci DPR Korea 6:5–31
4. Ogay V, Bae KH, Kim KW et al (2009) Comparison of the characteristic features of Bonghan ducts, blood and lymphatic capillaries. J Acupunct Meridian Stud 2:107–117
5. Lee BC, Yoo JS, Baik KY et al (2005) Novel threadlike structures (Bonghan ducts) inside lymphatic vessels of rabbits visualized with a Janus Green B staining method. Anat Rec B New Anat 286:1–7
6. Han HJ, Ogay V, Park SJ (2010) Primo-vessels as new flow paths for intratesticular injected dye in rats. J Acupunct Meridian Stud 3:81–88
7. Sung B, Kim MS, Lee BC et al (2008) Measurement of flow speed in the channels of novel threadlike structures on the surfaces of mammalian organs. Naturwissenschaften 95:117–124
8. Park SH, Lee BC, Choi CJ et al (2009) Bioelectrical study of Bonghan corpuscles on organ surfaces in rats. J Korean Phys Soc 55:688–693

9. Choi CJ (2009) Spontaneous action potential and basic electrical characteristics of cells in the Bonghan duct. Master's course, Graduate school of Seoul National University
10. Han TH, Lim CJ, Choi JH et al (2010) Viability assessment of primo-node slices prepared from surface of rat abdominal organs. J Acupunct Meridian Stud 3:241–248
11. Lee BC, Kim KW, Soh KS (2009) Visualizing the network of Bonghan ducts in the omentum and peritoneum by using Trypan blue. J Acupunct Meridian Stud 2:66–70
12. Shin HS, Johng HM, Lee BC et al (2005) Feulgen reaction study of novel threadlike structures (Bonghan ducts) on the surfaces of mammalian organs. Anat Rec B New Anat 284:35–40
13. Pillekamp F, Reppel M, Dinkelacker V et al (2005) Establishment and characterization of a mouse embryonic heart slice preparation. Cell Physiol Biochem 16:127–132
14. Spencer NJ, Henning GW, Dickson E et al (2005) Synchronization of enteric neuronal firing during the murine colonic MMC. J Physiol 564:829–847
15. Oike M, Creed KE, Onoue H et al (1998) Increase in calcium in smooth muscle cells of the rabbit bladder induced by acetylcholine and atp. J Auton Nerv Syst 69:141–147
16. Han TH, Lee K, Park JB et al (2010) Reduction in synaptic GABA release contributes to target-selective elevation of PVN neuronal activity in rats with myocardial infarction. Am J Physiol Regul Integr Comp Physiol 299:129–139
17. Choi JH, Lim CJ, Han TH et al (2011) TEA-sensitive currents contribute to membrane potential of organ surface primo-node cells in rats. J Membr Biol 239:167–175
18. Lee BC, Yoo JS, Ogay V et al (2007) Electron microscopic study of novel threadlike structures on the surfaces of mammalian organs. Microsc Res Tech 70:34–43
19. Gendelman HE, Ding S, Gong N et al (2009) Monocyte chemotactic protein-1 regulates voltage-gated K+channels and macrophage transmigration. J Neuroimmune Pharmacol 4:47–59

Chapter 35
Effects of Cholinergic Drugs on Membrane Potential of Cells in Organ Surface Primo Nodes

Sang-Hyun Park, Byung-Cheon Lee, Cheon-Joo Choi, Kwang-Sup Soh, and Pan Dong Ryu

Abstract The primo node (PN) was recently identified as a part of the primo network system in rats and rabbits. In this work, we studied the effects of acetylcholine, pilocarpine, and nifedipine, an antagonist of Ca^{2+} channels, on the membrane potential of cells in the organ-surface PN. Bath application of acetylcholine (15 μM) hyperpolarized the PN cells by 8.4±5.2 mV. A similar response was induced by pilocarpine, a muscarinic agonist. In addition, acetylcholine and pilocarpine changed the interval, duration, amplitude, frequency, and the half-width of the spikes for some cases. Nifedipine depolarized PN cells by 13.4±6.4 mV. Our results show that PN cells have cholinergic muscarinic receptors and voltage-dependent Ca^{2+} channels, indicating that the activity of the PN (i.e., contractility) can be under a cholinergic control.

1 Introduction

Many studies of the primo vessel (PV) and the primo node (PN) have been done in the course of investigating their properties since the first rediscovery of a PV in the blood vessels of rabbits and rats [1, 2]. Previous studies using light microscope revealed gross anatomical and histological property of the PV and the PN on the surfaces of internal organs of rabbits and rats [3, 4]. In addition, fine structures of the PV and PN were further characterized by the studies using various types of electron microscopy [5]. Both light and electron microscopic studies have revealed the structural characteristics of the PV and the PN, which are distinctively different

K.-S. Soh (✉)
Nano Primo Research Center, Advanced Institute of Convergence Technology, Seoul National University, Suwon 443-270, Korea
e-mail: kssoh1@gmail.com

K.-S. Soh et al. (eds.), *The Primo Vascular System: Its Role in Cancer and Regeneration*, DOI 10.1007/978-1-4614-0601-3_35, © Springer Science+Business Media, LLC 2012

from those of blood and lymphatic vessels [6]. Besides structural features, functional aspects also progressively emerged. Proteomics [7] and immuo-histochemical analysis [8] yielded supportive data for a PN being a possible niche of mesenchymal stem cells. A PN has chromaffin cells that produce and store catecholamine, which implies the medical significance of the PV as a hormonal pathway [9]. DNA-containing granules flowing in the PV were collected and examined by using electron microscopy [10] and atomic force microscopy [11]. These granules, the so-called primo micro cells, displayed active chaotic motion whose speed was affected by UV light of 360 nm [12], and the viscosity of the liquid in the PV was measured [13]. In the early 1960s, Bonghan Kim claimed that the PVs and the PNs formed a new circulatory system [14]. Therefore, the circulatory function of the primo system needed to be confirmed and understood. Measurement of the flow speed of the primo liquid in a PV was performed by injecting Alcian blue; the speed of travel was rather slow, 0.3 ± 0.1 mm/s [15]. The endothelial cells necessary for a flowing liquid system were also observed [6].

Exploring the excitability of the cells in a PN is important because the propelling power for the flow in PVs is supposed to come from the contractility of a PN. For this purpose, a necessary step is to study the electrophysiology of a PN. The resting was measured potential, spontaneous irregular bursting rhythms were indeed observed, and their biophysical characteristics were reported [16]. In this work, we investigated the changes in the bioelectric signals of PN cells after exposing to various drugs. In order to identify the cell type of the excitable cells in a PN, we studied the change in the membrane potential due to stimulation by drugs such as acetylcholine and pilocarpine. Since the contraction of muscle cells requires Ca^{2+} ions, we also tested the effects of a Ca^{2+} channel blocker, nifedipine.

2 Experiments

The electrophysiological system of this experiment was set up using the intracellular microelectrode recording method [17]. The bioelectrical signals from the rectal tissue, which were obtained by using the intracellular recording method, were used to calibrate the measurement system for our experiments [18]. The sample preparation and the experimental setup are reported elsewhere [16]. For the drug stimulation experiments, the sample needed to be adapted to a new environment and the effect of the insertion had to be removed, so we waited for more than 5 min for stabilization. The sample was treated with acetylcholine, pilocarpine, and nifedipine (15 mM, 300 μL, 37–39°C) that were dropped into the chamber (30 mL). For nifedipine, we used only 0.1% ethanol to dissolve the drug, which is lower than the maximum ethanol concentration with no harmful effect on the cells in cultures (0.2%) [19]. The position at which the drugs were dropped was 2 cm away from the sample. A grounded pipette was used to reduce electrical noise. The resting potential of the PNs from rats was measured to be 10.5 ± 8.4 mV. There were irregular bursts of spontaneously evoked spikes in the resting potential for an average duration of 17 s.

Detailed data and analysis of the resting potential and bursts without drug stimulation were recently reported [16]. These data were used as bases for the study of effects of drugs upon the bioelectrical signals of PNs.

3 Results

In this study, we tested a total of 91 cells in the PNs located on intestinal organs of rats with drugs. We obtained measurable responses in 6 of 12 cells treated with acetylcholine, 7 of 11 cells treated with pilocarpine, and 4 of 7 cells treated with nifedipine. The following results are based on these drug-responsive cells.

3.1 Hyperpolarization of Cells in Primo Nodes

3.1.1 Acetylcholine

Acetylcholine is one of the major neurotransmitters in the autonomic nervous system. It stimulates both nicotinic and muscarinic receptors and is commonly used to induce the contraction of intestinal smooth muscle. Here we examined the effects of acetylcholine (15 μM) on PN cells in vitro. The results are given in Table 35.1, and a representative experimental result is shown in Fig. 35.1. As shown in Fig. 35.1, acetylcholine hyperpolarized the membrane potentials of PN cells. In 6 cells, the resting potential decreased by 8.4 ± 5.2 mV upon bath application of acetylcholine, and the average time taken to be stabilized was 215 ± 232 s. In addition, acetylcholine also increased the bursting activity. The interval and the half-width of the spontaneous spikes changed dramatically by 0.51 ± 0.31 and 0.22 ± 0.13 s ($n=6$), respectively. These changes are very large in the sense that the original values are only 0.95 and 0.37 s ($n=6$), respectively ($p<0.05$ for the interval; $p<0.05$ for the half-width). The amplitude and the duration of the spikes in the burst changed significantly on a case-by-case basis even though their average values were small. Another notable change due to acetylcholine stimulation was the appearance of a high-frequency component at about 3.7 ± 1.7 Hz. Acetylcholine induced significant changes in the resting potential, the interval of the spikes, and the half-width of these spikes ($p<0.05$ by paired t-test)

3.1.2 Pilocarpine

Pilocarpine is a nonselective muscarinic acetylcholine receptor agonist working at concentrations of 10^{-8}–10^{-5} M for a rat [20] and was used to determine of the receptor types in PNs in this study. The effects of 15-μM pilocarpine stimulation on the PN were studied and the results are given in Table 35.2. Figure 35.2 illustrates a

Table 35.1 Effects of 15-μM acetylcholine stimulation on the resting potential and its spontaneous burst

	Resting potential				Spontaneous bursts of spikes												
Exp	V_R (mV)	V'_R	ΔV_R	T_R (s)	V_s	V'_s	ΔV_s	T_s (s)	T'_s	ΔT_s	D_s (s)	D'_s	ΔD_s	F_s (s)	F'_s	ΔF_s	HF (Hz)
1	−20.0	−20.7	−0.7	n/a	5.06	8.56	3.50	n/a	n/a	n/a	75	61	−14	0.38	0.26	−0.12	5.00
2	−18.0	−22.9	−4.9	75	0.36	0.31	−0.06	1.10	0.78	−0.32	10	568	558	0.43	0.22	−0.21	4.18
3	−21.0	−32.0	−11.0	609	0.15	0.98	0.83	0.74	0.57	−0.17	n/a	n/a	n/a	0.45	0.16	−0.29	1.00
4	−40.0	−46.5	−6.5	169	2.30	1.54	−0.76	1.13	0.21	−0.92	238	16	−222	0.21	0.06	−0.15	4.17
5	−25.0	−39.0	−14.0	20	3.14	8.29	5.15	0.80	0.40	−0.40	31	9	−22	0.25	0.15	−0.10	2.50
6	−33.0	−46.1	−13.1	203	0.30	0.43	0.13	0.98	0.22	−0.76	n/a	n/a	n/a	0.49	0.06	−0.43	5.50
Ave.	−26.2	−34.5	−8.4	215	1.89	3.35	1.47	0.95	0.44	−0.51	88	163	75	0.37	0.15	−0.22	3.73
Std.	8.6	11.2	5.2	232	1.98	3.96	2.33	2.89	0.24	0.31	103	2702	336	0.11	0.08	0.13	1.68

V_R, V'_R, ΔV_R: the resting potentials before and after stimulation, and the difference; T_R: the time required the resting potential to become stable after stimulation; V_s, V'_s, ΔV_s: the amplitude before and after stimulation, and the difference; T_s, T'_s, ΔT_s: the period of bursting spikes before and after stimulation, and the difference; D_s, D'_s, ΔD_s: the duration of bursting before and after stimulation, and the difference; F_s, F'_s, ΔF_s: The half-width of spikes in bursts before and after stimulation, and the difference; *HF*: high-frequency modes that appeared after stimulation

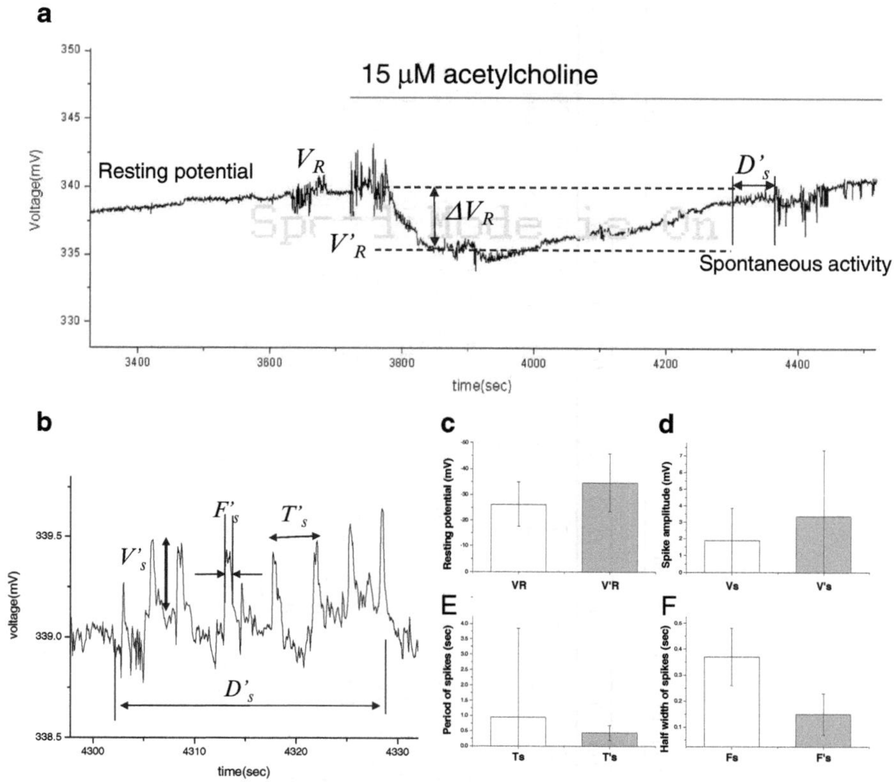

Fig. 35.1 Effects of acetylcholine (15 μM) on the resting potential and the spontaneous burst. (**a**) The resting potential dropped from V_R to V'_R, and the decrease was ΔV_R. The resting potential slowly returned to its original value by ΔV_R. (**b**) The amplitude (V'_s), period (T'_s), and half-width (F'_s) of the spikes in the burst show slight noticeable changes. (**c–f**) Mean values of the changes in the resting potential (**c**, $n=6$), the spike amplitudes (**d**, $n=6$), the spike duration (**e**, $n=6$), and the half-width of the spikes (**f**, $n=6$), (*asterisk* the paired *t*-test value $P<0.05$)

representative experimental result showing the effects of pilocarpine on the membrane potential of the PN cells. The effects of pilocarpine were tested with or without prior application of acetylcholine. In the cells tested pilocarpine only (Exp. 1, 2, and 3), the resting potential decreased by 7.5 ± 3.67 mV ($n=3$) upon stimulation with pilocarpine, and the average time taken for the decrease was 76 ± 50 s ($n=3$). The interval and the half-width of the spontaneous bursts changed dramatically by the amounts 0.50 ± 0.55 ($n=3$) and 0.4 ± 0.12 s ($n=3$), respectively. This result was similar to the acetylcholine effect. The amplitude was increased by 0.94 ± 0.88 mV ($n=3$), whereas the duration was changed little. Another notable change due to pilocarpine stimulation was the appearance of a high-frequency component at about 3.9 ± 1.1 Hz ($n=3$). In the experiments with prior application of acetylcholine (Exp. 4, 5, 6, and 7), upon stimulation with pilocarpine, the resting potential further decreased by 1.95 ± 27.5 mV ($n=4$) from the original decrease for acetylcholine,

Table 35.2 Effects of 15-μM pilocarpine stimulation on the resting potential and its spontaneous burst

	Resting potential				Spontaneous bursts of spikes												
Exp	V_R (mV)	V'_R	ΔV_R	T_R (s)	V_s	V'_s	ΔV_s	T_s (s)	T'_s	ΔT_s	D_s (s)	D'_s	ΔD_s	F_s (s)	F'_s	ΔF_s	HF (Hz)
1	−25.0	−30.0	−5.0	34	0.74	1.22	0.48	1.05	0.19	−0.86	7	9	2	0.40	0.11	−0.29	5
2	−45.2	−51.0	−5.8	62	0.68	1.07	0.39	0.77	0.69	−0.08	4	4	−1	0.61	0.15	−0.46	3.93
3	−70.9	−82.6	−11.7	131	0.00	1.96	1.96	–	1.32	–	–	4	–	–	0.11	–	2.83
Ave.	−47.0	−54.5	−7.5	76	0.47	1.42	0.94	0.91	0.73	−0.5	5.5	5.2	0.6	0.50	0.12	−0.4	3.9
Std.	23.0	26.5	3.66	50	0.41	0.48	0.88	0.20	0.57	0.55	1.68	2.87	1.76	0.15	0.02	0.12	1.09
4	−18.0	−17.5	0.5	48	0.34	1.00	0.66	1.07	0.15	−0.92	16	17	0	0.17	0.07	−0.10	4
5	−40.0	−40.3	−0.3	92	1.54	2.39	0.86	0.21	0.89	0.68	16	140	125	0.06	0.10	0.04	0
6	−40.0	−43.0	−3.0	12	2.09	1.50	−0.59	0.89	1.03	0.14	27	15	−12	0.53	0.34	−0.19	0
7	−33.0	−22.8	10.2	71	0.43	0.28	−0.15	0.22	0.38	0.16	–	–	–	0.06	0.15	0.09	2
Ave.	−32.8	−30.9	1.9	55	1.10	1.29	0.20	0.60	0.61	0.01	19	57	37	0.20	0.16	−0.04	1.50
Std.	10.4	12.6	5.8	34	0.86	0.89	0.68	0.45	0.42	0.67	6	71	75	0.22	0.12	0.13	1.91

The meaning of V_R, V'_R, ΔV_R, T_R, V_s, V'_s, ΔV_s, T_s, T'_s, ΔT_s, D_s, D'_s, ΔD_s, F_s, F'_s, ΔF_s, and HF were described in Table 35.1

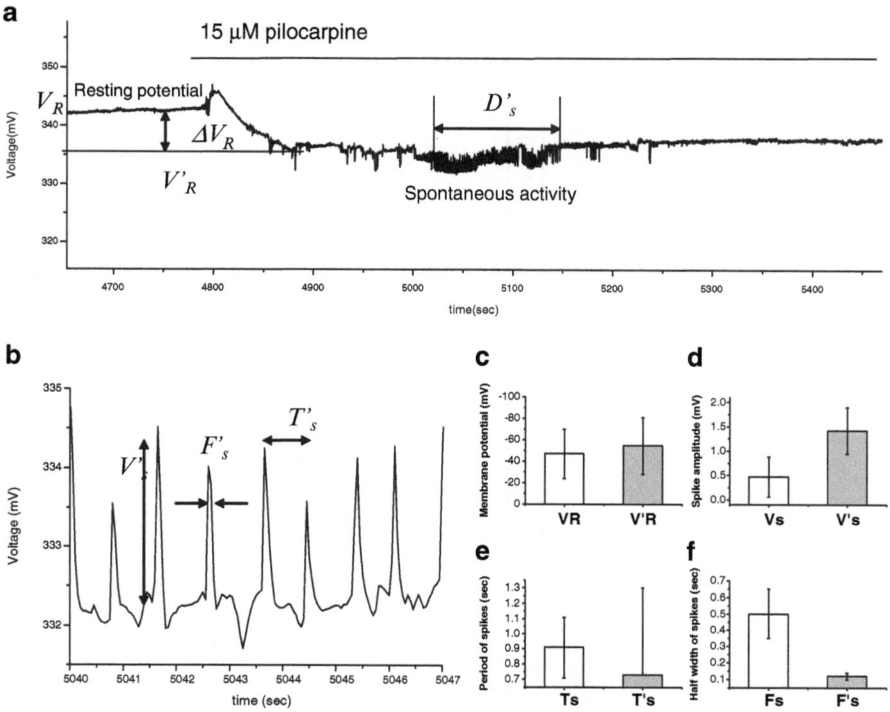

Fig. 35.2 Effects of pilocarpine (15 μM) on the resting potential and the spontaneous bursts. (**a**) The resting potential dropped from V_R to V'_R, and the decrease was ΔV_R. The resting potential slowly became stable. The duration of the spontaneous burst (D'_s) increased dramatically. (**b**) The amplitude (V'_s), the period, and the half-width are presented. (**c**) The amplitude of spikes increased, the period (T'_s), in the spikes decreased, and the half-width (F'_s) of the spikes in the burst decreased noticeably ($n=3$)

and the average time taken for the decrease was 55 ± 34 s ($n=4$). The spontaneous bursting activity was changed remarkably in some cases, but the change was rather irregular. For example, V_s of the Exp. 4 increased, but V_s of the Exp. 6 decreased after stimulation with pilocarpine. Also, the T_s of Exp. 4 decreased by 0.92 s, while the T_s of Exp. 5 increased by 0.68 s after pilocarpine treatment.

3.2 *The Existence of Ca^{2+} Channels in Primo-Node Cells*

Nifedipine is an L-type Ca^{2+} channel blocker working at concentrations of 10^{-8}–10^{-5} M [20] for a rat. Nifedipine is a blocker, and we investigated the effects of blocking by observing the changes evoked in the spikes by using pilocarpine or acetylcholine. The effects of 15 μM nifedipine on the PN were studied and the results are given in Table 35.3. Figure 35.3 illustrates a representative experimental result. Exp. 1

Table 35.3 Effects of 15-μM nifedipine stimulation on the resting potential and its spontaneous burst

	Resting potential				Spontaneous bursting of spikes											
Exp	V_R (mV)	V'_R	ΔV_R	T_R (s)	V_s	V'_s	ΔV_s	T_s (s)	T'_s	ΔT_s	D_s (s)	D'_s	ΔD_s	F_s (s)	F'_s	ΔF_s
1	−21.7	−2.2	19.5	n/a	2.34	1.97	−0.37	5.73	n/a	n/a	188	n/a	n/a	0.61	0.20	−0.41
2	−7.9	−3.3	4.6	176	1.41	0.26	−1.15	0.93	1.89	0.96	30	16	−14	0.39	0.24	−0.15
3	−23.7	−10.0	13.7	184	2.40	0.90	−1.50	0.22	0.39	0.17	32	18	−14	0.08	0.20	0.12
4	−27.8	−11.9	15.9	44	1.33	0.21	−1.13	0.95	0.93	−0.02	32	19	−13	0.37	0.38	0.01
Ave.	−20.3	−6.9	13.4	135.0	1.87	0.84	−1.04	1.96	1.07	0.37	31	18	−14	0.36	0.26	−0.11
Std.	8.6	4.8	6.4	78.6	0.58	0.82	0.48	2.54	0.76	0.52	1	2	1	0.22	0.09	0.23

The meaning of V_R, V'_R, ΔV_R, T_R, V_s, V'_s, ΔV_s, T_s, T'_s, ΔT_s, D_s, D'_s, ΔD_s, F_s, F'_s, ΔF_s, and HF were described in Table 35.1

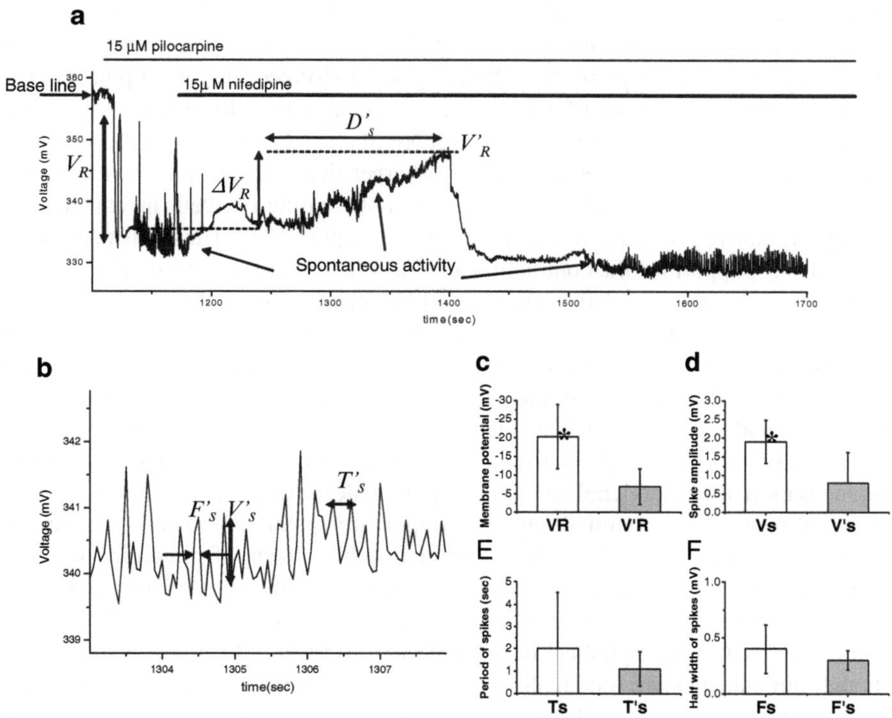

Fig. 35.3 Effects of nifedipine (15 μM) on the resting potential and the spontaneous burst. (**a**) The resting potential increased from V_R to V'_R, and the increase was ΔV_R. The resting potential slowly returned to its original value by ΔV_R. (**b**) The amplitude (V'_s), period (T'_s), and half-width (F'_s) of the spikes in the burst did not show noticeable changes, but sharp spikes appeared. (**c–f**) Mean values of the changes in the resting potential (**c**, $n=4$), the spike amplitudes (**d**, $n=4$), the spike duration (**e**, $n=4$), and the half width of the spikes (**f**, $n=4$) (*asterisk* the paired *t*-test value $P<0.05$)

shows the effect on a sample of nifedipine treatment without prestimulation, Exp. 2 shows that for a washed sample, Exp. 3 shows the effect for a sample pretreated by pilocarpine, and Exp. 4 shows the effect for a sample prestimulated with acetylcholine. In all the experiments, the resting potential increased upon application of nifedipine, and the spontaneous bursting increased dramatically in the amplitude. We observed a significant change in the resting potential and the spike amplitude ($p<0.05$ in both parameters by paired *t*-test)

4 Discussion

The most prominent feature of the acetylcholine effect was a decrease in the resting potential (hyperpolarization). Noticeable changing responses to the cholinergic drugs stimulus were in the frequency and the half-width of the spikes. The frequency and the half-width were changed more 50% than before the stimulation, respectively. These changes in the membrane potential imply a hyperpolarizing of the cells due to acetylcholine stimulation. In order to further check the receptors mediating the hyperpolarization of the cells, we performed the pilocarpine experiment. The resting potential of a PN decreased again upon pilocarpine stimulation, which confirmed our expectation. Furthermore, the spikes in the bursts in the pilocarpine stimulated case became strikingly sharper than these in the nonstimulated cases. Our data suggest that PN has vascular smooth muscle-like cells in that the cells are hyperpolarized by activation of muscarinic receptors [21]. Thus, this result provides strong support for the contracting and the relaxing movements of a PN, which is necessary for the circulatory function of the primo system. More evidence for smooth muscle-like movement of the PN is the Ca^{2+} ion channel test. Nifedipine is a Ca^{2+} ion channel blocker. Thus, we expect the drop in the resting potential to become very small if nifedipine is used; indeed, as shown in Table 35.3, cases 1 and 2 show, V'_s *of* −2.2 and −3.3 mV, which are about 10% of the average V_R. We expect the resting potential changes (ΔV_R) due to acetylcholine or pilocarpine to be blocked if nifedipine is used after acetylcholine or pilocarpine. Indeed, we observed that the resting potential increased substantially upon nifedipine treatment, as shown in the experiments 3 and 4 in Table 35.3. One of the noticeable effects of nifedipine was the change in the shape of the spikes. They became very sharp. Nifedipine is a blocker, so this change reflects blocking effects. These phenomena were very noticeable in some cases, but more statistical data are needed to make a firm conclusion. Even though our statistics are too small to make a firm conclusion, a tendency for nifedipine to block the Ca^{2+} ion channel was clearly observed. In conclusion, the present study shows the presence of cholinergic hyperpolarization and voltage-dependent Ca^{2+} channels in PN cells. The cells were hyperpolarized by acetylcholine and pilocarpine. The basic test of contractibility was done with the Ca^{2+} ion channel blocker nifedipine. These results support the cells in the PN being similar to vascular smooth muscles, not cardiac, skeletal, gastro-intestinal muscle, according to data tables on the reference materials [20–22].

Acknowledgments This work was supported by Industrial Strategic technology development program (10035137-2010-01) funded by the Ministry of Knowledge Economy (MKE, Korea). K. Soh thanks the Association of Korean Oriental Medicine for support.

References

1. Jiang X, Kim HK, Shin HS et al (2002) Method for observing intravascular Bonghan duct. J Kor Orient Prevent Med Soc 6:162–166
2. Lee BC, Baik KY, Cho SM et al (2003) Comparison of intravascular Bonghan ducts from rats and mice. J Kor Orient Prevent Med Soc 7:47–53
3. Shin HS, Johng HM, Lee BC et al (2005) Feulgen reaction study of novel threadlike structures (Bonghan ducts) on the surfaces of mammalian organs. Anat Rec B New Anat 284(1):35–40
4. Lee BC, Kim KW, Soh KS (2009) Visualizing the network of Bonghan ducts in the omentum and peritoneum by using Trypan blue. J Acupunct Meridian Stud 2(1):66–70

5. Lee BC, Yoo JS, Ogay V et al (2007) Electron microscopic study of novel threadlike structure on the surfaces of mammalian organs. Microsc Res Tech 70(1):34–43
6. Ogay V, Bae KH, Kim KW et al (2009) Comparison of the characteristic features of Bonghan ducts, blood and lymphatic capillaries. J Acupunct Meridian Stud 2:107–117
7. Lee SJ, Lee BC, Nam CH et al (2008) Proteomic analysis of Bonghan liquid and duct on the surface of rabbit intestine. J Acupunct Meridian Stud 1:1–13
8. Kim MS, Hong JY, Hong S et al (2008) Bong-Han corpuscles as possible stem cell niches on the organ-surfaces. J Kor Inst Herb Acupunct 11(1):5–12
9. Kim J, Ogay V, Lee BC et al (2008) Catecholmine producing novel endocrine organ: Bonghan system. Med Acupunct 20:97–102
10. Ogay V, Baik KY, Lee BC et al (2006) Characterization of DNA-containing granules flowing through the meridian-like system on the internal organs of rabbits. Acupunct Electrotherap Res 31(1–2):13–31
11. Kwon J, Baik KY, Lee BC et al (2007) Scanning probe microscopy study of microcells from the organ surface Bonghan corpuscle. Appl Phys Lett 90(17):17903
12. Sung B, Ogay V, Yoo JS et al (2005) UV-A-induced activation of Bonghan granules in motion. J Intl Soc Life Info Sci 23(2):297–300
13. Sung B, Kim MS, Corrigan A et al (2009) In situ microextraction method to determine the viscosity of biofluid in threadlike structures on the surfaces of mammalian organs. Phys Rev E 79:1
14. Kim BH (1963) On the Kyungrak system. J Acad Med Sci DPR Korea 90:1–41
15. Sung B, Kim MS, Lee BC et al (2008) Measurement of flow speed in the channels of novel thread-like structures on the surfaces of mammalian organs. Naturwissenschaften 95:117–124
16. Park SH, Lee BC, Choi CJ et al (2009) Bioelectrical study of Bonghan corpuscles on organ surfaces in rat. J Kor Phys Soc 55(2):688–693
17. Stamford JA (1992) Monitoring neuronal activity: a practical approach. IRL Press at Oxford University Press, Oxford
18. Kuriyama H, Mekata F (1971) Biophysical properties of the longitudinal smooth muscle of the guinea-pig rectum. J Physiol 212(3):667–683
19. Herembert T (1995) Mechanism of action of the inhibitory effect of nifedipine on the growth of cultured aortic cells from spontaneously hypertensive and normotensive rats. Br J Pharmacol 114:1703
20. Urakawa N, Karaki H (1998) Smooth muscle experiment instruction. MEDRANG INFORANG, Seoul
21. Lee WJ (2008) Pharmacology, 6th edn. MEDRAN INFORANG, Seoul
22. Bkaily G (1994) Ionic channels in vascular smooth muscle. R.G. Landes, Austin

Chapter 36
Apoptotic Cardiomyocyte Beating Frequency Detected with Optical Intensity Fluctuation Spectrometer

Svetlana Norina, Byung Cheon Lee, Jungdae Kim, and Ku Youn Baek

Abstract Dynamical spectroscopy of intensity fluctuations allows measurements of the dynamic scattering of light and motility detection of biomicroparticles in suspensions and tissues. Spectroscopy of intensity fluctuations allows the detection of complex rhythms of the rat heart in vitro. The optical portable spectrometer used detected Fourier spectra and real-time rhythms in rat cardiomyocytes, the beating in the infarct model of the rat heart. The findings indicated that single cardiomyocyte beating was superimposed on the autowave heart tissue rhythm and associated with the apoptotic process. A dependence was observed of the superposition of individual cardiomyocyte beatings at 8–22 Hz on their metabolic state and autooscillations of isolated rat heart tissue at 1–2 Hz.

1 Introduction

Current technologies based on spectroscopy of intensity fluctuations [1–5] allow precise microdynamical measurements of a small number of cells in suspensions and tissues. In the present study, the measuring system detected microdynamical events in rat heart tissue. A spectrometer allowed the detection of Fourier spectra and real-time rhythms of rat cardiomyocytes, the beating in the infarct model of the rat heart. The findings indicated that single cardiomyocyte beating was superimposed on the autowave heart tissue rhythm and was associated with the apoptotic process. The superposition of individual cardiomyocyte beatings at 8–22 Hz was observed to depend on the metabolic state and autooscillations of the isolated rat

S. Norina (✉)
Physics Dept., Lomonosov Moscow State University,
Moscow, Russia
e-mail: n.svetlana@ymail.com

K.-S. Soh et al. (eds.), *The Primo Vascular System: Its Role in Cancer and Regeneration*,
DOI 10.1007/978-1-4614-0601-3_36,

heart tissue at 1–2 Hz. The beating and the modulation processes when disturbed cardiomyocytes appeared in heart tissue were observed, and spontaneous beating was superposed with the heart rhythm.

2 Methods

The cardiomyocyte beat frequencies were photoelectrically counted with two photodiodes connected in a difference scheme. The fluctuations of the photoelectric current, which correspond to the cell beating, were amplified and converted with an analog-to-digital device to the LabView system, adjusted to document the intensity fluctuations, the real-time parameters, and the temperature measurements. Microscopic images of cardiomyocytes and rat heart tissue were projected simultaneously onto the photodiode planes coinciding with the ocular observation planes of the microscope. The rhythmic events counted every 2 s were documented on a six-graph display containing real-time dependencies: frequency, temperature, amplitude, intensity via time, frequency via temperature, and the Fourier-spectral dependencies. The effects of Ca^{2+} ions (present or absent in PBS saline solution), and the maximal motile frequency were measured from the signals maintained for 2 s and were plotted vs. time. Cardiomyocytes from 4- to 6-week-old rats were observed through a Olympus microscope (Dongwon, Korea and Olympus X-51, Japan) after collagenase treatment of perfused hearts to cause initial or enormous apoptosis in the tissue regions of the isolated rat hearts. Cardiomyocytes were sedimented in PBS. Cardiomyocyte cells were kept in cold PBS or in cold cell lysis buffer (50-mmol/L NaCl, 5-mmol/L $MgCl_2$, 20-mmol/L Hepes, pH 8.2, and 1-mmol/L dithiothreitol), and cells were homogenized slightly or considerably.

Dynamical spectroscopy of intensity fluctuations allows measurements of the dynamic scattering of light and motility detection of biomicroparticles in suspensions and tissues. A portable device was used to measure visually targeted structures of specimens for analyses of their Fourier spectra and for dynamical process documentation in the spectral range from 1 to 1,000 Hz and permitted motility analysis with high spatial resolution of moving structures up to 0.7 μm. An autocorrelation function of the photocurrent generated from a coherence area can be represented by a relation including two shifted currents induced in a differential scheme with two photodiodes, and the fluctuations of the light intensity due to the beats of the microobject were used to study the photodynamical changes and motile biological activity.

Light from the incident lamp was focused onto a spot in the investigated biosample. The computer image analyzer processes a sequence of video images (frame-by-frame analysis). The temporal fluctuations of the scattered intensity are detected by the two photo-receivers. The electrical signal was amplified, recorded, and processed by using a computer with the LabView Software. Using the LabView software, we measured the mean quadratic amplitude of the cell fluctuations, the frequency spectrum, and the temporal correlation function at two temporal moments. The x-axis in Fig. 36.1 (b–e, at lower graphs) shows the time of the input array for 5,000 points corresponding to 2 s.

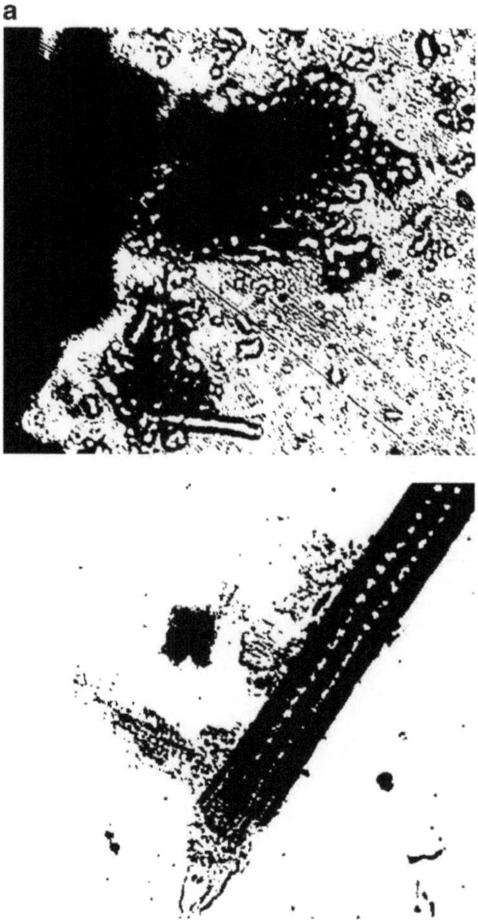

Fig. 36.1 (**a**) Rat heart tissue (*upper*) and cardio-myofibrile (*lower*) in apoptosis; (**b, c**) Fourier spectra of beating (*upper*) and live rhythms (real-time graphs) during 2 s (*lower*) of cardiomyocytes on the surface of the broken rat heart; the cell beating is seen as superposed with a low-frequency rat heart tissue rhythm (1–2 Hz) at the beginning of apoptotic changes and over 8 h; (**d, e**) Fourier spectra and real-time graphs for close to stopping a rhythm (**d**) and death (**e**) of cardiomyocytes. Frequency is double value on *upper* spectra due to 2 s-real-time input signal detection

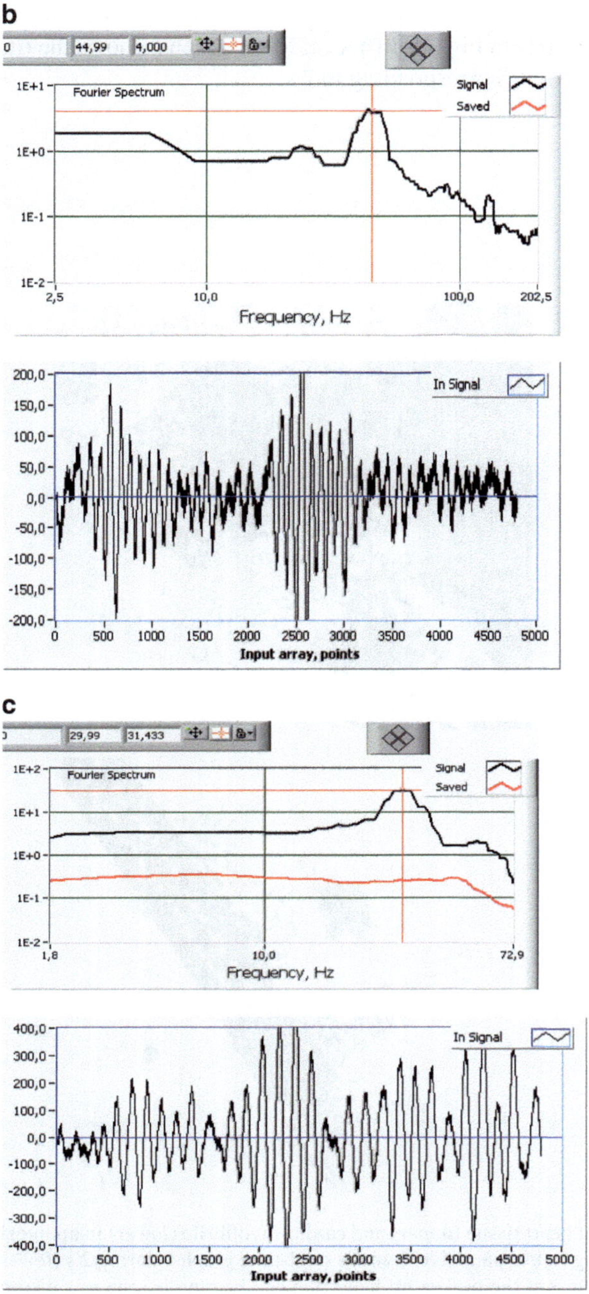

Fig. 36.1 (continued)

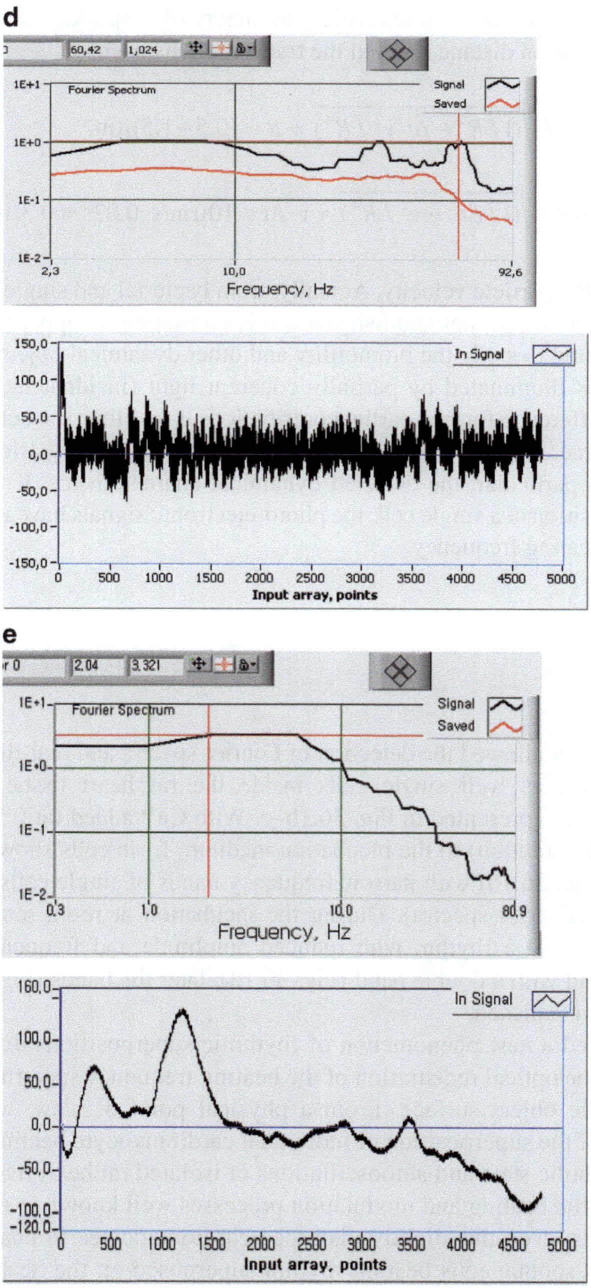

Fig. 36.1 (continued)

One can estimate the characteristic parameters of a speckle motion for single cells, the correlation distance r_c, and the translation distance r_t:

$$r_c = 1/\sqrt{(1/R^2 - \Delta t^2 v^2 / R^4)} \approx R \approx (0.5-1.5)\mu\text{m},$$

$$r_t = v/\sqrt{(1/\Delta t^2 - v^2 / R^4)} \approx v \cdot \Delta t \approx 10\mu\text{m/s} \cdot 0.03\text{s} \approx 0.3\mu\text{m},$$

where v is the particle velocity. Actually, from bacterial and single protozoa cilia sizes to invertebrate animal and fish sizes, it is convenient to apply this fluctuation detection method to study the biomotility and other dynamical objects. If a scattering medium is illuminated by partially coherent light (incident lamp) and if the path-length difference for two partial contributions is less than the coherence length of a light source, then the partial coherence does not essentially affect the speckle image and, in particular, the detected dynamical characteristics. If the detector is placed in a position at a single cell, the photo-electronic signals have a characteristic shape at the beating frequency.

3 Results

The spectrometer allowed the detection of Fourier spectra and real-time rhythms of rat heart tissue, as well single cells inside the rat heart tissue (Fig. 36.1a). Measurements are presented in Fig. 36.1b–e. With Ca^{2+} added (at 0.5% concentration in the PBS solution) to the incubation medium, fresh cells showed immediate oscillations (Fig. 36.1b) with narrow frequency bands of single cells with maxima at 22–24 Hz (Fourier spectra). During the incubation at room temperature, this rhythm changed to a rhythm with reduced amplitude and frequency of beating (Fig. 36.1c) and with a double band (Fig. 36.1d); later the bands (Fig. 36.1e) in the Fourier spectra vanished.

We observed a new phenomenon of rhythmic superpositions due to the independence of the optical registration of the beating frequency spectrum used on the distance to the object surface. From a physical point of view, we observed a dependence of the superposition of individual cardiomyocyte beatings at 8–22 Hz on their metabolic state and autooscillations of isolated rat heart tissue at 1–2 Hz. We observed the beating and modulation processes well known in radio technics. In other words, even the slightly disturbed cardiomyocytes in heart tissue have appeared as a spontaneous beating rhythm superposed on the heart rhythm as a whole. Immediately after collagenase treatment or mechanical disturbance, the individual cardiomyocytes started to beat with their own frequency. Under videomicroscope observation of individual cardiomyocytes (Fig. 36.1a), cells that preserved their rod shapes did not show beating. In contrast, apoptotic cardiomyocytes first initiated beating when they still had normal rod shapes and continued beating until the cells maximally shriveled to form a round shape. Cardio-myofibrillar packing (Fig. 36.1a, right) was easily spoiled by apoptotic treatments, Z-disks were disturbed, and single cells and aggregates became separated. Under a video microscope, cells that kept the form of a rod within 8 h following processing did not show beating. On the contrary, apoptotic cardiomyocytes first began beating when they still had the normal forms of a rod and continued their beating until the cell had a round form.

4 Conclusion

Spectroscopy of intensity fluctuations allowed the detection of complex rhythms of the rat heart in vitro; the disturbed heart tissue generated its own heart rhythms at 1–2 Hz; simultaneously in apoptotic areas of tissue, cardiomyocytes produced spontaneous beating rhythms at 8–22 Hz, which depended on the metabolic state. The Fourier spectra obtained showed the superposition of two rhythms and provided evidence of apoptotic processes in rat heart tissue and in individual cardiomyocytes.

These results indicate that beating is associated with the apoptotic process. Cardiomyocyte contraction or relaxation is governed by changes in the Ca^{2+} concentration. In an adult cardiomyocyte, the appearance of beating is an early sign of apoptosis, which is followed by morphological changes and DNA fragmentation (appearance of apoptotic bodies). The maximal beating frequencies are related to the start of the apoptotic process. Because both apoptosis and beating are ATP-energy-dependent, rapid or slow beating may reflect the functional and metabolic state of cardiomyocytes.

References

1. Norina S, Kim J, Soh KS (2007) Gradient magnetic field effects on separation, Fourier-spectra and motion of chromatin particles and cells. Int J Appl Electromagn Mech 25(1–4):419–427
2. Norina S, Kim J, Yang JM et al (2007) Spectroscopy of intensity fluctuations of cell dynamics, tissue reflectance and auto-fluorescence. In: R. Magjarevic and J.H. Nagel (eds) World congress on medical physics and biomedical engineering 2006. Springer, Heidelberg, pp 1324–1326
3. Maruyama R, Takemura G, Tohse N et al (2006) Synchronous progression of calcium transient-dependent beating and sarcomere destruction in apoptotic adult cardiomyocytes. Am J Physiol Heart Circ Physiol 290:H1493–H1502
4. Zimnyakov DA, Tuchin VV (2003) Speckle correlometry. In: Tuan Vo-Dinh Chief (ed) Biomedical photonics handbook. SPIE Press, Bellingham, pp 14-1–14-20
5. Rastopov S (2004) US Patent Application 20040070756

Chapter 37
PKA Activation in Cardiac Myocytes Affects the Voltage Dependence of Na-K ATPase Pump and Na-Ca Exchange Currents Differently

Chin Ok Lee and David C. Gadsby

Abstract The nature of voltage dependence of Na-K pump and Na-Ca exchange currents was investigated in isolated guinea pig ventricular myocytes in the absence and presence of forskolin, the adenylyl cyclase activator. Whole-cell membrane currents of myocytes were measured by using wide-tipped, perfused pipettes. The steady-state Na-K pump current–voltage (*I–V*) relationship was obtained by subtracting the *I–V* relationship determined in the presence of 0.5 mM strophanthidin to inhibit the pump from that determined just before application of strophanthidin. The steady-state Na-Ca exchange outward *I–V* relationship was obtained by subtracting the *I–V* relationship determined at 0 mM $[Ca^{2+}]_o$ from that determined at 1 mM $[Ca^{2+}]_o$. As membrane voltage was depolarized from −100 to +40 mV, the Na-K pump current increased steeply at negative potentials and then saturated at positive potentials, whereas the Na-Ca current increased exponentially. The voltage dependence of Na-K pump exchange current was not altered by the presence of forskolin. However, the increase of Na-Ca exchange current by forskolin (roughly 60% at +20 mV) became more pronounced as membrane voltage was depolarized from −100 to +100 mV. The results indicate that activation of adenylyl cyclase did not alter the voltage dependence of Na-K pump current while it appeared to alter the voltage dependence of Na-Ca exchange current.

1 Introduction

The Na-K pump plays a major role in keeping the intracellular Na ion concentration low, thereby maintaining the large transmembrane Na ion gradient. The Na-Ca exchanger activity depends on that large transmembrane gradient of low intracellular

C.O. Lee (✉)
Laboratory of Cardiac and Membrane Physiology, Rockefeller University, New York, NY, USA
e-mail: leech@mail.rockefeller.edu

K.-S. Soh et al. (eds.), *The Primo Vascular System: Its Role in Cancer and Regeneration*,
DOI 10.1007/978-1-4614-0601-3_37, © Springer Science+Business Media, LLC 2012

and high extracellular Na ion concentrations. In the so-called "forward" mode, the Na-Ca exchanger extrudes Ca ions from cells using the inward driving force of the transmembrane Na ion gradient. Ca extrusion by Na-Ca exchanger in cardiac muscle cells maintains the low sarcoplasmic Ca ion concentration which is essential for normal cellular functions. Thus, Na-Ca exchanger activity is closely related to Na-K pump activity in the regulation of intracellular Ca ion concentration in cardiac myocytes.

Na-K pumps move three Na ions out of and two K ions into a cell at the expense of a single molecule of ATP, thus generating outward electrical current across the cell membrane. In the "reverse" mode, Na-Ca exchangers move three or four Na ions out in exchange for one Ca ion into a cell, thereby also generating outward electrical current across cell membrane. Na-K pump and Na-Ca exchange currents are known to be voltage-dependent [1, 2], and in both cases, the outward currents reflect a net outward movement of Na ion. The characteristic of voltage dependence of Na-K pump current appears different from those of Na-Ca exchange current. As the membrane is depolarized, Na-K pump current is first increased and then saturates, whereas Na-Ca exchange current increases exponentially. This difference in the change of Na-K pump and Na-Ca exchange currents with voltage likely reflects differences in the Na-K pump and Na-Ca exchanger mechanisms.

The β_1 adrenergic receptor plays important roles in regulation of ventricular function. It is well known that activation of β_1 adrenoceptor increases intracellular Ca ion concentration, $[Ca^{2+}]_i$, via G protein–cAMP–PKA (cAMP-dependent protein kinase) pathway which increases open probability of L-type Ca^{2+} channels [3]. However, the effects of β-adrenergic stimulation on Na-K pump current or Na-Ca exchange have been controversial (see Sect. 4). Effects of isoprenaline on the Na-K pump current in guinea pig ventricular myocytes were reported to depend on $[Ca^{2+}]_i$; the Na-K pump current decreasing in a voltage-independent manner at low $[Ca^{2+}]_i$, but increasing at negative, but not positive, voltages at higher $[Ca^{2+}]_i$ [4]. In rat cardiac myocytes, however, isoprenaline increased Na-K pump current at both negative and positive potentials, independent of $[Ca^{2+}]_i$ [5]. Effects of β-adrenergic stimulation on Na-Ca exchange are also controversial. For example, isoprenaline was initially reported to increase Na-Ca exchange current in guinea pig myocytes over the full voltage range tested [6], but later work found no effect on the Na-Ca exchange current in either rabbit or guinea pig myocytes [7, 8].

Our study focused on the characteristics of voltage dependence of Na-K pump and Na-Ca exchange currents without and with activation PKA to test whether activation of PKA could alter their voltage dependence in cardiac myocytes. The present study has two major aims. We first examined the different voltage dependent characteristics of Na-K pump and Na-Ca exchange currents. We then investigated effects of PKA activation on the voltage dependence of Na-K pump and Na-Ca exchange currents. The results have been presented in abstract form [9].

2 Methods

2.1 *Isolation of Myocytes*

Single ventricular myocytes were isolated by collagenase digestion of guinea pig hearts with a Langendorf perfusion system [1]. Guinea pigs of either sex (300–500 g) were anesthetized with pentobarbital (~100 mg/kg, i.p.). The heart was quickly excised, and the aorta cannulated for retrograde coronary perfusion with oxygenated normal Tyrode's solution at 36°C. After about 4 min, the perfusate was switched to a Ca^{2+} free Tyrode's solution, and then to Ca^{2+}-free solution containing 1 mg/mL Sigma Type I collagenase. Following about 10 min collagenase digestion, the enzyme was washed out with Ca^{2+}-free Tyrode's solution for about 4 min. Then the heart was perfused with KB (see below) solution for about 4 min and was cut into small chunks. The cells were filtered and stored at 4°C in KB solution until used.

2.2 *Solutions*

Tyrode's solution used during the myocyte isolation and superfusion contained (mM): 145 NaCl, 5.4 KCl, 1.8 $CaCl_2$, 1 $MgCl_2$, 5.5 glucose, and 5 HEPES (pH 7.4 with NaOH). The modified Ca^{2+}-free Tyrode's solution used in most experiments to establish control conditions contained (mM): 145 NaCl, 5.4 KCl, 1 $MgCl_2$, 2 $BaCl_2$, 1 $CdCl_2$, 5.5 Glucose, and 5 HEPES (pH 7.4 with NaOH). For the experiments of Na-K pump, the Tyrode's solution in the pipette was exchanged for standard pipette solution contained (mM): 50 NaOH, about 76 CsOH, 90 aspartic acid, 20 TEACl, 3 $MgCl_2$, 5.5 dextrose, 10 EGTA, 10 Mg ATP, 5 Tris2-creatine phosphate, 5 pyruvic acid, 10 HEPES, and $CaCl_2$ 4.0 or 5.6 mM (free Ca^{2+} 50 or 100 nM, respectively), and (pH 7.4). For the experiments of Na-Ca exchanger, the Tyrode's solution in the pipette was exchanged for standard pipette solution which contained (mM), 20 NaOH, 8 aspartic acid, 5 pyruvic acid, 29.4 EGTA/NMG, 20.6 CaEGTA/NMG, 20 TEACl, 2 $MgCl_2$, 10 MgATP, 5 Tris-CrP, 0.1 Tris-GTP, 40 HEPES, 5.5 glucose, and pH 7.4. Stock solutions of strophanthidin (0.5 M in DMSO) and forskolin (10 mM in DMSO) were diluted to the desired final concentration immediately before use. KB solution was used to wash out collagenase and to store myocytes: it contained (mM), 0.5 EGTA, 10 KH_2PO_4, 25 KCl, 20 taurine, 70 L-glutamic acid, 20 glucose, and 10 HEPES, adjusted to pH 7.4 with KOH.

2.3 *Whole-Cell Current Recording*

Whole-cell currents were recorded via wide-tipped pipettes with resistance of about 1 M ohm when filled Tyrode's solution as described [1]. After obtaining a giga-ohm seal near the center of the cell by gentle suction, the pipette contents were exchanged for standard pipette solution and the membrane was then ruptured by more vigorous suction. The extracellular Tyrode's solution was switched to modified Tyrode's solution. The holding potential was set to 0 mV to inactivate voltage-gated Na^+ and Ca^{2+} channels. Whole-cell currents were recorded in response to 40 ms voltage pulses to potentials from −100 to +40 or +100 mV in 20 mV steps. The steady-state current level at test potential was determined by averaging points over 25 ms near the end of each clamp pulse. The steady-state current was plotted against the test potential.

3 Results

3.1 Effect of PKA Activation on Voltage Dependence of Na-K Pump Current

Figure 37.1 illustrates the effect of forskolin, the adenylyl cyclase activator, on voltage dependence of the Na-K pump current measured in a ventricular myocyte. Figure 37.1a shows changes of membrane current during exposures of the myocyte to strophanthidin (Stro), the specific Na-K pump inhibitor, and/or forskolin at 0 mV holding potential. The membrane current reached a stable level of about 490 pA within about 1 min after break-in by rupture of the membrane patch. The first vertical line indicated by "a" marks an application of 40 ms voltage step pulses from the 0 mV holding potential to test potentials between −100 and +40 mV. The Na-K pump current was determined as cardiac steroid (Stro)-sensitive current. Application of 0.5 mM Stro after the vertical line "a" caused a rapid decrease of outward holding membrane current. During the application of Stro, another series of 40 ms voltage step pulses was applied as indicated by the vertical line "b." After washing off the Stro, the membrane current returned to the control level it had before Stro application. An application of 5 μM forskolin produced a rapid increase of membrane current, partly due to activation of cardiac CFTR chloride channels (e.g., [8]), and 40 ms voltage step pulses were applied as indicated by the vertical line "c." Then application of 0.5 mM Stro caused a rapid decrease of membrane current in the presence of forskolin. During the application of Stro, voltage step pulses were applied as indicated by the vertical line "d." After washing off Stro, the membrane current returned to a level similar to that before forskolin application. After removal of forskolin, the membrane current eventually (in 2–3 min) recovered to the level before forskolin application. After the recovery of membrane current, voltage step pulses were applied in the absence and presence of Stro as indicated by vertical lines "e" and "f," respectively.

Figure 37.1b shows plots of membrane current (I) against membrane voltage (V) which were obtained from the experiment like Fig. 37.1a. Figure 37.1b shows I–V relationships averaged from five myocytes. As membrane voltage was depolarized, the Na-K pump currents were increased in the negative voltage range and then saturated in the positive voltage range, both in the absence (filled circles) and

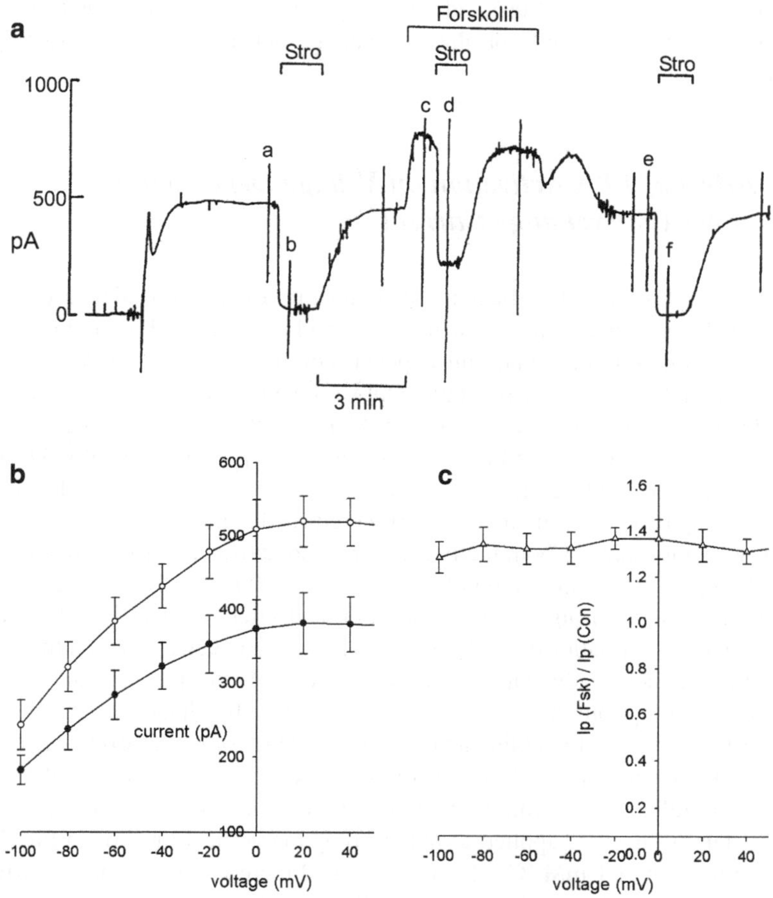

Fig. 37.1 Effect of forskolin on Na-K pump current. (**a**) Chart record of changes in whole-cell current at 0 mV holding potential was obtained from a myocyte. *Bars* above the current record indicate periods of myocyte exposure to 0.5 mM strophanthidin (Stro) and 5 μM forskolin. The *vertical lines* labeled a, b, c, d, e, and f reflect the application of voltage step pulses (−100 to +40 mV in 20 mV steps) to obtain current–voltage data shown in (**b**). (**b**) Voltage dependence of averaged Na-K pump currents: effect of forskolin on current–voltage relationship of Na-K pump. The averaged amplitude of the Na-K pump currents, from five myocytes, in the presence (*open circles*) and absence (*filled circles*) of forskolin is plotted against voltage. (**c**) The averaged ratio, from five myocytes, of the Na-K pump current in the presence of forskolin to that in the absence of forskolin is plotted against voltage

presence (open circles) of forskolin. The *I–V* relationships show that forskolin increased the strophanthidin-sensitive Na-K pump current over the full voltage range tested. In five myocytes examined with 100 nM $[Ca^{2+}]_i$, forskolin increased the strophanthidin-sensitive current by 35 + 7% (mean + SE) at the membrane voltage of 0 mV. Figure 37.1c shows that the ratios of strophanthidin-sensitive currents in the presence (open circles) and in the absence (filled circles) of forskolin at each test potential, Ip (Fsk)/Ip (Con), did not change over the voltage range tested. This indicates that forskolin did not alter the voltage dependence of the Na-K pump currents.

3.2 *Effect of PKA Activation on Voltage Dependence of Na-Ca Exchange Current*

Figure 37.2 illustrates the effect of forskolin on voltage dependence of reverse Na-Ca exchange current measured in a ventricular myocyte. Figure 37.2a shows changes of membrane current during exposures of the myocyte to 1 mM $[Ca^{2+}]_0$ and/or forskolin at 0 mV holding potential. Outward Na-Ca exchange current was activated by first loading the cell with 20 mM intracellular Na^+ and then suddenly increasing $[Ca^{2+}]_0$. As seen in Fig. 37.2a, raising $[Ca^{2+}]_0$ from nominally 0 to 1 mM rapidly increased holding membrane current to a new stable outward level. After washing away $[Ca^{2+}]_0$, the membrane current returned to the control level. A steady-state *I*–*V* relationship for both the background and activated currents was obtained by applying 40 ms voltage step pulses (−100 to +100 mV in 20 mV steps) before (vertical line "a"), during (vertical line "b"), and after (vertical line "c") raising $[Ca^{2+}]_0$. The *I*–*V* relationship of the $[Ca^{2+}]_0$-sensitive current was obtained by subtracting the net steady-state current (an average of currents "a" and "c") in the absence of $[Ca^{2+}]_0$ from that ("b") in its presence at each voltage.

To test the effect of forskolin on voltage dependence of Na-Ca exchange current, 5 μM forskolin was applied. Its application produced an increase of membrane current, largely reflecting activation of chloride currents (as in Fig. 37.1a), and 40 ms voltage step pulses were applied as indicated by the vertical line "d" in Fig. 37.2a. Then application of 1 mM $[Ca^{2+}]_0$ caused a rapid increase of membrane current in the presence of forskolin. During the application of $[Ca^{2+}]_0$, voltage pulses were applied as indicated by the vertical line "e." After washing off the $[Ca^{2+}]_0$, another set of voltage pulses was applied as indicated by "f." After removal of forskolin, membrane current recovered to the level before forskolin application. The *I*–*V* relationship of the $[Ca^{2+}]_0$-activated current was obtained by subtracting the net steady-state current (an average of currents "d" and "f") in the absence of $[Ca^{2+}]_0$ from that ("e") in its presence at each voltage.

Figure 37.2b shows the *I*–*V* relationships averaged from five myocytes. It shows that the $[Ca^{2+}]_0$-activated currents in the absence (filled circles) and the presence (open circles) of forskolin were all outward and increased steadily as the membrane potential was made more positive. These properties of the steady-state $[Ca^{2+}]_0$-activated current are consistent with its being entirely generated by the Ca influx-Na efflux mode of the Na-Ca exchanger [2, 6–8]. The $[Ca^{2+}]_0$-sensitive Na-Ca exchange currents increased exponentially as membrane potential was depolarized from −100 to +100 mV. The results of Fig. 37.2b show that the increase of the Na-Ca exchange current by forskolin appeared to become more pronounced as membrane potential was depolarized from −100 to +100 mV. The Na-Ca exchange current was increased

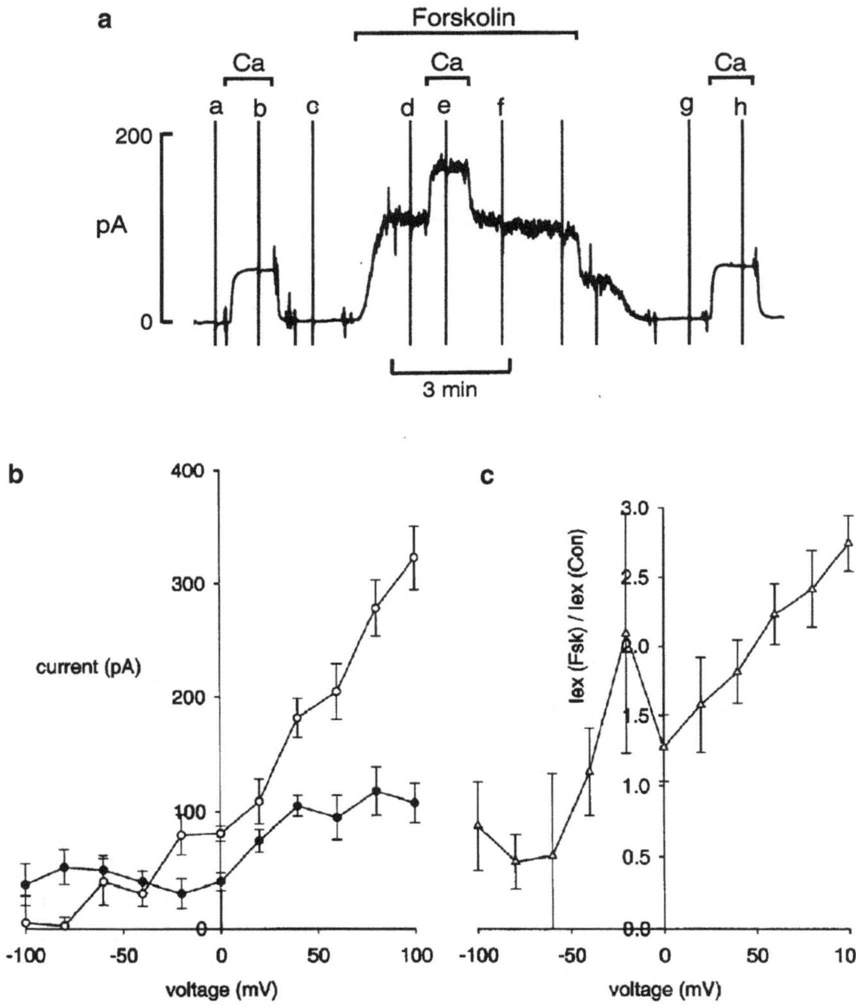

Fig. 37.2 Effect of forskolin on outward Ca-activated Na-Ca exchange current. (**a**) Chart record of changes in whole-cell current from a myocyte at 0 mV holding potential. *Bars* above the current record indicate periods of myocyte exposure to Ca^{2+} (1 mM) and forskolin (5 μM). The concentration of Ca^{2+} inside the myocyte (pipette) was 50 nM. The *vertical lines* labeled a, b, c, d, e, f, g, and h show the timing of application of voltage step pulses. Note that the whole-cell current increase by Ca application before exposure to forskolin looks similar to that after recovery from exposure to forskolin. (**b**) The averaged amplitude of outward Na-Ca exchange currents from five myocytes in the presence (*open circles*) and absence (*filled circles*) of forskolin is plotted against voltage. (**c**) The ratios of outward Na-Ca exchange currents in the presence, Iex (Fsk), and absence, Iex (Con) of forskolin were averaged over the five myocytes and were plotted against voltage

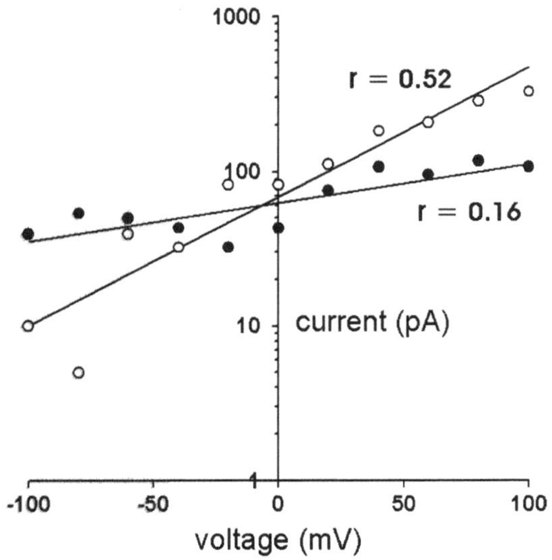

Fig. 37.3 Semilogarithmic plots of averaged Na-Ca exchange currents, taken from Fig. 37.2b, against membrane voltage in the absence (*filled circles*) and presence (*open circles*) of forskolin. The slope of a current–voltage relation is represented by its *r* value

by roughly 50% at +20 mV, but was more than doubled at +100 mV. Figure 37.2c shows the average ratios of the Na-Ca exchange currents in the presence, Iex (Fsk), and in the absence, Iex (Con), of forskolin. The ratios increased as membrane potential was depolarized from −100 to +100 mV. This is in contrast to the unchanged ratios of the Na-K pump currents (Fig. 37.1c). This increase in the ratios of the Na-Ca exchange currents implies that forskolin altered the voltage dependence of Na-Ca exchange current.

Figure 37.3 shows semilogarithmic plots of the averaged Na-Ca exchange currents taken from Fig. 37.2b against membrane voltage in the absence (filled circles) and the presence (open circles) of forskolin. The slopes (*r*) of the current–voltage relationships represent the steepness of voltage dependence of the Na-Ca exchange currents. The alteration of voltage dependence is indicated by the increase of the slope (*r*) of the current–voltage relationship which was caused by forskolin (see Sect. 4).

4 Discussion

4.1 Working Relation Between Na-K Pump and Na-Ca Exchange

The Na-K pump generates a large gradient of Na ions across the cell membrane. The potential energy of the Na ion gradient is essential for many cellular functions such as action potentials, neuromuscular transmission, synaptic transmission of neurons, regulation of cell volume, and membrane transport of other ions and organic molecules. In addition to maintaining the Na ion concentration gradient, the Na-K pump generates outward electrical current across the cell membrane by steady exchange of three cytosolic Na ions for two extracellular K ions [10, 11]. The electrical current generated by the Na-K pump influences the cell membrane voltage. Thus, the Na-K pump plays important roles in the generation of chemical and electrical gradients across the cell membrane.

The large electrochemical gradient of Na ion across cell membrane provides the energy source for membrane transport of other ions and organic molecules. Such membrane transport includes Na-Ca exchange, Na-H exchange, Na-glucose transport, and Na-amino acid transport. The Na-Ca exchange activity depends on the transmembrane electrochemical gradient of Na ion. The Na-Ca exchange can operate in either forward mode or reverse mode. In the forward mode, the Na-Ca exchanger moves three Na ions into and moves one Ca ion out across the cell membrane. This mode generates net inward current and lowers cytosolic Ca ion concentration. In the reverse mode, the exchanger moves three Na ions out of and one Ca ion into a cell. This mode generates net outward current and loads cells with Ca ions. In the resting state of cardiac muscle cells, the large transmembrane Na gradient generated by the Na-K pump favors the forward mode which maintains the normal low intracellular Ca ion concentration. Therefore, the interplay of Na-K pump and Na-Ca exchanger activity is important for regulation of intracellular Ca ion concentration. It has been shown that the interaction of Na-K pump and Na-Ca exchange is the basis for the mechanism of digitalis action in cardiac muscle [12].

4.2 Voltage Dependence of Na-K Pump and Na-Ca Exchange Currents

Both Na-K pump and Na-Ca exchange activity generate electrical currents across the cell membrane [13–17]. Our experiments were designed so that Na-K pumps and Na-Ca exchangers would both generate outward currents, whose current–voltage (*I–V*) relationships could be compared.

The current–voltage (*I–V*) relationships from −100 to +40 mV (Fig. 37.1) show that the Na-K pump current first increased steeply in the negative voltage range, and then became saturated and showed little change at positive membrane voltages. This confirms previous reports (e.g., [1]). Such voltage dependence may have some electrophysiological significance. Thus, during the plateau phase of the cardiac action potential, following Na entry through voltage-gated channels, the Na-K pumps would be maximally activated. The steepness of the pump current increase in the negative voltage range was smaller at normal physiological intracellular Na ion concentrations (5–10 mM), so that changes in membrane potential might have weaker effects on pump current amplitude [1].

Information on the mechanisms of Na-K pump and Na-Ca exchanger can be gained by studying their *I–V* relationships [1, 2]. This study compared the voltage dependence of the Na-K pump current with that of Na-Ca exchange current. To isolate the Na-Ca exchange current, we pharmacologically blocked other currents likely to interfere with the measurements. After first loading the cell with 20 mM Na^+, outward Na-Ca exchange current was activated by suddenly increasing external Ca ion concentration, $[Ca^{2+}]_o$. The $[Ca^{2+}]_o$-activated current was entirely outward (positive) at all membrane voltages. The outward direction of this current and its approximately exponential increase in magnitude as the membrane voltage was made more positive are consistent with its being entirely generated by the Ca influx/ Na efflux mode (reverse mode) of the Na-Ca exchange, as previously reported [2]. The influence of membrane potential on outward Na-Ca exchange current and on outward Na-K pump current therefore seems different both in the negative range of membrane voltages and in the positive voltage range. In the negative voltage range, changes in membrane potential seem to affect the Na-Ca exchange current little, whereas they have large effects on the Na-K pump current. In contrast, in the positive voltage range, changes in membrane potential have a large effect on the Na-Ca exchange current, whereas they have little effect on the Na-K pump current.

The exponential function of the *I–V* relation of outward Na-Ca exchange can be represented by the equation [2, 17]:

$$I_{\text{Na-Ca}} = a\exp(r\ \text{EF/RT})$$

where a scales the exchange current magnitude and r is the steepness factor of the voltage dependence of the exchange current determined as the slope of a semilogarithmic plot of the voltage dependence of the exchange current (Fig. 37.3). In this study, the r value obtained from the averaged data was roughly 0.2 under the control condition with 20 mM $[Na^+]_i$, 150 mM $[Na^+]_o$, 50 nM $[Ca^{2+}]_i$, and 1 mM $[Ca^{2+}]_o$, comparable to the r value of about 0.3 reported previously under similar conditions [2]. The magnitude of the outward Na-Ca exchange current in guinea pig ventricular myocytes depends on $[Na^+]_i$ (pipette Na^+ concentration) and is half-maximally activated at about 20 mM $[Na^+]_i$ (M. Noda, R.N. Shepherd, M. Nakao, D.C. Gadsby, unpublished data). At normal physiological $[Na^+]_i$ (5–10 mM), the outward Na-Ca exchange current would therefore appear to vary less with changes in membrane potential.

4.3 *Effects of PKA Activation on Voltage Dependence of Na-K Pump and Na-Ca Exchange Currents*

Effects of PKA activation on Na-K pump and Na-Ca exchanger have been controversial. Gao et al. [4] found that, in guinea pig ventricular myocytes at low $[Ca^{2+}]_i$ (15 nM), the Na-K pump current was decreased by isoprenaline in a nonvoltage-dependent manner; but, at high $[Ca^{2+}]_i$ (1.4 μM), isoprenaline shifted the Na-K pump *I–V* relationship to more negative potentials, increasing the pump current at negative voltages while leaving the pump current unchanged at positive voltages. However, Kockskamper et al. [18] reported that the Na-K pump current in guinea pig ventricular myocytes was increased by forskolin at nanomolar (15 nM) and subnanomolar (0.013 nM) $[Ca^{2+}]_i$, although voltage dependence was not examined. Dobretsov et al. [5] reported that isoprenaline increased Na-K pump current in rat myocytes with no effect on the *I–V* curve shape, and in a manner that was not $[Ca^{2+}]_i$-dependent.

In this study, the ratios of the Na-K pump currents in the presence and absence of forskolin Ip (Fsk)/Ip (Con) were not voltage-dependent, indicating that PKA activation did not alter the Na-K pump's voltage dependence. A simple possible interpretation is that PKA activation may result in an increase in the number of active Na-K pumps in the cell membrane without changing their functional characteristics (but cf. [5]). How might PKA activation increase the number of functioning Na-K pumps? β-Adrenergic stimulation of the Na-K pump has been proposed to be mediated by phosphorylation of phospholemman, a member of the FXYD family of proteins that includes the Na-K pump γ-subunit [19]. It is possible, therefore, that phospholemman phosphorylation may somehow activate inactive Na-K pumps.

Effects of PKA activation on the Na-Ca exchanger have been similarly disparate and hence controversial, with stimulation of the Na-Ca exchanger, or inhibition, or no effect being reported in different studies. Increases in Na-Ca exchanger current in response to isoprenaline or forskolin in ventricular myocytes from guinea pig [6] or from pig [20] were interpreted as reflecting stimulation of the phosphorylation of Na-Ca exchange protein by PKA. Similarly, PKA activation by forskolin in *Xenopus* oocytes expressing the cardiac isoform of the rat Na-Ca exchanger increased the current attributed to Na-Ca exchange [21]. On the other hand, β-adrenergic stimulation by isoprenaline was reported to not significantly affect Na-Ca exchange current in guinea pig, mouse, or rat ventricular myocytes [8]. In addition, Na-Ca exchange current in rabbit cardiac myocytes was not altered by PKA activation [7]. Also, activation of PKA by 8-Br-cAMP had no effect on Na-Ca exchange current in human embryonic kidney cells expressing dog Na-Ca exchanger, but inhibited the frog Na-Ca exchanger [22]. However, in HEK293 cells coexpressing the rat cardiac Na-Ca exchanger with phospholemman, phosphorylation of the phospholemman by PKA upon exposure to forskolin decreased the exchange current [23].

The reasons for these discrepant findings are not known. One difficulty concerns the methods for isolating the Na-Ca exchanger current from other components of whole-cell current. Because there is no readily available specific inhibitor of the Na-Ca exchanger (in contrast to the cardiac steroids for inhibition of the Na-K pump), complications from nonspecificity of the methods used (in several cases, application of millimolar Ni^{2+}) might have influenced interpretation of some of the experimental results.

We found the shape of the *I–V* relationship of outward Na-Ca exchange current to be altered in the presence of forskolin. The increase of outward Na-Ca exchange current by forskolin became more pronounced as membrane voltage was made more positive (Fig. 37.2). Thus, the slope of the *I–V* relationship of Na-Ca exchange current in a semilogarithmic plot increased from ~0.2 to ~0.5 (Fig. 37.3) during PKA activation by forskolin. These results indicate that the increase of Na-K pump current by PKA activation is voltage-independent, whereas the increase of outward

Na-Ca exchange current by PKA activation is voltage-dependent. The latter suggests that PKA acts by altering the functional characteristics of active Na-Ca exchangers, but whether the Na-Ca exchanger itself is phosphorylated by PKA, or some interacting molecule is phosphorylated by PKA, is still not clear.

Acknowledgments This work was supported by National Institutes of Health grant HL36783. We thank Peter Hoff for technical assistance.

References

1. Gadsby DC, Nakao M (1989) Steady state current-voltage relationship of the N^+-K^+ pump in guinea-pig ventricular myocytes. J Gen Physiol 94:511–537
2. Kimura J, Miyamae S, Noma A (1987) Identification of sodium-calcium exchange current in single ventricular cells of guinea-pig. J Physiol 348:199–222
3. McDonald TF, Pelzer S, Trautwein W, Pelzer DJ (1994) Regulation and modulation of calcium channels in cardiac, skeletal and smooth muscle cells. Physiol Rev 74:365–507
4. Gao J, Mathias RT, Cohen IS, Shi J, Baldo GJ (1996) The effects of β-stimulation on the Na+-K+ pump current-voltage relationship in guinea-pig ventricular myocytes. J Physiol 494: 697–708
5. Dobretsov M, Hastings SL, Stimers JR (1998) Na^+-K^+ pump cycle during β-adrenergic stimulation of adult rat cardiac myocytes. J Physiol 507:527–539
6. Perchenet L, Hinde AK, Patel KCR, Hancox JC, Levi AJ (2000) Stimulation of Na/Ca exchange by the β-adrenergic/protein kinase A pathway in guinea-pig ventricular myocytes at 37 C. Pflugers Arch 439:822–828
7. Ginsberg KS, Bers DM (2005) Isoproterenol does not enhance Ca-dependent Na/Ca exchange current in intact rabbit ventricular myocytes. J Mol Cell Cardiol 39:972–981
8. Lin X, Jo H, Sakakibara Y, Tambara K, Kim B, Komeda M, Matsuoka S (2006) β–Adrenergic simulation does not activate Na+/Ca^{2+} exchange current in guinea pig, mouse, and rat ventricular myocytes. Am J Physiol Cell Physiol 290:C601–C608
9. Lee CO, Gadsby DC (2007) The nature of voltage dependence of Na-K pump and Na-Ca exchange current in cardiac myocytes. Biophysical J 51:461a
10. Rakowski RF, Gadsby DC, De Weer P (1989) Stoichiometry and voltage dependence of the sodium pump in voltage-clamped, internally dialyzed squid axon. J Gen Physiol 93:903–941
11. Rakowski RF, Gadsby DC, De Weer P (1997) Voltage dependence of the Na/K pump. J Membr Biol 155:105–112
12. Lee CO (1985) 200 Years of digitalis: the emerging central role of the sodium ion in the control contractile force in cardiac muscle. Am J Physiol Cell Physiol 249:C367–C378
13. De Weer P, Gadsby DC, Rakowski RF (1988) Voltage dependence of Na/K pump. Annu Rev Physiol 50:225–241
14. Lauger P, Apell HJ (1986) A microscopic model for the current-voltage behavior of the Na/K pump. Eur Biophys J 13:309–321
15. Kimura J, Norma A, Irisawa H (1986) Na-Ca exchange current in mammalian heart cells. Nature 319:596–597
16. Kimura J, Miyamae S, Norma A (1987) Identification of sodium-calcium exchange current in single ventricular cells of guinea-pig. J Physiol 384:199–222
17. Noble D (1986) Sodium-calcium exchange and its role in generating electrical current. In: Nathan RD (ed) Cardiac muscle: the regulation of excitation and contraction. Academic Press, Orlando, FL, pp 171–200

18. Kockskamper J, Erlenkamp S, Glitsch HG (2000) Activation of the cAMP-protein kinase A pathway facilitates Na^+ translocation by the Na^+-K^+ pump in guinea-pig ventricular myocytes. J Physiol 532:561–574
19. Despa S, Bossuyt J, Han F, Ginsberg KS, Jia LG, Kutchai H, Tucker AL, Bers DM (2005) Phosphoslemman-phosphorylation mediates the β-adrenergic effects on Na/K pump function in cardiac myocytes. Circ Res 97:252–259
20. Wei SK, Ruknudin A, Hanlon SU, McCurley JM, Schulze DH, Haigney MC (2003) Protein kinase A hyperphosphorylation increases basal current but decreases β–adrenergic responsiveness of the sarcolemmal Na^+-Ca^{2+} exchanger in failing pig myocytes. Circ Res 92:897–903
21. Ruknudin A, He S, Lederer WJ, Schulze DH (2000) Functional differences between cardiac and renal isoforms of the rat Na^+-Ca^{2+} exchanger NCX1 expressed in *Xenopus* oocytes. J Physiol 529:599–610
22. He LP, Cleemann L, Soldatov NM, Morad M (2003) Molecular determinants of cAMP-mediated regulation of the Na^+-Ca^{2+} exchanger expressed in human cell lines. J Physiol 548: 677–689
23. Zhang XQ, Ahlers BA, Tucker AL, Song J, Wang J, Moorman JR, Mouney JP, Carl LL, Rothblum L, Cheung JY (2006) Phospholemman inhibition of the cardiac Na^+/Ca^{2+} exchanger. J Biol Chem 281:7784–7792

Chapter 38
Bioimaging of Stem Cells, Live Tissue, and Whole Animals Using Diversity-Oriented Fluorescence Library Approach

Young-Tae Chang

Abstract Embryonic stem cells (ESC) and induced pluripotent stem cells (iPSC) hold a tremendous potential for biomedical research. However, their application in areas such as regenerative medicine or drug discovery is hindered by their heterogeneity and differentiation. Hence, the development of selective probes for the detection and isolation of stem cells is of great interest. We have employed combinatorial chemistry to develop several diversity-oriented fluorescence libraries and successfully applied them for ESC and iPSC imaging probe, compound of designation yellow 1 (CDy1). In further characterization of the fluorescent probe in iPSC, we found CDy1 can detect the iPS-forming cells much earlier than Oct4-GFP (green fluorescent protein) reporter. Thus, this new tool will allow us to study the early-stage events of the cell destiny changes from somatic cell to iPS. Considering the possible connection of primo vascular system (PVS) to stem cells, CDy1 may serve as a new probe for PVS study.

1 Introduction

There has been increasing interest in the use of ESC and iPSC in biomedical research and clinical therapy. The recent success of induced pluripotent stem cell (iPSC) generation using elderly amyotrophic lateral sclerosis patient's skin fibroblasts and its differentiation into motor neurons exemplifies how stem cells can be used for the treatment of specific individual patients with complex diseases. However, despite the general enthusiasm about the multiple applications of stem cells, their practical use both in research and disease therapy has been hampered by the heterogeneity of

Y.-T. Chang (✉)
Department of Chemistry and Medicinal Chemistry Program, National University of Singapore, 3 Science Drive 3, Singapore, Singapore 117543
e-mail: chmcyt@nus.edu.sg

K.-S. Soh et al. (eds.), *The Primo Vascular System: Its Role in Cancer and Regeneration*, DOI 10.1007/978-1-4614-0601-3_38,

stem cells and their unpredictable proliferation and differentiation. Therefore, the development of tools and technologies that may facilitate the isolation, identification, and characterization of stem cells is one of the most demanding requisites in the field of stem cell research and applications. We have employed combinatorial chemistry to develop several diversity-oriented fluorescence libraries (DOFL) and successfully applied them for the discovery of bioimaging probes for a number of targets such as muscle cells, β-amyloid plaque, DNA, RNA, GTP, human serum albumin, chymotrypsin, glutathione, and heparin as described previously [1]. Recently, this approach was applied to identify pancreatic alpha cells and demonstrated that the small molecule probe is working as well as antibody for specific staining in live cells. We applied the same approach to selectively stain embryonic stem cell (ESC) and iPSC, and the result is described here.

2 Results and Discussion

Among our DOFL we screened the rosamine library to discover a novel fluorescent compound that selectively stains ESC and iPSC. We named this compound as "compound of designation yellow 1" (CDy1, $\lambda_{ex}/\lambda_{em}$ = 535/570 nm) (Fig. 38.1) [2].

For high-throughput screening, we incubated mouse ESC (mESC) and mouse embryonic fibroblast (MEF) feeder cells with 280 rosamine compounds at a concentration of 500 nM in 384-well microplates. After 0.5, 24, and 48 h, tetramethylrhodamine isothiocyanate (TRITC) fluorescence and bright-field images were taken using an ImageXpress^{MICRO} (Molecular Devices) imaging system. From this primary screening, 20 compounds that stained mESC consistently with stronger intensity than MEF were manually selected. As a secondary screening, we incubated mESC and MEF separately with each of the hit compounds and analyzed them using flow cytometry. CDy1 was identified as the most selective compound for mESCs among the 20 hits.

Having found that CDy1 selectively stains ESC, we applied the dye to iPSC which was generated from MEF of transgenic mice that express green fluorescent protein (GFP) under the control of the Oct4 promoter. The reprogramming was performed in a 6-well culture dish by retroviral introduction of four transcriptions factors, Oct4, Sox2, Klf4, and c-Myc, and iPSC generation was verified by GFP expression, an alkaline phosphatase assay, and immunostaining of SSEA-1 at 17 days postinfection (dpi). We found CDy1 also selectively stains the iPSC colony. When the 155 colonies grown in a 6-well plate cells were treated with CDy1 at 17 dpi, 101 colonies (65%) were both CDy1- and GFP-positive, 26 (17%) were CDy1-only positive, 4 (3%) were GFP-only positive, and 24 (15%) were negative for both CDy1 and GFP, despite the fact that the morphology of the colonies was indistinguishable.

In a cell culture treated with CDy1 at an earlier time point of iPSC generation (10 dpi), an increasing number of CDy1-stained colonies started to show a GFP signal. To perform a more systematic analysis, we stained the cells with CDy1 at 2 dpi, when iPSC was not distinguishable by any means and tracked the CDy1 and GFP

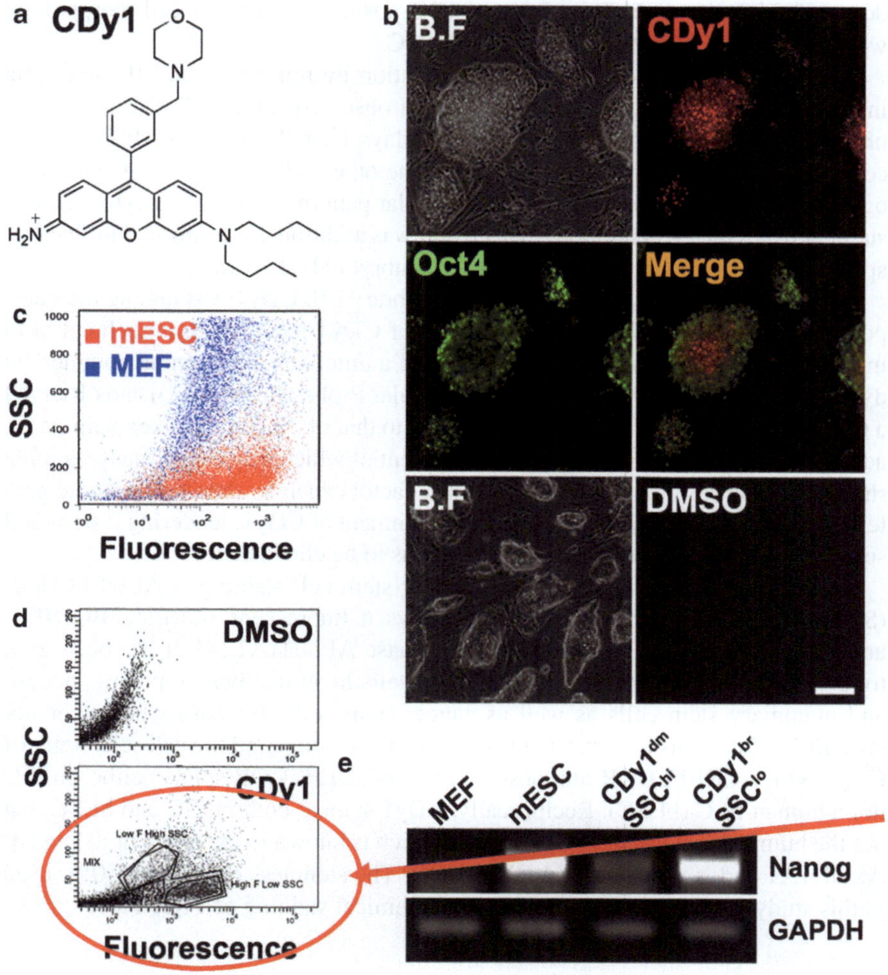

Fig. 38.1 Selective staining of mouse ESC (mESC) by compound of designation yellow 1 (CDy1). (**a**) Chemical structure of CDy1. (**b**) *Top and middle*: CDy1-stained mESC were immunostained with anti-Oct4 antibody, *bottom*: DMSO was used as a negative control. *B.F* Bright field. *Scale bar*, 100 μm. (**c**) Flow cytometry dot plot image of CDy1-stained mESC and mouse embryonic fibroblast (MEF). Images of pure cell populations were overlaid. (**d**) Flow cytometry dot plot image of mESC and MEF mixture. *Top*: mESC and MEF mixed cells incubated with DMSO, *bottom*: mESC and MEF mixed cells incubated with CDy1. (**e**) Nanog expression analysis using RT-PCR. SSC^{low} $CDy1^{bright}$ and SSC^{high} $CDy1^{dim}$ cells were sorted from mESC and MEF mixture after CDy1 staining (reproduced with permission from Im et al. [2]. Copyright 2010 Wiley)

signals by daily acquisition of cell images using an ImageXpress^{MICRO} system. At 10 dpi, when small colonies started to appear, we selected 342 CDy1-positive colonies that were not expressing GFP and monitored their GFP expression up to 25 dpi. More and more colonies started to express GFP during this period, and 338 (99%) out of the 342 tracked colonies were GFP-positive at 25 dpi. During this period, any detectable differences in the number of GFP-positive colonies or cell morphology were not observed compared to untreated iPSC.

In contrast, we induced mESC differentiation by removing the LIF (leukemia inhibitory factor) from the culture media and observed some cells that were morphologically distinguishable from mESC 3 days later. Most of the differentiated cells were not stained by CDy1, whereas some other cells having mESC morphology were stained by the dye, and showed similar patterns for immuno-cytochemical staining using the Oct4 antibody. This result was additionally confirmed in lineage-specific cells differentiated from mESC by embryoid body formation.

In our previous study with the rosamine library [3], CDy1 was among the compounds targeting mitochondria. To examine if CDy1 localizes in mitochondria in mESC, we costained the cells with CDy1 and a mitochondria-staining commercial dye (MitoTracker Deep Red 633 from Molecular Probes (Invitrogen)) and observed a CDy1 staining pattern that was very similar to that of the MitoTracker staining. In addition to the mitochondrial membrane potential which sequesters many cationic rhodamine and rosamine compounds, other factors such as stem-cell-specific proteins appear to play roles in the entry and retainment of CDy1, rendering it stem-cell selective. A more detailed mechanism remains to be elucidated.

Among the few fluorescent dyes used for stem-cell staining is ALDEFLUOR (STEMCELL Technologies), which employs a fluorescent substrate BODIPY-aminoacetaldehyde for aldehyde dehydrogenase ALDH1A1 [4]. It has been used to identify and isolate certain types of stem cells including hematopoietic, neural, and mammary stem cells as well as cancer stem cells. Because whether or not ALDEFLUOR stains ESC has not been known, we compared the cell selectivity of CDy1 with ALDEFLUOR and observed that ALDEFLUOR stains neither mESC nor a human ESC (hESC). Reciprocally, CDy1 stained both mESC and hESC, but not the human lung cancer cell line H522 which is known to express a high level of ALDH1A1 and is stained by ALDEFLUOR. The stemness of hESC BG01V used in this study was verified by immuno-cytochemical staining of TRA-1-60.

3 Conclusions

We have developed a novel bioimaging probe, CDy1, for ESC and iPSC detection. The experimental results presented herein strongly demonstrate that CDy1 can be used for the identification and isolation of live ESC and iPSC without the aid of a genetic reporter system at an earlier stage of the reprogramming and during the ESC differentiation. To our knowledge, this is the first ESC- or iPSC-selective fluorescent probe that has been reported [2]. As primo vascular system (PVS) may have some functional connection to stem cell systems, this new probe, CDy1, would be a useful tool not only for stem-cell research, but potentially also for PVS.

Acknowledgments This work was supported by a Young Investigator Award (R-143-000-353-101) granted to Y.-T.C. from the National University of Singapore and intramural funding from the A*STAR Biomedical Research Council.

References

1. Lee JS, Kim YK, Vendrell M et al (2009) Diversity-oriented fluorescence library approach for discovery of sensors and probe. Mol Biosyst 5:411–421
2. Im CN, Kang NY, Ha HH et al (2010) A fluorescent rosamine compound selectively stains pluripotent stem cells. Angew Chem Int Ed Engl 49:7497–7500
3. Kim YK, Ha HH, Lee JS et al (2010) Control of muscle differentiation by a mitochondria-targeted fluorophore. J Am Chem Soc 132:576–579
4. Storms RW, Trujillo AP, Springer JB et al (1999) Isolation of primitive human hematopoietic progenitors on the basis of aldehyde dehydrogenase activity. Proc Natl Acad Sci USA 96:9118–9123

Chapter 39
The Clinical Application of Optical Spectroscopy in Monitoring Tissue Oxygen Supply Following Cancer Treatment

David K. Harrison

Abstract This paper describes a number of clinical applications of non-invasive visible wavelength lightguide spectroscopy (VLS) for monitoring tissue oxygen saturation (SO_2) in patients following (or in one case during) cancer treatment. The first study involved measurements on patients receiving standard fractionated radiotherapy to the breast for conservation therapy of breast carcinoma. The aim was to investigate whether VLS could detect changes in breast skin early on in the course of radiotherapy treatment that might be predictive of acute or delayed severe skin reactions. In the second study described here, VLS was applied to the continuous monitoring of transverse rectus abdominis myocutaneous flap tissue oxygenation for the first 72 h post-surgery in order to investigate whether changes in tissue SO_2 can be used to warn of possible vascular occlusion. In the third study, we applied the technique peri-operatively, for the measurement of colonic mucosal and serosal tissue oxygenation in order to attempt to define safe margins for colon resection in patients with bowel cancer. In all of the above studies, VLS has proved to be a robust and sensitive method for measuring tissue oxygenation. It may, therefore, prove to be a valuable tool for investigating the role of the primo-vascular system in studies of tumour pathophysiology.

1 Introduction

In the early 1990s, we developed the technique of visible wavelength lightguide spectroscopy (VLS) to predict flap viability following lower limb amputation in critical limb ischaemia [1]. At the same time, we used VLS to study tissue oxygen supply in inflammatory reactions induced experimentally in forearm skin [2].

D.K. Harrison (✉)
Institute of Cellular Medicine, Newcastle University, Newcastle upon Tyne NE1 7RU, UK

Heilig-Kreuz-Strasse 19, 39030 St Lorenzen, Italy
e-mail: d.k.harrison@ncl.ac.uk

K.-S. Soh et al. (eds.), *The Primo Vascular System: Its Role in Cancer and Regeneration*, DOI 10.1007/978-1-4614-0601-3_39, © Springer Science+Business Media, LLC 2012

Briefly, VLS makes use of a light source transmitted along an optical fibre, or emitted by a light-emitting diode (LED), mounted in an optode at the surface of the tissue being interrogated. A receiving lightguide fibre (or fibre bundle), also mounted within the optode, captures the back-scattered light and transmits it to a spectrophotometer. The absorption spectrum of the remitted light is analysed and the tissue SO_2 derived from an algorithm that makes use of the characteristics of fully oxygenated and deoxygenated haemoglobin spectra [3].

Although we have not applied VLS directly to the study of tumour oxygenation, it has been used in a number of studies related to the consequences of cancer treatment. The results of these studies, although published separately in detail elsewhere [4–7], will be summarised here in the context of the potential application of VLS to investigate the role of the primo-vascular system in cancer.

2 Breast Radiotherapy

A range of acute skin reactions, ranging from mild erythema to moist desquamation, can be seen in patients receiving standard fractionated radiotherapy to the breast for conservation therapy of breast carcinoma. In a number of cases, these reactions can cause considerable discomfort and seriously affect the patient's quality of life [8]. The application of VLS in cancer patients was, therefore, to investigate whether VLS could detect changes in breast skin early on in the course of radiotherapy treatment and to what extent these early changes may be able to predict the later occurrence of severe acute or delayed reactions.

An EMPHO II micro-lightguide spectrophotometer (BGT, Überlingen, Germany) was used to measure the SO_2 profiles in the skin along a central line of the treated breast from lateral to medial and proximal to distal within the marked treatment area. Histograms, consisting of 100 measurement points, were also recorded centrally at the medial and lateral extremities of the treatment area.

Radiation treatment consisted of a dose of 45 Gy administered in 20 fractions over 4 weeks. Measurements were carried out in ten patients prior to the first fraction, after 5, 10 and 20 fractions and, in eight patients, after 3 months.

Mean SO_2 values (±Standard Deviation, SD) for (a) areolar, lateral and medial areas, and (b) the lateral-medial and proximal-distal measurements on the treated breast prior to treatment and after 5, 10, 20 fractions and 3 months are shown in Fig. 39.1. Figure 39.1a shows the mean values from all SO_2 values measured, representing about 1,000 measurement points at the medial and lateral sites and over 200 in the areolar areas. Figure 39.1b shows the mean values for the lateral-to-medial and proximal-to-distal SO_2 measurements. Figure 39.1a shows that the SO_2 values in the areolar and lateral areas were only significantly increased after 20 fractions (week 4). However, the SO_2 values at the medial site increased significantly after 5 fractions (week 1), were not significantly different than control at 10 fractions (week 2), but were further increased after 20 fractions (week 4), and were even higher after 3 months (week 13). Significant increases were observed in the lateral-to-medial and proximal-to-distal SO_2 measurements at all stages of treatment. While starting to fall

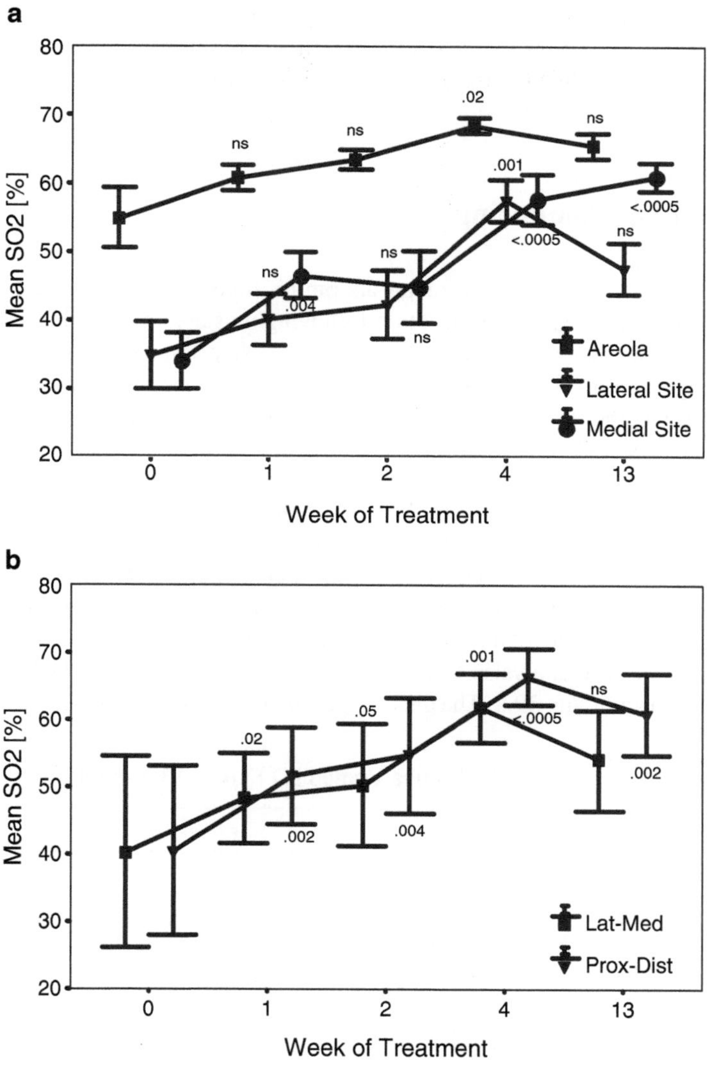

Fig. 39.1 Mean SO_2 values (±SD) for (**a**) areolar, lateral and medial areas, and (**b**) the lateral-medial and proximal-distal measurements on the treated breast at each stage of the study. Note that to avoid superimposition, all but the lateral site data points are offset from their true *x*-axis values

at 3 months, the proximal-to-distal SO_2 values remained significantly higher than control values.

It is interesting to note the differences between the lateral and medial sites with the SO_2 at the medial site continuing to increase at the 3-month stage. This is consistent with the fact that the most severe skin reactions tend to occur medially and may reflect continuing erythema.

There were no severe acute skin reactions among this group. It is therefore not possible to determine at this stage which area of measurement may be most predictive of a severe reaction occurring or what the warning level of SO_2 might be. However, the current results demonstrate that VLS is sufficiently sensitive to detect significant changes in skin SO_2, in all of the areas of the breast investigated, after only five fractions of the radiation dose.

3 Free Flap Monitoring

Free flaps, such as the transverse rectus abdominis myocutaneous (TRAM) flap, are frequently used in breast reconstruction following mastectomy. While the majority of transplants are successful, the consequences of flap failure can be devastating. Early recognition of vascular problems in the early post-operative stage, and recognition as to whether the problem is arterial or venous in origin, is vital if a failing flap is to be salvaged.

In 2006, we reported the results from 14 patients who had been monitored using VLS (RM200, Whitland Research, Whitland, UK) to detect early vascular compromise following free TRAM flap breast reconstruction [6]. Tissue oxygen saturation was compared with clinical observation and laser Doppler (LD) flowmetry (Moor Instruments DRT 4, Axminster, UK). An update on this study including a further eight patients monitored up to 70 h post-operatively has recently been published [9].

Further combined analysis of the relationship between SO_2 and LD flux from 13 patients is shown in Fig. 39.2. The data suggest that the sensitivity of SO_2 monitoring

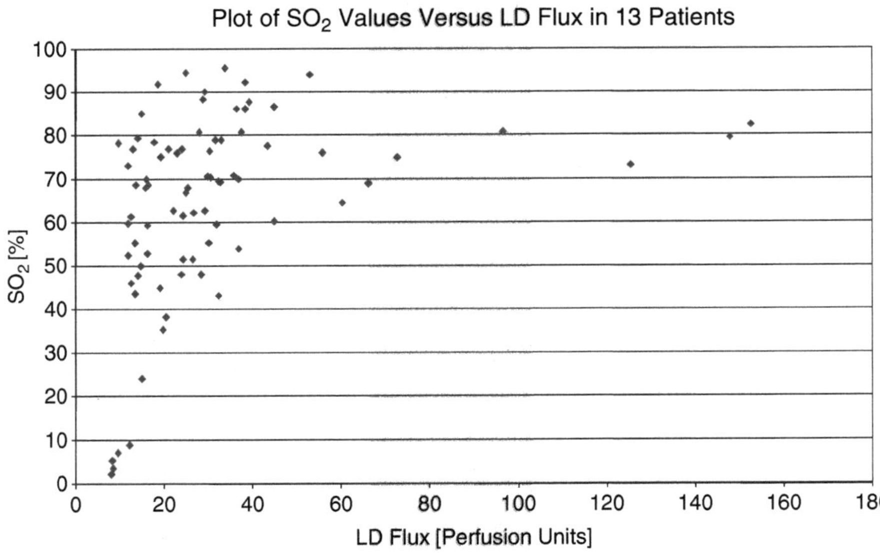

Fig. 39.2 Plot of SO_2 values vs. LD flux in all patients in whom both probes were sited in zone 1 of the flap ($n=13$). The rapid, early rise in SO_2 (y-axis) relative to LD (x-axis) suggests a high degree of sensitivity of SO_2 to detect changes in saturation at low flow rates where clinical determination of viability is frequently difficult

is superior to that of LD flux at low flow rates where determination of viability is particularly crucial.

As with breast radiotherapy, while a critical alarm value for SO_2 has yet to be established, the results show an encouraging correlation between the tissue SO_2 readings and both clinical outcome and laser Doppler. The increased sensitivity of the SO_2 measurement at low levels of perfusion when compared with LD (Fig. 39.2) reflects the improved potential VLS may have for free flap monitoring.

4 Colonic Oxygenation During Bowel Resection

Sufficiency of tissue blood flow and tissue oxygenation is known to be essential for anastomotic healing following colorectal resections both in animal experiments and human clinical studies. Poor tissue blood flow is associated with higher anastomotic failure rates. LD flowmetry has previously been used in a number of clinical studies to measure tissue blood flow. The aim of our study was to explore the use of VLS in assessing bowel SO_2 in the basal and ischemic states.

Nine patients undergoing elective left-sided colon resections were included in the study. SO_2 values were measured peri-operatively with the RM200 SO_2 monitor in the serosal and the mucosal layers of the descending colon, sigmoid colon and the rectum on four occasions; at the beginning of the operation (baseline), after mobilisation of the sigmoid, after ligation of the inferior mesenteric artery (IMA) and after complete devascularisation of the sigmoid. The mucosal measurements were obtained by using a catheter probe through a flexible sigmoidoscope. Detailed results of the study have been published elsewhere [7].

Of interest was the failure of the serosal SO_2 to demonstrate hypoxia. The serosal SO_2 in the sigmoid was not affected by mobilisation and ligation of IMA. Even after complete devascularisation, the sigmoid serosal SO_2 decreased from baseline only by 7–79%. We postulated that the failure of the serosal SO_2 to fall to a lower level may be due to slow utilisation of oxygen in the serosa together with the maintenance of the SO_2 by diffusion of atmospheric oxygen into the serosa. When the latter was prevented using an impermeable film, the serosal SO_2 fell to a level that reflected tissue ischaemia.

Mucosal SO_2 measurements using VS can accurately diagnose an ischaemic state of the colon and therefore may be a useful investigation in patients with suspected colonic ischaemia. If reliable serosal measurements are required, then the diffusion of atmospheric oxygen into the tissue must be prevented.

5 Summary

In all of the above studies, visible wavelength range optical spectrophotometry has shown itself to be a robust and sensitive method for measuring tissue oxygenation.

It may, therefore, prove to be a valuable tool for future investigations into the role of the primo-vascular system in studies of tumour pathophysiology – in particular into whether the primo-vascular system contributes in any way to oxygen supply to tissue.

Acknowledgments The author is most grateful for the collaboration of Phyllis Windsor, David Newton, and Eileen Harrison (breast radiotherapy); Jenny Caddick, Matt Erdmann and Daya Singh (TRAM flaps); Iain Bain, Daya Singh, Gerry Stansby and Yan Yiannakou (colonic oxygenation).

References

1. Harrison DK, McCollum PT, Newton DJ et al (1995) Amputation level assessment using light-guide spectrophotometry. Prosthet Orthot Int 19:139–147
2. Newton DJ, Harrison DK, McCollum PT (1996) Oxygen extraction rates in inflamed human skin using the tuberculin reaction as a model. Int J Microcirc Clin Exp 16:118–123
3. Harrison DK, Evans SD, Abbot NC et al (1992) Spectrophotometric measurements of haemoglobin saturation and concentration in skin during the tuberculin reaction in normal human subjects. Clin Phys Physiol Meas 13:349–363
4. Harrison DK, Harrison EM, Newton DJ, Windsor PM (1999) Changes in microvascular blood flow patterns and oxygen saturation in breast skin during radiotherapy. Adv Exp Med Biol 471:481–486
5. Harrison DK, Harrison EM, Newton DJ, Windsor PM (2001) Laser Doppler imaging, thermographic imaging and tissue oxygen saturation measurements detect early skin reactions during breast radiotherapy. In: Tuchin VV (ed) Saratov fall meeting 2000: optical technologies in biophysics and medicine II, SPIE Proceedings Series. SPIE, Bellingham, 4241:435–444
6. Caddick J, Raine C, Harrison D et al (2006) Lightguide spectrophotometry to monitor free TRAM flaps. Adv Exp Med Biol 578:351–356
7. Singh DB, Stansby G, Bain I, Harrison DK (2009) Intraoperative measurement of colonic oxygenation during bowel resection. Adv Exp Med Biol 645:261–266
8. Mendelson EB (1993) Radiation changes in the breast. Semin Roentgenol 28:344–362
9. Harrison DK (2011) Clinical applications of tissue oxygen saturation measurements. Adv Exp Med Biol (in press)

Part VI
Acupuncture

Chapter 40
Oriental Medicine in Japan, Lymphology and the Primo Vascular System

Moriya Ohkuma

Abstract Recently, the treatment of various diseases by Oriental medicine has become very popular. The author would like to introduce Japanese Oriental medicine and his experience in the treatment of lymphedema with oral Oriental drugs which may be useful for doctors in other countries.

1 History of Japanese Oriental Medicine

Ancient Chinese medicine which originated in BC 97 was brought to Japan, mostly in the seventh century. In AD 500, Lian wrote a famous book called Shen Non Ben Cao Jing Ji Zhu which collected together 365 drugs. Hanoka Seishuu developed this medicine in a different way in Japan in the seventeenth century (edo era).

2 Integrated Medicine (Table 40.1)

1. The constitution of the individual is emphasized. Not only one organ but the entire body as a whole, physical body and spirit are harmonized.
2. So long as there is pathological symptom (even though laboratory tests are normal), they select and prescribe drugs.
3. Different drugs are given to the same diseases and also the same drugs are given to different diseases.

M. Ohkuma (✉)
Department of Dermatology, Sakai Hospital, Kinki University, School of Medicine, Osaka, Japan
e-mail: lymph_ohkuma@r8.dion.ne.jp

K.-S. Soh et al. (eds.), *The Primo Vascular System: Its Role in Cancer and Regeneration*, DOI 10.1007/978-1-4614-0601-3_40, © Springer Science+Business Media, LLC 2012

Table 40.1 Comparison of oriental medicine with western medicine

Western medicine	Oriental medicine
Analyzing	Synthesizing
Etiology dependent	Symptom-dependent
Synthesized drug	Natural material
Single substance	Multiple components
Treating disease	Also prophylaxis and better QOL
Some adverse effects	Rare (1/10,000 in incidence)
Expensive	Less expensive

4. Prophylaxis is also considered.
5. Drugs contain multiple natural materials including from plants (stem, root, skin, fruit, etc.), animal, insect, and even stone.
6. The treatment is more economical and the less expensive [1].

3 Characteristics (Table 40.1)

This is characterized by

1. Multiple pharmacological effects happen in the body at the same time, having an influence on each other. 1 + 2 is not 3, but 4 or 5.
2. If there is only one substance in a large amount, it is not comfortable for the tissue and the tissue excretes, destroys, neutralizes, transports it away via the lymphatics, or produces an antibody. However, if there are multiple substances in moderate amounts, the tissue does not produce its antibody and there is less rejection. This is why the drugs are suitable for long term prescription.
3. If a drug has caused any adverse reaction, such as hepatic, renal, or hematological toxic effects, it has been excluded from the collection after a long history of prescription.
4. After a long history of prescription, the specialists know that so and so drug fits to so and so symptom. They know a large number of references of prescriptions reported in the past.
5. Hospital expense is less [1].

4 Criteria and Ideas Applied for Diagnosis and Treatment

1. Vital energy (energy of living), red body fluid, and transparent body fluid.
2. Five organs (liver, spirit, spleen, lung, and kidney) form a unit in which spirit and body are combined.

3. Yin-yang (relativity of disease condition), hypofunction and hyperfunction, alternating episodes of chills and fever, superficies and interior.
4. Six stages of diseases; large sun (initial stage of yang disease with ectodermal symptoms), small sun (a stage of yang disease with mesodermal symptoms), yan-mini (a stage of yang disease with endodermal symptoms), large yin (initial stage of yin disease), small yin (second stage of yin disease), terminal stage of yin disease.
5. Recognition of transition of disease.

5 Diagnosis and Treatment (Acupuncture and Moxibustion Omitted)

There are two methods:

A. Therapy is based on diagnosis and treatment by Oriental medicine.

   1. Signs collected from the patient by four examination methods, inspection (including tongue), hearing and smelling, questioning, and palpation (mainly pulse and abdomen).
   2. Their analyzing and ordering.
   3. Diagnosis.
   4. Selecting formulations (prescriptions).

B. Therapy is based on diagnosis by western medicine, but treated with oriental drugs.

6 Drugs

They are mostly made of plants (seed, leaf, stem, stalk and root). Shells, insects, and stones are also used. Three hundred drugs are available in Japan and some of them are capulated. Usually they are powders, made from the raw materials by boiling in water, freeze-dried, and packed in an aluminum envelope. All procedures are done under sterile conditions. About 20 pharmaceutical companies exist in Japan. One hundred and forty-seven drugs have been approved by the medical insurance and they are commonly numbered, although the producing company may be different. Seventy-two percent of Japanese medical doctors prescribe these oriental drugs. Only medically qualified doctors are allowed to prescribe. If there is any adverse effect, the doctor must report it to the Japanese Ministry of Health & Labor; this is obligatory. These drugs are suitable for prescription over a long period. Generally speaking, the incidence of adverse effects is about 1/10,000 as compared with western drugs.

7 Social Activities

There are two Societies of Oriental medicine in Japan. One is the Japanese Society of Wakanyaku which has 914 members and the other is the Japanese Society for Toyo Igaku which has 8,637 members. They both have annual domestic and also sometimes international meetings. They also publish their own journals, partly in English. There are 2,427 specialists who have been approved after passing examinations, publishing papers, and oral presentations in the scientific meetings. It is reevaluated each 5 years.

8 Trial Treatment of Lymphedema by Physiotherapy Combined with Oral Oriental Drugs

8.1 *Materials and Methods*

I. Five cases of lymphedema were treated with physiotherapy using pulse magnetic fields, vibration, and hyperthermia [2]. The patients have become edema free after the treatment. Edema has never recurred even after release of compression.
II. Eight cases received the same treatment, with success, but edema recurred after release of compression.
III. Fourteen patients were treated successfully in the same way but combined with oral Oriental drugs Tj. 41 and Tj. 48. No edema reappeared even after release of compression.
IV. Eight patients treated in the same way with the same results as in III but edema recurred.
V. Five patients were treated with only oral Oriental drugs as a control.
VI. The same control of five cases was treated with compression alone.

I–VI. All the patients received compression with elastic and less elastic bandages both during and between the treatments. Disappearance of edema after the treatment is judged by the difference in circumference of the extremity measured 10 cm above and beneath the patella edge or olecranon being less than 0.5 cm for unilateral lymphedema of the extremity. The criterium for the bilateral type is made by subjective observation by the patients as well as by macroscopic observation by the doctor. Clinical data such as duration of edema, mycotic infection as well as the PPD skin test were compared in each group.

8.2 *Results*

The results are summarized in Table 40.2. The control V is not more effective than VI and the cases have been left out. The control VI is effective to some extent, but no patient became edema free.

Table 40.2 Summary of clinical results of lymphedema treated by the physiotherapy alone (13 cases) and combined with oral oriental drugs (22 cases)

	Physiotherapy alone		Physiotherapy with oriental drugs	
Treatment	No recurrence	Recurrence	No recurrence	Recurrence
Recurred after release of compression	5/13	8/13	14/22	8/22
Duration since onset	1–2 Months	1 Month to 10 years	7 Months to 17 years	6 Months to 19 years
Duration until recurrence		2–10 Months		2–7 Months
Observed after release	5–10 Years	10 Months	1–8 Years	2 Month
PPD skin test reaction	Moderate	Weak	Strong	Weak
PPD test after treatment	Intensified		Very much intensified	

8.3 *Discussion*

Tj. 41, hochuekkito contains *Glycyrrhizae radix*, *Asteragali radix*, *Atractyloides lanceae rhizoma*, *Ginseng radix*, *Augelicae radix*, *Bupleuri radix*, *Zizyphi fructus*, *Aurantil nobilis pericarpium*, *Cimicigugae rhizona,* and *Zingiberis rhizoma*. The main effect of this drug is to increase NK cells [3]. Tj. 48, juzendaihotou contains the same first five components as Tj. 41 as well as *Cinamoni cortex*, *Rehmanniae radix*, *Paeoniae radix*, *Cnidii rhizoma,* and *Poria*. Its main pharmacological effect is to stimulate macrophages [4]. They are associated with innate immunological reaction. However, why this increased innate immunity gives a good result is not known. The PPD skin test is related to delayed immunological reaction. It is also not known why positive tests give good results. Old Japanese people (secondary lymphedema is often seen after cancer operation) have, in most of the cases, a positive PPD skin test. It forms a parameter to know about a delayed immunological reaction. In all the text books, lymphedema is said to be an incurable disease. However, this should be changed: lymphedema can be cured perfectly.

8.4 *Summary and Conclusion*

1. Secondary lymphedema patients treated with physiotherapy using magnetic fields, vibration, and hyperthermia combined with the oral Oriental drugs, Tj. 41 and Tj. 48 show a better rate of perfect healing even with long intervals between the onset of the edema and the start of the treatment (it is all right even if the treatment is given after many years).
2. The difference between the patients with perfect and temporary healing is stronger with the PPD skin test after the treatment in the former.

9 Primo Vascular System

It is interesting to compare this system with the lymphatic system.

It has no valves, multiple small lumina which are not covered by endothelial cells, no vascular smooth muscle cell except those in the skin, a high content of oxygen (lymph has lower oxygen content than venous blood), open peripheral end (the initial lymphatic always has endothelial cells and its end is covered by the endothelial cells), negative immunohistochemical staining for LIVE-1 as well as Prox-1, etc. It sometimes exists in another vascular lumen. This morphological and functional comparison will give us good information about the primo vascular system [5–7].

References

1. Akase A (2000) Hospital expense in case of flu syndrome. J Orient Med 50:656–663
2. Ohkuma M (2002) Treatment of peripheral lymphedema by concomitant application of magnetic fields, vibration and hyperthermia: a preliminary report. Lymphology 35:87–90
3. Kuroiwa A et al (2004) Function of NK cell. Int Immunopharmacol 4:313–324
4. Saiki I (2000) Improved homeostatic condition in balance by increased phagocytosis and antibody production. Biol Pharmacol Bull 23:677–688
5. Lee B-C et al (2010) Characteristic features of a nerve primo-vessel suspected in rabbit brain ventricle and central canal. J Acupunct Meridian Stud 3:75–80
6. Wang G-J et al (2010) Meridian studies in China: a systematic review. J Acupunct Meridian Stud 3:1–9
7. Jia Z-F et al (2010) Fluorescent nanoparticles for observing primo vascular system along sciatic nerve. J Acupunct Meridian Stud 3:150–155

Chapter 41
From the Anatomical Discovery of Meridians and Collaterals to Fasciaology Theory

Yu Bai, Lin Yuan, Yong Huang, Chun-lei Wang, Jun Wang, Jin-peng Wu, Jing-xing Dai, Dong-fei Li, Chun Yang, Mei-chun Yu, Hui-ying Yang, Hui Tao, Ou Sha, and David Tai Wai Yew

Abstract The theory of meridians and collaterals is the basis of acupuncture in the traditional Chinese medicine; however, their anatomical existence is still unknown. The aim of this study was to investigate the anatomical basis of acupoints and meridians. Based on the digital datasets of Virtual Chinese human (VCH) bodies, 3-dimensional (3D) structures of virtual meridians and fascia connective tissue-gathering areas were constructed, and they were compared with each other. The shortest distances between virtual acupoints and constructed fascia connective tissues were measured. 3D structures of fascia connective tissue-gathering areas were also constructed based on CT and MRI images of living human bodies, and also compared with meridians. 3D structures of fascia connective tissue-gathering areas in the VCH bodies showed a pattern of beads-on-strings. Furthermore, the distances between the fascia strings and virtual meridians were close. More fascia connective tissue areas were constructed, more string-like structures were found. When 3D structures of all fascia connective tissues through over the body were constructed, a body-shaped connective tissue network appeared. 3D structures of fascia connective tissue-gathering areas constructed from both CT and MRI images also appeared beads-on-string patterns and colocalized with traditional Chinese meridians. The fascia network all over the body is the anatomical basis of acupoints and meridians in the traditional Chinese medicine. The histological composition of Meridian is the nonspecific connective tissue (including loose connective tissue and fat tissue). Therefore, we put forward a new approach for division of anatomical discipline. According to this method, the human body consists of two major systems: one is the supporting and storing system and the other is the functional system. The anatomical discipline which based is on this division method is named fascial anatomy. The discipline which studies the supporting and storing system and the mutual relationship between this system and the functional system, the latter is made of differentiated cells, is named fasciaology.

L. Yuan (✉)
Department of Anatomy, Southern Medical University, Guangzhou 510515, China
e-mail: yuanl@fimmu.com

K.-S. Soh et al. (eds.), *The Primo Vascular System: Its Role in Cancer and Regeneration*, 305
DOI 10.1007/978-1-4614-0601-3_41,

1 Introduction

The theory of meridian system is one and it is the core of acupuncture, moxibustion, and massage in the traditional Chinese medicine [1]. Meridians are in fact a collection of acupoints, which may be clarified as major highways, meridians, on a map converging on and exiting from a large metropolis, acupoint [2, 3]. The meridian system is composed of 12 principal meridians and 8 extra meridians (collaterals). Principal meridians provide the linking between different organs, and extra meridians communicate with principal meridians. The meridian map as above has been well established by the ancient Chinese, and the treatment efficacy of acupuncture, moxibustion, and massage based on the meridian system is not in doubt [4] However, meridians are still invisible to scientists, since no anatomical evidence has been discovered.

Many efforts have been made to prove the existence of acupoints and meridians, which have been described as different substances, such as neurovascular bundles, neuromuscular attachments, sensory nerve endings, perivascular space, and perineurial vessels [5–10]. Recently, more and more researchers have suggested that a correspondence may exist between acupuncture meridians and connective tissues [11–14]. However, there is an argument about the types of connective tissues. Omura et al. [15] linked the acupuncture points mainly to dense connective tissue, while Langevin and Yandow [16] viewed the meridians as the network form by interstitial connective tissue. Furthermore, all above studies were experimented in parts of the body.

The virtual human technique is a new methodology, and visualization of 3D human renderings allows an accurate anatomy and function mapping and a quick analysis of structure-to-function relationships [17–23]. Therefore, the data of VCH bodies were employed in this project. The aim of this study was to investigate the anatomical structures of acupoints and meridians three-dimensionally in VCH bodies.

2 Methods

2.1 VCH Image Datasets

The digital image datasets, including a set of VCH Male 1 (VCH-M1) and a set VCH Female 1 (VCH-F1), were collected as prepared [24]. Briefly, a 176 cm tall male cadaver and a 155 cm tall female cadaver were perfused with a red filling material through the femoral artery before freezing and embedding. The red filling material is a mixture composed of gelatin, cinnabar, and starch. The whole cadavers were sectioned by a JZ1500A vertical milling machine at intervals of 0.2 mm. Images of all sequential sections were captured with a FujiFinePixS2Pro camera and saved as Tagged Image File Format (TIFF). The digitized VCH male and female image datasets were established [25].

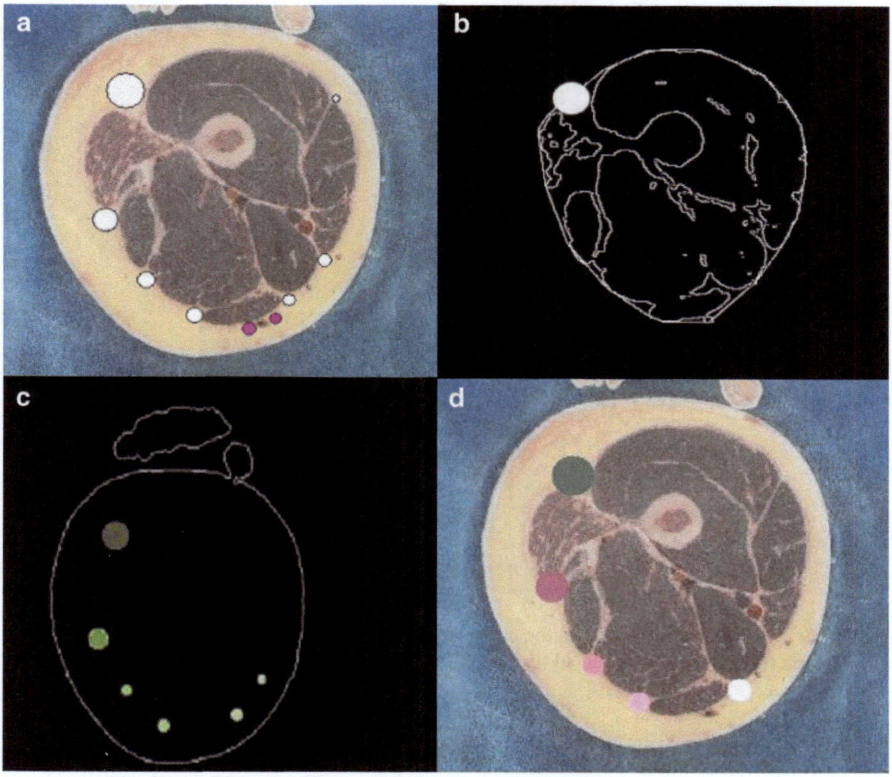

Fig. 41.1 Marking procedure. The connective tissue-gathering areas in a VCH image were marked in (**a**). The muscle tissue was extracted, and the connective tissue areas were automarked by the computer (**b**, **c**, separately). (**d**) Showed the marked VCH image

2.2 Construction of 3D Structures of Connective Tissues

100 of the upper extremity and 100 of the lower extremity, all at the interval of 5 mm, were sequentially selected from the image database of VCH-M1 [26]. Using four marker points on the original images, image registration was performed and followed by confirmation with a standard edge detection algorithm (Sobel). During the localization, the parameters were adjusted according to the actual images. After removal of the backgrounds in the images in three ways, three groups of images were obtained (Fig. 41.1). The images were then compared with the standard human body tomograms to identify connective tissues. Green circles were inscribed in the centers of thick connective tissue areas, which were usually triangular or polygonal in shape (Fig. 41.2a, c). The diameter of a green circle was half of the maximum diameter of a thick connective tissue area (Fig. 41.6a). The maximum diameter of a thick connective tissue is the vertical distance from the point in intermuscular septum between two muscles and subcutaneous thick connective tissue to the

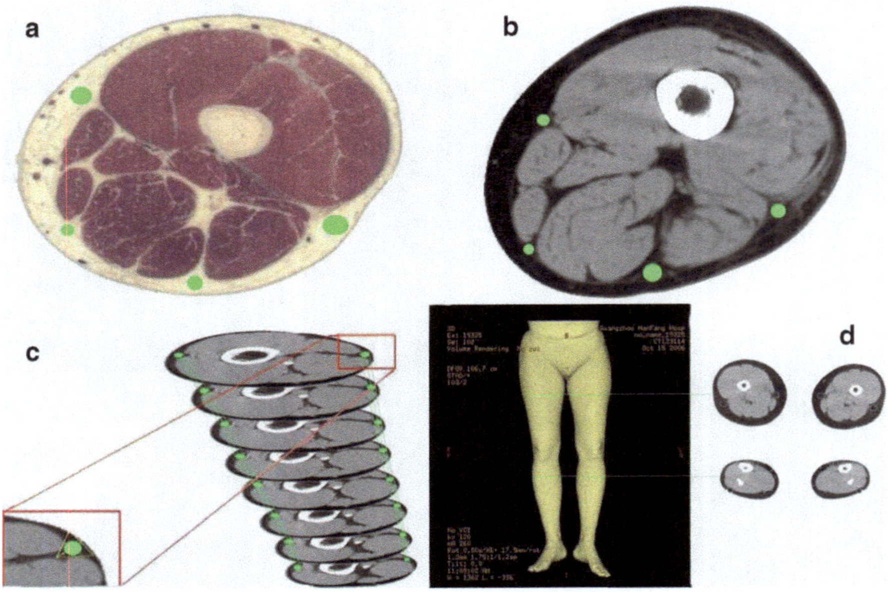

Fig. 41.2 *Circles* were inscribed in the connective tissues in VCH (**a**) and CT (**b**) images. Then the 3D structures were reconstructed (**c, d**)

dermis. Subsequent image segmentation and diminution by bicubic interpolation resulted in three groups of images, each with a size of 810×390 pixels. The images were conversed into gray color. Finally, a 3D model was then reconstructed.

2.3 Marking Procedure

The connective tissue-gathering areas in a VCH image were marked in Fig. 41.1a. The muscle tissue was extracted, and the connective tissue areas were automarked by the computer (Fig. 41.1b, c, separately). Figure 41.1d shows the marked VCH image.

Circles were inscribed in the connective tissues in VCH (Fig. 41.2a) and CT (Fig. 41.2b) images. Then the 3D structures were reconstructed (Fig. 41.2c, d).

2.4 Comparison of 3D Connective Tissue Strings with Traditional Meridians

The 3D connective tissue structures were rendered from the green circles inscribed in the VCH images close to the anterior and posterior midlines of the trunk, as well as close to the medial and the lateral lines of the upper and lower extremities.

The courses of the constructed connective strings were compared with those of meridians in TCM, including Ren and Du meridians in the trunk, Large Intestine meridian of Hand-Yangming (LI) and Triple-warmer meridian of Hand-shaoyang in upper limbs, as well as Spleen meridian of Foot-taiyin (SP) and Urinary Bladder meridian of Foot-taiyang (BL) in lower limbs. The vertical distances between the connective tissue strings and respective above meridians were also measured.

2.5 *Study on CT Images of Living Human Bodies*

Twelve lower limbs of volunteers were scanned by GE Lightspeed 16 row spiral CT. Circles were added into the connective tissue-gathering areas in the CT images similar to those in VCH images (Fig. 41.2b, c). Their 3D structures were then reconstructed through 3D volume rendering. The outputs data were transformed from DICOM to JPEG format. The correlation between acupoints and fascia converging areas was analyzed. We measure the space between traditional meridians and the reconstructed fascia converging lines. First, ten acupoints in every six standard meridian lines were selected, then the vertical distance from the ten acupoints in every traditional meridians to the reconstructed fascia converging lines in the vicinity were measured [26].

2.6 *Study on MRI Images of Living Human Bodies*

MRI images of volunteers were randomly selected to carry out 3D connective tissue reconstruction. Traditional Chinese meridians and acupoints were first marked on one upper limb of the volunteers, and a plastic tracer filled tube was placed along the marked meridians. GE 3.0 MRI was used to scan the limbs. The fascia converging areas in MRI images were marked and their 3D structures were reconstructed by MIMICS11.02.

3 Results

3.1 *Datasets of VCH Bodies*

A total of 9,232 and 8,556 TIFF images were obtained separately from the male and female cadavers. The database of VCH-M1 was 161.56 GB, and that of VCH-F1 was 149.7 GB.

3.2 *3D Constructions of VCH Connective Tissues*

The reconstructed fascia strings were similar to the distribution of traditional meridians (Fig. 41.3). The reconstructed fascia strings showed beads-on-string structures in

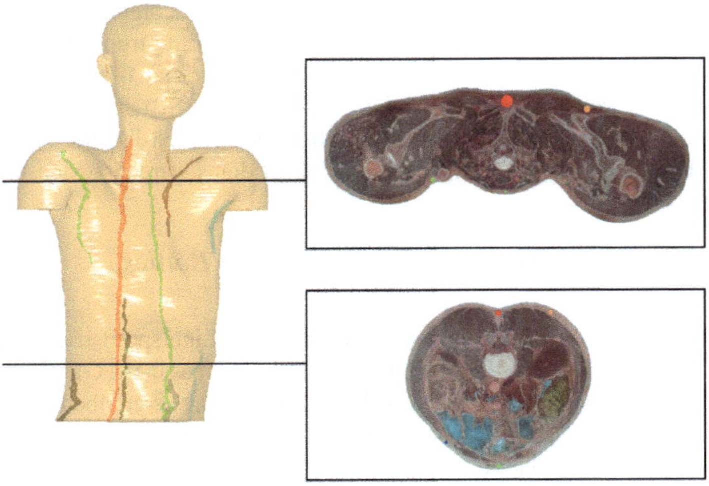

Fig. 41.3 Reconstructed fascial lines (fascia gathered area) were separately showed in the trunk in the VCH body

certain parts of limbs and the body trunk. For example, the reconstructed connective tissue string in the back, in a beads-on-string pattern, was very similar to the distribution of Du meridian, and the string in the abdomen was very similar to the distribution of Ren meridian (Fig. 41.3). This may be due to the regional anatomical structure, the vertebral column. The green connective tissue strings in upper limbs were comparable with the classical Large Intestine meridian of Hand-yangming and Triple Energizer meridian of Hand-shaoyang that were indicated in red (Fig. 41.4). The connective tissue strings in lower limbs were also comparable with Kidney meridian of Foot-shaoyin and Gallbladder meridian of Foot-shaoyang (Fig. 41.5). The vertical distance of these fascia connective strings were very close to the locations of meridians. The results showed a strong consistency between the intermuscular connective tissues and distributions of meridians.

Reconstructed fascial lines (fascia gathered area) were separately shown in the trunk, upper limbs, and lower limbs in the VCH body (Figs. 41.3–41.5 separately).

The results of analysis: select six standard meridian lines to compare with traditional meridians. The vertical distance between traditional meridians and the reconstructed fascial lines at ten different acupoints that were selected on respective lines were measured. The vertical distance to the classic description of the meridian near body middle line was taken as a positive value, away from body middle line as a negative value, and the unit is mm (Table 41.1).

When more connective tissues were measured and their 3D structures were rebuilt, more fascia strings were found. In other words, more 3D structures of connective tissues were reconstructed, more meridians appeared. When the fascia connective tissues of the whole body were measured and their 3D structure was rebuilt, a complete fascia network was figured out (Fig. 41.6). The connective tissues

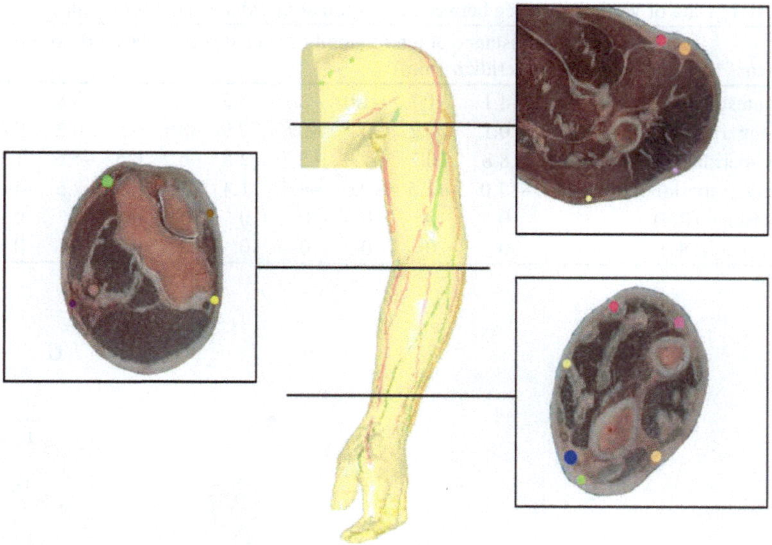

Fig. 41.4 Reconstructed fascial lines (fascia gathered area) were separately showed in the upper limbs in the VCH body

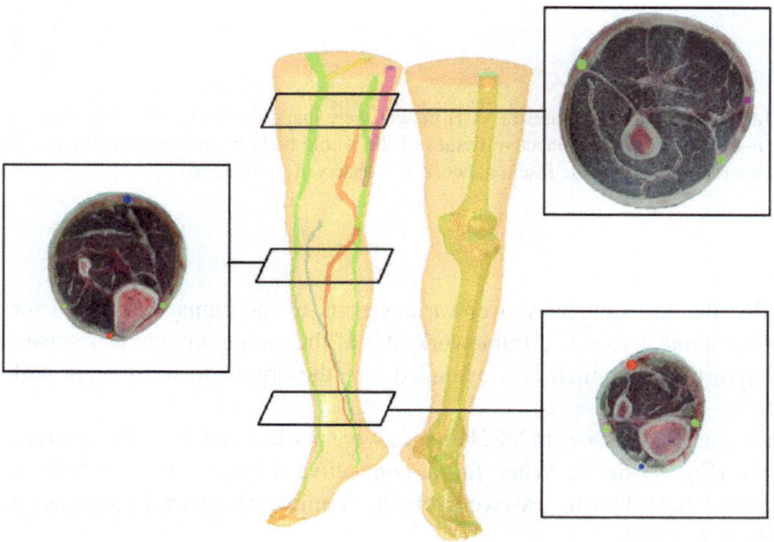

Fig. 41.5 Reconstructed fascial lines (fascia gathered area) were separately showed in the lower limbs in the VCH body

Table 41.1 Data of vertical distance between cognominal meridians pairs (unit: mm)

Meridians	Distance of ten randomly selected points between two kinds of meridians(mm)									
Large intestine meridian (LI)	14.1	1.7	−1.0	2.1	−3.2	1.8	1.2	3.8	0.4	3.6
Sanjiao meridian (SJ)	0.1	0.2	0.2	−0.8	−2.9	−8.1	9.5	0.2	0.6	3.0
Kidney meridian (KI)	18.8	10.2	−7.9	−5.7	2.5	−6.7	6.5	−5.6	1.1	9.0
Gallbladder meridian (GB)	−17.0	−11.5	−9.6	−4.5	1.4	−1.1	0.5	2.5	−1.6	4.1
Ren meridian (RN)	0	0	0	0	0	0	0	0	0	0
Du meridian (DU)	0	0	0	0	0	0	0	0	0	0

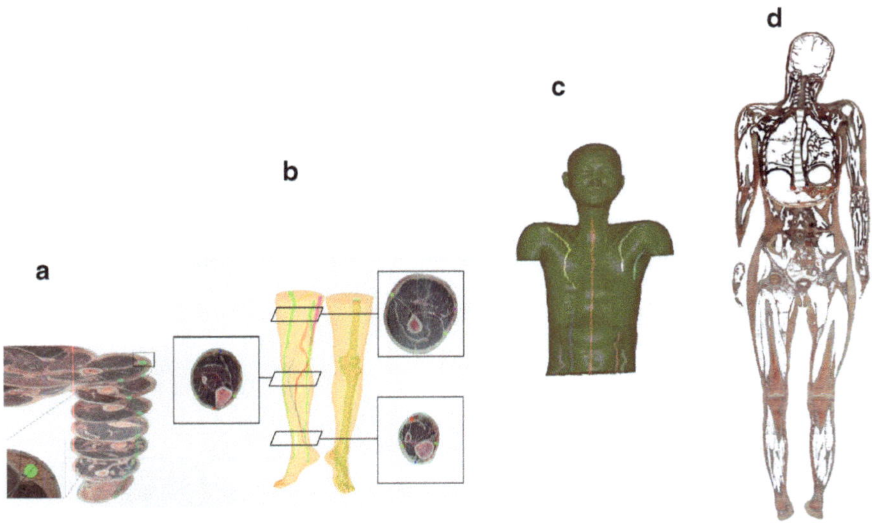

Fig. 41.6 The connective tissues in VCH images were marked and their 3D structures were rendered (**a–c**). When fascia connective tissues of the whole body were marked and their 3D structures were rebuilt, a complete fascia network was observed as in the (**d**)

were distributed extensively over various parts of the human body and formed a complete connective tissue framework. All of the human organs were coated with connective tissues, which also extended into the organs to form septa within the organs.

The connective tissues in VCH images were marked and their 3D structures were rendered (Fig. 41.6a–c). When fascia connective tissues of the whole body were marked and their 3D structures were rebuilt, a complete fascia network was observed as in the Fig. 41.6d.

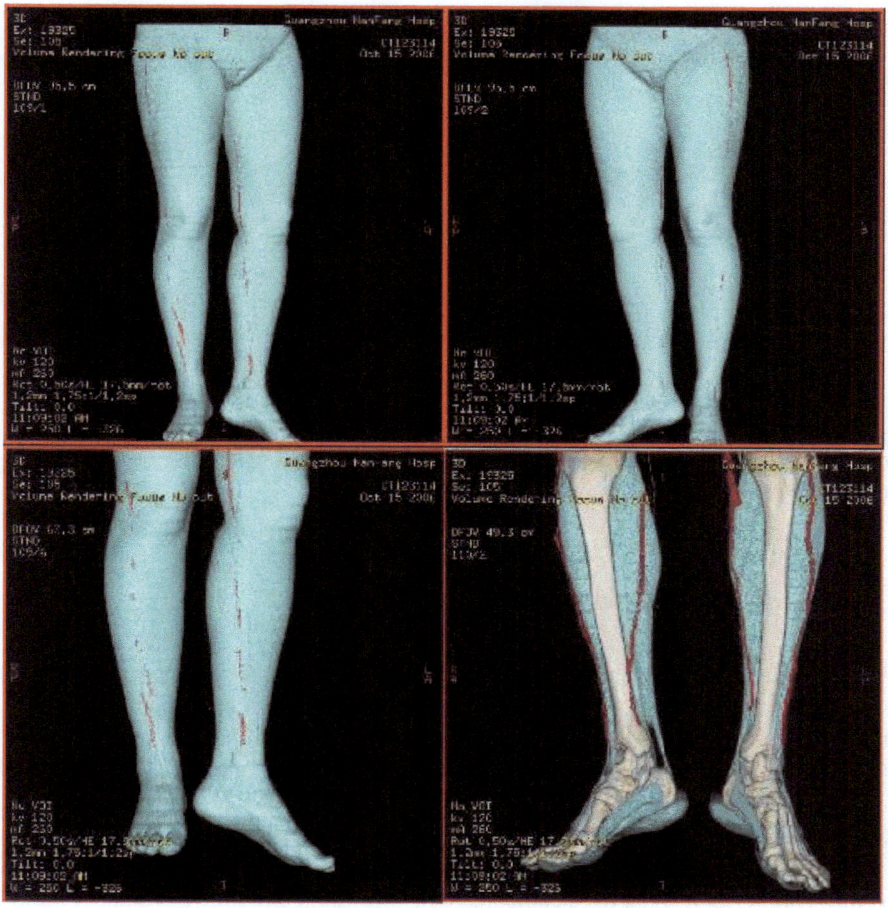

Fig. 41.7 Based on CT images of lower limbs, the 3D structures of gathering connective tissues were rebuilt.

3.3 Study on CT Images

The distribution of the rebuilt fascia meridians was similar to that of the traditional ones. When the classical meridians were compared with connective tissue string rebuilt from CT images, the distribution of the rebuilt fascia meridians was similar to that of the traditional ones (Fig. 41.7).

Based on CT images of lower limbs, the 3D structures of gathering connective tissues were rebuilt.

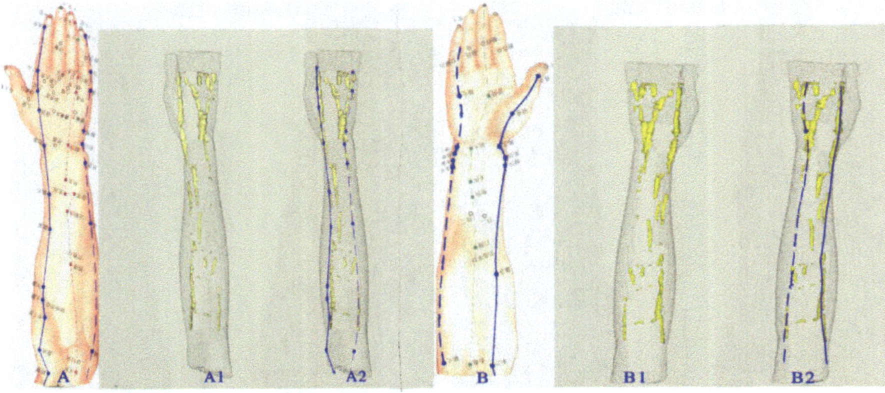

Fig. 41.8 Based on MRI images of upper limbs, the constructed connective tissue strings, in *yellow* (*A1*, *B1*), were compared with *blue* large intestine meridian of Hand-yangming (*A*, *A2*) and lung meridian of Hand-taiyin (*B*, *B2*)

3.4 Study on MRI Images

When the classical meridians were compared with the rebuilt connective tissue string from MRI images, the distribution of the rebuilt fascia meridians was similar to that of the traditional ones. The results were similar with those from VCH study (Fig. 41.8). The constructed connective tissue strings were indicated in yellow and the meridians, including Large Intestine meridian of Hand-yangming (Fig. 41.8A, A2) and Lung meridian of Hand-taiyin (Fig. 41.8B, B2), were in blue.

Based on MRI images of upper limbs, the constructed connective tissue strings, in yellow (Fig. 41.8A1, B1), were compared with blue Large Intestine meridian of Hand-yangming (Fig. 41.8A, A2) and Lung meridian of Hand-taiyin (Fig. 41.8B, B2).

All the results showed that connective tissues permeated the whole body and formed a network. Acupoints are located in the connective tissues either between muscles or between the muscle and bones [27, 28]. Fascia has some kind of self-regulating function to adapt to the body pressure, and massage therapy can affect its tension, viscoelasticity, and structure [29–32]. Fascia is also responsible for maintaining structural integrity of the body. Fascia connective tissue is rich in blood vessels, nerves, lymph and provides the matrix that allows for intercellular communication; it may play an important role in surveillance and regulation of the body [33, 34].

4 Discussion

Our research shows that the histological composition of Meridians has been regarded as nonspecific connective tissues, including loose connective tissue and fat tissue [25, 26]. Therefore, we propose that the effective sites of acupuncture should be fascial connective tissues, including the cells and tissues inside these sites, such as nerve ending, capillaries, fibroblasts, undifferentiated mesenchymal cells, lymphocytes, etc.

Previous work has shown that connective tissues permeated the whole body and formed a network [12, 16]. Acupoints are located in the connective tissues either between muscles or between the muscle and bones. The fascia has a self-regulating function to adapt to the body pressure, and massage therapy can affect its tension, viscoelasticity, and structure [29, 30]. It is also responsible for maintaining structural integrity of the body. The fascia connective tissue is rich in blood vessels, nerves, lymph and provides the matrix that allows for intercellular communication. It may play an important role in surveillance and regulation of the body [27].

If we trace the origin of fascia during biological evolution, we can find that the fascia network is homologous with the extracellular matrix of a single germ layer organism, the middle lamella of a two germ layer organism, and the mesenchyma of a three embryonic layer organism. When the structures are reversely displayed and modeled, the components of an organism during each period can be summarized into two major systems, which are (i) the supporting and storing system containing adipose and loose connective tissues and (ii) the functional system which constituted of specialized cells. In other words, the fascial network, mesenchyme, and ECF are homologous, and their common function is to sustain stability of the internal environment of an organism. The mesoderm further evolves into organs and systems with specific functions, including locomotor system, urinary system, reproductive system, circulatory system, and so on. The residues of the mesenchyme then differentiate into connective tissues which are distributed throughout the body. The above theory is the basis of fascial anatomy [26, 27].

Consequently, two new definitions have been introduced as below: (i) Fascial anatomy: this is a new theory in anatomy. According to fascial anatomy, each living organism is composed of two major systems. One is the supporting-storing system, which is constituted of the network of unspecialized connective tissues. The other is the functional system, which contains all kind of differentiated cell that are surrounded by the supporting-storing system. (ii) Fasciaology: the research field studying the supporting-storing system and the mutual relationships between the above two major systems in a living organism has been named fasciaology. The theory of fascial anatomy and fasciaology highlights the significance of fascial research, which is important not only in the field of TCM, but also in other biomedical research and clinical therapies. TCM theories may be scientifically interpreted in fasciaology.

On the basis of this new approach, the functional system does not refer to the traditional nine major systems in systemic anatomy, including musculoskeletal system, nervous system, endocrine system, cardiovascular system, immune system, respiratory system, urinary system, digestive system, and reproductive system. Instead, it refers to all the cells, tissues, and organs which were wrapped and segmented by the supporting-storing system. Furthermore, the supporting-storing system provides physical support for the structural integrity and physical stability of each individual organ [31, 32]. The functional system is composed of various committed stem cells and functional cells. These cells are different from the mesoderm and folded endoderm and/or ectoderm cell. The functional cells work together to fulfill living activities. When the supporting-storing system wears out, the body will die. If the living condition of an organism is regarded as a blazing candle, then the fascial system, the network of unspecialized connective tissues containing adipocytes and stem cells, can correspond to the stick of the candle, and the functional system can correspond to the flame of the candle. The supporting-storing system provides energy and cell reserve for the functional system, just like the stick of the candle providing "fuel" to the flame. When the stick depletes, the flame will extinguish. So does the human body. When the supporting-storing system wears out, the body dies [27].

In TCM, there are many kinds of physical treatments including acupuncture, Gua Sha, plum blossom needle, and so on. They all stimulate the fascia network so as to regulate the metabolism and functions of functional cells. Since fascia connective tissue is distributed throughout body, acupoints may exist in every part of the body. Moreover, we believe that Guan Sha therapy is also targeted at the dermal dense connective tissue all over the body. The difference between clinically so-called acupoints and nonacuppoints, as well as between main acupoints and supplementary acupoints, is more the intensity of biological reactions rather than structural components. Nonmeridian extra acupoints are the areas rich in fascial connective tissues, and they are located outside meridians. The histological composition of the target areas of acupuncture can be classified into five types: (1) Dermal dense connective tissue, (2) Subcutaneous loose connective tissue, (3) loose connective tissue in intermuscular septum, (4) loose connective tissue around neurovascular tract, (5) loose connective tissue of hilus and tunica of visceral organ [26].

The locations and depths of these areas are different in all individuals, as is sensitivity. Therefore, the therapeutic effects are different. The ancient records of meridians and acupoints as well as herbal medicine of TCM lack a scientific basis, so they can only serve as references, but should not be strictly followed as textbooks. Fascial research may provide the evidence for the accurateness of acupoints.

5 Conclusions

The anatomical basis of acupoints and meridians is the fascial network throughout the body. The histological composition of meridians is nonspecific connective tissues, including loose connective tissue and fat tissue. Acupoints are the sites that produce strong biological reactions when stimulated. The nonspecific connective tissue network has been proposed as the supporting-storing system composed in the hypotheses of fascial anatomy and fasciaology. The hypotheses are involved in the research of the mechanism of acupuncture and TCM, evolutionary biology, holistic therapies, integrative medicine, and alternative/complementary medicine.

Acknowledgments The work described in this paper was supported by the following funding agencies: The National Basic Research Program (also called 973 Program) (Grant No.2007CB512705). National Natural Science Foundation of China (Grant No.30801464).

References

1. Birch SJ, Felt RL (1999) The theoretical basis of acupuncture: fundamental concepts and explanatory models. In: Understanding acupuncture. Churchill Livingstone, Edinburgh, pp 110–113
2. Bensoussan A (1991) The nature of the meridians. In: The vital meridian: a modern exploration of acupuncture. Churchill Livingstone, Melbourne, pp 51–53
3. Sutherland JA (2000) Meridian therapy: current research and implications for critical care. AACN Clin Issues 11:97–104
4. Board of Science and Education, British Medical Association (2000) The evidence base of acupuncture. In: Acupuncture: efficacy, safety and practice. Harwood Academia Publishers, Amsterdam, pp 11–21
5. Bossy J (1984) Morphological data concerning the acupuncture points and channel network. Acupunct Electrother Res 9:79–106
6. Dung HC (1984) Anatomical features contributing to the formation of acupuncture points. Am J Acupunct 12:139–143
7. Ciczek LSW, Szopinski J, Skrzypulec V (1985) Investigations of morphological structures of acupuncture points and meridians. J Tradit Chin Med 5:289–292
8. Ma W, Tong H, Xu W, Hu J, Liu N, Li H, Cao L (2003) Perivascular space: possible anatomical substrate for the meridian. J Altern Complement Med 9:851–859
9. Hashimoto PH (2005) The perineurial vessel: a possible candidate for the structural basis of the meridian (Jing-Luo) in Chinese medicine. Anat Sci Int 80:177–180
10. Yung KT (2005) Birdcage model for the Chinese meridian system: part VI. Meridians as the primary regulatory system. Am J Chin Med 33:759–766
11. Ho MW, Knight DP (1998) The acupuncture system and the liquid crystalline collagen fibers of the connective tissues. Am J Chin Med 26:251–263
12. Langevin HM, Churchill DL, Wu J, Badger GJ, Yandow JA, Fox JR, Krag MH (2002) Evidence of connective tissue involvement in acupuncture. FASEB J 16:872–874
13. Ahn AC, Wu J, Badger GJ, Hammerschlag R, Langevin HM (2005) Electrical impedance along connective tissue planes associated with acupuncture meridians. BMC Complement Altern Med 5:10
14. Konofagou EE, Langevin HM (2005) Using ultrasound to understand acupuncture. Acupuncture needle manipulation and its effect on connective tissue. IEEE Eng Med Biol Mag 24:41–46
15. Omura Y, Takeshige C, Shimotsuura Y, Suzuki M (1988) Imaging of the stomach, and localization of the stomach meridian & its acupuncture points in a human cadaver by the use of the indirect "Bi-Digital O-Ring Test Imaging Technique". Acupunct Electrother Res 13:153–164
16. Langevin HM, Yandow JA (2002) Relationship of acupuncture points and meridians to connective tissue planes. Anat Rec (New Anat) 269:257–265
17. Ackerman MJ, Spitzer VM, Scherzinger AL (1995) The Visible Human data set: an image resource for anatomical visualization. Medinfo 8:1195–1198
18. Moore KL, Agur AMR (1995) Introduction to clinical anatomy. In: Essential clinical anatomy. Lippincott Williams & Wilkins, New York, p 8
19. Spitzer V, Ackerman MJ, Scherzinger AL (1996) The visible human male: a technical report. J Am Med Inform Assoc 3:118–130
20. Ckerman MJ, Yoo T, Jenkins D (2001) From data to knowledge–the Visible Human Project continues. Medinfo 10:887–890

21. Caon M (2004) Voxel-based computational models of real human anatomy: a review. Radiat Environ Biophys 42:229–235
22. Huang Y, Yuan L, He ZQ, Wang CL (2006) Study on the meridians and acupoints based on fasciaology: an elicitation of the study on digital human being. Zhongguo Zhen Jiu 26:785–788
23. Robb RA (1999) 3-D visualization in biomedical applications. Annu Rev Biomed Eng 1:377–399
24. Zhong SZ, Yuan L, Tang L, Huang WH, Dai JX, Li JY, Liu C, Wang XH, Li H, Luo SQ, Qin D, Zeng SQ, Wu T, Zhang MC, Wu KC, Jiao PF, Lu YT, Chen H, Li PL, Gao Y, Wang T, Fan JH (2003) Research report of experimental database establishment of digitized virtual Chinese No.1 female. Di Yi Jun Yi Da Xue Xue Bao 23:196–200
25. Yuan L, Tang L, Huang WH (2003) Construction of dataset for virtual chinese male No. 1. J First Mil Med Univ 23:520–523
26. Wang CL, Yuan L, Wang J, Jiao PF (2007) Contrast study on the line course of fascia meridians made by three dimensional reconstruction and classical meridians in human body. Chin J Anat 30:340–343
27. Wang CL, Wu JP, Wang J (2008) An interpretation on the essence of meridians and acupuncture mechanism from fasciaology view. Chin J Basic Med Tradit Chin Med 14:312–314
28. Langevin HM, Churchill DL, Cipolla MJ (2001) Mechanical signaling through connective tissue: A mechanism for the therapeutic effect of acupuncture. FASEB J 15:2275–2282
29. Robert S (2003) Fascial: plasticity – a new neurobiological explanation: part 1. J Bodyw Mov Ther 7:11–19
30. Robert S (2003) Fascial: plasticity – a new neurobiological explanation: Part 2. J Bodyw Mov Ther 7:104–116
31. Thomas WM (2004) Structural integration-developments in Ida Rolf's 'Recipe'-I. J Bodyw Mov Ther 8:131–142
32. Swartz MA, Tschumperlin DJ, Kamm RD (2001) Mechanical stress is communicated between different cell types to elicit matrix remodeling. Proc Natl Acad Sci USA 98:6180–6185
33. Scheip R, Klinger W, Lehmann-Horn F (2005) Active fascial contractility: fascia may be able to contract in a smooth muscle-like manner and thereby influence musculoskeletal dynamics. Med Hypotheses 65:273–277
34. Langevin HM, Churchill DL, Fox JR, Garra BS, Krag MH (2001) Biomechanical response to acupuncture needling in humans. J Appl Physiol 91:2471–2478

Chapter 42
An Evidence-Based Review of Acupuncture as an Adjunctive Therapy in Comprehensive Cancer Care

Christopher Zaslawski

Abstract The use of acupuncture as an adjunctive therapy for cancer has been increasing over several decades. While there has only been one Cochrane review completed in this area, with another at the protocol stage, several high-quality RCTs have been published in the last decade. These studies have focused on the effect of acupuncture on managing many of the disabling symptoms associated with radiation therapy, chemotherapy or surgical intervention. These include conditions such as nausea, pain control, xerostomia and fatigue. This chapter will report on the current evidence for using acupuncture for these conditions as well as some of the problems associated with designing and implementing a rigorous acupuncture clinical trial.

1 Introduction

During the last two decades, there has been increasing use of Complementary and Alternative Medicine (CAM) therapies including acupuncture as an adjunctive therapy for cancer-related conditions. Several studies have shown variable rates of CAM use for cancer ranging from 34 to 91% in the USA [1]. Surveys conducted in Asian populations have also shown high use of CAM, for example, a recent study of CAM usage among a sample of Singaporean cancer patients was 55%. Independent predictors were Chinese ethnicity, tertiary education, age greater than 65 years and previous CAM use [2]. When viewed as a specific modality, acupuncture use among cancer patients ranges between 1.2 and 31% [3].

C. Zaslawski (✉)
College of Traditional Chinese Medicine, Department of Medical and Molecular Biosciences, University of Technology, Sydney, Australia
e-mail: Chris.Zaslawski@uts.edu.au

K.-S. Soh et al. (eds.), *The Primo Vascular System: Its Role in Cancer and Regeneration*, DOI 10.1007/978-1-4614-0601-3_42,

Acupuncture has been suggested for use in several areas of cancer care as an adjunctive treatment. These include pain, fatigue, chemotherapy-induced nausea and vomiting, xerostomia, anxiety and depression and cancer-related hot flushes (prostate and breast cancer). It is only since 2005 that systematic reviews have been published which critically evaluate the efficacy of acupuncture for these conditions. Earlier this year, a systematic review of the current systematic reviews, a meta-review, had been published [4]. This review evaluated and summarised the current state of evidence for seven areas where acupuncture has been used as an adjunct therapy for cancer. The authors concluded that there was strong evidence supporting the use of acupuncture for chemotherapy-induced nausea and vomiting; however, it was unclear for the areas of adverse events in breast cancer patients and chemotherapy-induced leukopenia, while the areas of hot flashes in prostate cancer as well as breast cancer, cancer pain and xerostomia all failed to show evidence of efficacy.

While three of the reviews were relatively current being published in 2009, the other four were dated with two having been published in 2005, one in 2006 and the remaining one in 2007. In order to update these four systematic reviews, a PubMed search was conducted in September 2010 to ascertain whether there had been further clinical trials published in these areas and whether the results of these clinical trials may affect the conclusion made by the authors of these early systematic reviews.

2 Nausea and Vomiting

Chemotherapy-induced nausea and vomiting remain a problem for many cancer patients despite ongoing development of antiemetics for this debilitating side effect such as the 5-HT(3) inhibitors and dexamethasone. Some of the earliest studies were conducted by Dundee in the 1980s and showed promising results with the most up-to-date Cochran review published in 2006 [5]. This review included 11 trials comprising 1,247 pooled subjects. The reviewers found that acupuncture modalities combined (including manual, electro and acupressure stimulation) statistically reduced the incidence of acute vomiting ($p=0.04$), but not acute or delayed nausea severity compared to control. The review included studies published up to 2003.

Since 2003, several studies have been published. A recent PubMed search conducted found a further six studies related to acupuncture treatment of chemotherapy-related nausea [6–11]. While all reported significant effects when comparing the acupuncture with various controls, a review for trial quality was not undertaken, so the results from these six studies need to be interpreted cautiously. Given this increase in published papers in the area, an updated systematic review would seem warranted. Furthermore, two recent studies while not directly related to chemotherapy-related nausea also support the use of acupuncture for post-operative nausea [12, 13].

3 Cancer Pain

The treatment of cancer pain is another area where acupuncture has been purported to be of value. A systematic review was published in 2005 which included only seven studies [14]. The authors reported that "one high quality randomised clinical trial of ear acupuncture showed statistically significant pain relief" with two being unblinded and four being uncontrolled. Most studies were flawed due to "inadequate study design, poor reporting of results, small sample size and overestimation of the results". This led them to conclude that "the notion that acupuncture may be an effective analgesic adjunctive method for cancer patients is not supported by the data currently available from the majority of rigorous clinical trials".

More recently, an updated systematic review from China was published [15]. This review included studies up to 2008. Their review only included RCTs and involved not only an English literature search, but also a Chinese database and several Chinese medical journals. In total, seven published RCTs were reviewed involving a total of 634 patients, with the quality of one of the included trials scored as high. They reported that a "high-quality trial showed that auricular acupuncture therapy was significantly superior to placebo in pain alleviation while the other six low-quality trials with non-placebo showed that acupuncture therapy had some positive effects". This led them to conclude that "acupuncture is effective for pain relief". However, they also added the caveat that "the poor quality of the majority of the trials reduces the reliability of the conclusion" and the often heard request that more high-quality RCTs are needed to verify the effects of acupuncture for cancer-related pain.

Another recent review evaluated 9 years of published research [16]. This study concluded that there was an evident lack of level I evidence regarding the use of acupuncture as a cancer pain treatment modality and that "because the majority of evidence is level III or higher; therefore, causality cannot be inferred".

Again an updated search using PubMed was conducted and four new published studies were found, three in 2010 [17–19] and one in 2008 [20]. Of the 2010 studies, the first looked at whether acupuncture reduces pain and dysfunction in patients with cancer following neck dissection ($n=58$) [17], the second evaluated whether there was an analgesic effect of transcutaneous electrical acupoint stimulation (TEAS) in assisting general anaesthesia (GA) for radical operation of breast carcinoma ($n=60$) [18], while the final report evaluated whether acupuncture improved aromatase inhibitors-induced arthralgia in women with early-stage breast cancer ($n=38$) [19].

The three recent 2010 studies all found significant reduction in pain following the administration of acupuncture. However, the 2008 study ($n=162$) which evaluated the use of a specially developed acupuncture technique (the insertion of small intradermal needles preoperatively and retained for 4 weeks) for a surgical oncology population reported no significant reduction in pain or analgesic use after thoracotomy compared to the sham intervention [20].

Finally, two studies published in 2007 which used acupuncture massage or scalp acupuncture found positive outcomes for pain reduction in both studies. The first study involved 60 cases of radical operation of intestinal cancer which were randomly divided into two groups, with the first group of 30 receiving scalp acupuncture plus epidural analgesia (scalp acupuncture group), while the other 30 were the active control epidural analgesia group [21]. VAS scores at 6, 12, 24 and 48 h in the scalp acupuncture group were lower than the epidural analgesia group, with significant differences at 6 and 12 h between the two groups. The second study was an RCT ($n = 138$) which assessed the effect of massage and acupuncture added to usual care vs. usual care alone in post-operative cancer patients [22]. They reported that providing massage and acupuncture in addition to usual care resulted in significantly reduced pain and depressive mood scores among post-operative cancer patients when compared with usual care alone.

4 Xerostomia

Another area where there has been additional published research is for the condition of xerostomia. Xerostomia or dry mouth can occur following radiotherapy for head and neck cancers which damages the salivary gland. The only systematic review in this area was published in 2005 and included only three studies [23]. Of those three studies, one was of high quality and it presented indifferent results. The other two studies were both assessed as low quality with one presenting positive results and one reporting indifferent results. The author concluded that there was "no evidence for the efficacy of acupuncture in the management of xerostomia… there is a need for future high quality randomised controlled trials". However, this review only assessed clinical studies up to September 2003, and similar to the areas of nausea and pain, several further studies have now been published.

A 2010 study involved 60 patients with a history of neck dissection being randomly allocated to either an acupuncture treatment group or a usual care group who received a combination of physical therapy, analgesia, and/or anti-inflammatory drugs, per patient preference or physician recommendation [17]. At completion of the 4-week study period, the acupuncture group reported statistically greater improvement in self-reported xerostomia as measured by the Xerostomia Inventory. A small study published in 2008 consisted of 12 patients with radiation-induced xerostomia who were randomised into two groups (verum and sham acupuncture) for 6 weeks (12 sessions) of acupuncture [24]. They reported that both groups showed a slight increase in whole salivary flow rates, but there was no significant difference between the two groups. The verum acupuncture did, however, markedly increase unstimulated salivary flow rates, and a significant improvement was observed for the score for dry mouth in the verum acupuncture group for the xerostomia questionnaire.

Several small uncontrolled studies have also been published. A 2008 uncontrolled pilot study which included 12 patients reported statistically significant improvement for salivary flow rates on both objective (resting salivary flow rate) and subjective evaluations (visual analogue scale for salivary production) following 12 sessions of acupuncture over 6 weeks [25]. Another 2009 study of 19 radiation-induced xerostomia patients received eight sessions of acupuncture. A xerostomia inventory and patient benefit questionnaire scores were significantly improved after acupuncture at weeks 4 and 8 compared to baseline [26].

The only other study found was a 2010 Phase II randomised trial of acupuncture-like transcutaneous electrical nerve stimulation for the prevention of radiation-induced xerostomia [27]. Sixty patients were randomised to receive the TENS or normal mouth care. No significant difference was found for either total saliva production or a visual analogue scale xerostomia symptom score. While the numbers of participants were adequate and methodology rigorous (random allocation and an active control), the question remains about whether acupuncture-like TENS has any relationship to needle acupuncture. While these recent additional studies which reported positive outcomes were not without methodological flaws, two were uncontrolled pilot studies and the randomised studies involved small subject numbers, they do support the need for larger studies.

5 Chemotherapy-Induced Neutropenia

The last area of interest concerns the effect of acupuncture on chemotherapy-induced neutropenia. The only systematic review on this condition was published in 2007 [28]. This review evaluated the published research between 1979 and 2004. The authors searched both English and Chinese databases which resulted in 33 reviewed articles, of which 682 patients from 11 eligible trials were included in analyses. All trials were published in non-PubMed listed journals from China. While they reported a significant increase in white blood cells on average, they noted that "the inferior quality and publication bias present in these studies may lead to a false-positive estimation". They further stated that "meta-analysis based on these published trials should be treated in an exploratory nature only".

Interestingly, the main author of this review was also the chief investigator in the only English language study published in the period between 2004 and 2010 [29]. This pilot study was well designed with 21 subjects with ovarian cancer who had neutropenia while undergoing myelosuppressive chemotherapy and were randomly allocated to either a verum acupuncture group or an invasive shame acupuncture group. All patients received at least eight sessions of manual/electro acupuncture or the invasive sham acupuncture. They reported "consistent trends of higher WBC counts in active acupuncture relative to sham acupuncture in a group of patients". They did, however, note that because of small numbers the results they observed may simply be due to chance and that a larger trial is warranted. While this later study may not change the conclusion of the 2007 systematic review on chemotherapy-induced leucopenia, as is common with many systematic review conclusions, more rigorous studies are needed with larger sample sizes.

6 Conclusion

So what can be concluded about the evidence for the use of acupuncture for these cancer-related conditions when more recent studies are considered? First, there seems to be increasing evidence for the use of acupuncture for chemotherapy-induced nausea and vomiting. While the 2006 systematic review was positive for the acupuncture, the more recent clinical trials reported today also support the previous 2006 conclusion that there is a "biological effect of acupoint stimulation". The more recent cancer pain studies also show significant outcomes for acupuncture. This may alter the 2005 study conclusions which were negative for acupuncture. The situation for xerostomia does not seem to have improved with only some studies showing positive outcomes and the others not. Again small samples sizes and failure to control mean the results need to be interpreted cautiously. Finally, while only one additional study was found for the effect of acupuncture on leukopenia, it did find significance for acupuncture in increasing white blood cell count but again a small sample size of 21 precludes a definitive result.

Future acupuncture studies in these areas should consider recruiting sufficient numbers of subjects to ensure an adequate sample size power and the use of the use of CONSORT [30] and STRICTA [31] to ensure good trial design and reporting.

References

1. Dean-Clower E, Doherty-Gilman AM et al (2010) Acupuncture as a palliative therapy for physical symptoms and quality of life for advanced cancer. Integr Cancer Ther 9:158
2. Chow WH, Chang P et al (2010) Complementary and alternative medicine among Singapore cancer patients. Ann Acad Med Singapore 39(2):129–135
3. Gansler T, Kaw C et al (2008) A population-based study of prevalence of complementary methods use by cancer survivors: a report from the American Cancer Society's studies of cancer survivors. Cancer 113(5):1048–1057
4. Ernst E, Lee MS (2010) Acupuncture for palliative and supportive cancer care: a systematic review of systematic reviews. J Pain Symptom Manage 40(1):e3–e5
5. Ezzo JM, Richardson MA et al (2006) Acupuncture-point stimulation for chemotherapy-induced nausea or vomiting. Cochrane Database Syst Rev (2):CD002285
6. Wang XQ, Yu JL et al (2010) Electroacupoint stimulation for postoperative nausea and vomiting in patients undergoing supratentorial craniotomy. J Neurosurg 22(2):128–131
7. Yang Y, Zhang Y et al (2009) Electroacupuncture at Zusanli (ST 36) for treatment of nausea and vomiting caused by the chemotherapy of the malignant tumor: a multicentral randomized controlled trial. Zhongguo Zhen Jiu 29(12):955–958
8. Gottschling S, Reindl TK et al (2008) Acupuncture to alleviate chemotherapy-induced nausea and vomiting in pediatric oncology – a randomized multicenter crossover pilot trial. Klin Padiatr 220(6):365–370
9. Nystrom E, Ridderstrom G et al (2008) Manual acupuncture as an adjunctive treatment of nausea in patients with cancer in palliative care – a prospective, observational pilot study. Acupunct Med 26(1):27–32
10. Chao LF, Zhang AL et al (2009) The efficacy of acupoint stimulation for the management of therapy-related adverse events in patients with breast cancer: a systematic review. Breast Cancer Res Treat 118(2):255–267

11. Gardani G, Cerrone R et al (2007) A progress study of 100 cancer patients treated by acupressure for chemotherapy-induced vomiting after failure with the pharmacological approach. Minerva Med 98(6):665–668
12. Sima L, Wang X (2009) Therapeutic effect of acupuncture on cisplatin-induced nausea and vomiting. Zhongguo Zhen Jiu 29(1):3–6
13. Fu J, Meng ZQ et al (2006) Clinical observation on electric stimulation of Yongquan (KI 1) for prevention of nausea and vomiting induced by cisplatin. Zhongguo Zhen Jiu 26(4):250–252
14. Lee H, Schmidt K et al (2005) Acupuncture for the relief of cancer-related pain – a systematic review. Eur J Pain 9:437–444
15. Peng H, Peng HD et al (2010) Efficacy of acupuncture in treatment of cancer pain: a systematic review. Zhong Xi Yi Jie He Xue Bao 8(6):501–509
16. Hopkins Hollis AS (2010) Acupuncture as a treatment modality for the management of cancer pain: the state of the science. Oncol Nurs Forum 37(5):E344–E348
17. Pfister DG, Cassileth BR et al (2010) Acupuncture for pain and dysfunction after neck dissection: results of a randomized controlled trial. J Clin Oncol 28(15):2565–2570
18. Yu JM, Qu PS et al (2010) Observation on the analgesic effect of transcutaneous electrical acupoint stimulation for breast radical carcinoma operation. Zhen Ci Yan Jiu 35(1):43–46
19. Crew KD, Capodice JL et al (2010) Randomized, blinded, sham-controlled trial of acupuncture for the management of aromatase inhibitor-associated joint symptoms in women with early-stage breast cancer. J Clin Oncol 28(7):1154–1160
20. Deng G, Rusch V et al (2008) Randomized controlled trial of a special acupuncture technique for pain after thoracotomy. J Thorac Cardiovasc Surg 136(6):1464–1469
21. He BM, Li WS et al (2007) Effect of previous analgesia of scalp acupuncture on post-operative epidural morphine analgesia in the patient of intestinal cancer. Zhongguo Zhen Jiu 27(5):369–371
22. Mehling WE, Jacobs B et al (2007) Symptom management with massage and acupuncture in postoperative cancer patients: a randomized controlled trial. J Pain Symptom Manage 33(3):258–266
23. Jedal E (2005) Acupuncture in xerostomia – a systematic review. J Oral Rehabil 32:392–396
24. Braga FP, Sugaya NN et al (2008) The effect of acupuncture on salivary flow rates in patients with radiation-induced xerostomia. Minerva Stomatol 57(7–8):343–348
25. Cho JH, Chung WK et al (2008) Manual acupuncture improved quality of life in cancer patients with radiation-induced xerostomia. J Altern Complement Med 14(5):523–526
26. Garcia MK, Chiang JS et al (2009) Acupuncture for radiation-induced xerostomia in patients with cancer: a pilot study. Head Neck 31(10):1360–1368
27. Wong RK, Sagar SM et al (2010) Phase II randomized trial of acupuncture-like transcutaneous electrical nerve stimulation to prevent radiation-induced xerostomia in head and neck cancer patients. J Soc Integr Oncol 8(2):35–42
28. Lu W, Hu D et al (2007) Acupuncture for chemotherapy-induced leukopenia: exploratory meta-analysis of randomized controlled trials. J Soc Integr Oncol 5(1):1–10
29. Lu W, Matulonis UA et al (2009) Acupuncture for chemotherapy-induced neutropenia in patients with gynecologic malignancies: a pilot randomized, sham-controlled clinical trial. J Altern Complement Med 15(7):745–753
30. Schulz KF, Altman DG et al (2010) CONSORT 2010 statement: updated guidelines for reporting parallel group randomised trials. BMC Med 8:18
31. MacPherson H, Altman DG et al (2010) Revised STandards for Reporting Interventions in Clinical Trials of Acupuncture (STRICTA): extending the CONSORT statement. Acupunct Med 28(2):83–93

Chapter 43
Thermal Characteristics of Moxibustion and its Implication to Primo Vascular System

Seung-Ho Yi, Moo-Won Park, and Hye-Jung Lee

Abstract In traditional Chinese medicine (TCM), moxibustion is one of the main therapeutic tools by delivering heat to specific areas like meridian points. Heat stimulation by moxa cones are varied with their physical dimension such as mass, shape, density, and ingredients. Practically, it is not easy to control temperature or heat amount of moxibustion in clinical settings, which might cause inconsistency of moxibustion treatment. In this study, we measured temperature variation of moxa cones fabricated with two different moxa flosses while changing air flow. The results clearly showed the significant difference between them in terms of max temperature, duration, and the amount of heat generated. Radiation from a burning cone could have some therapeutic effects. Elemental and thermal analysis demonstrated the considerable difference in those components and heat generation. Our results suggest some guidelines for thermal stimuli on the meridian system or primo-vessels with moxibustion. Since the moxibustion has been used for heat stimuli, it could impact the behavior of primo-vessels and Bonghan corpuscle, which requires further study on thermal properties of those entities.

1 Introduction

Primo-vessels or Bonghan ducts (BHDs) have been claimed to be another integrated circulation system that is totally different from the known blood, nerve, and lymphatic system in the living body based on several results [1]. The existence of them has recently been getting interests from scientific bodies due to possible significant

S.-H. Yi (✉)
Acupuncture and Meridian Science Research Center,
Kyung Hee University, Seoul 130-701, Korea

Department of Oriental Medicine, Kyung Hee University, Seoul 130-701, Korea
e-mail: shyi@khu.ac.kr

K.-S. Soh et al. (eds.), *The Primo Vascular System: Its Role in Cancer and Regeneration*,
DOI 10.1007/978-1-4614-0601-3_43, © Springer Science+Business Media, LLC 2012

impacts on basic biology and medical science for clinical applications [2]. The study of BHDs started by Bonghan Kim, a North Korean medical doctor, in the mid of 50th and its history was somewhat dramatic with respect to its flourishing and vanishing in the scientific society. Based on his discovery, yet to be verified, he hypothesized that his BHD is the anatomico-histological entities of the meridian system on which the traditional Chinese medicine (TCM) has been established. In his original article, he claimed the BHD is the Kyungrak (meridian) system and the Bonghan corpuscle (BHC) is the Kyunghyul (acupoint) [3]. He executed morphological, physiological, bioelectrical, biochemical, and histo-chemical studies of the system. According to the bioelectrical study, the new system has distinctive bioelectric characteristics compared with the nervous and skeletal-muscle tissues. Other physical stimuli evoked by needle (acupuncture) and moxacautery (moxibustion) and chemical stimulus also were proven to induce the electric changes.

The TCM has administrated three types of interventions such as acupuncture, moxibustion, and therapeutic herb. Their functions were described in classical textbooks like Yellow Empire's Internal Medicine, but needed to be proven based on rigorous basic studies as well as clinical trials. Acupuncture seems to generate mechanical stimuli by inserting needles into the designated skin or tissues while moxibustion is likely to deliver thermal stimuli via burning moxa cones. The fact that the BHD and BHC response to these traditional physical stimuli implies the therapeutic methods of the TCM might aim to stimulate them. So far, intensive results were reported on the bioelectric properties of them with electrical stimuli [4]. Some results on needle stimuli also were presented. However, no significant results with thermal stimuli were introduced yet despite its importance in the TCM.

In this study, we focused on the thermal properties of moxibustion as a possible tool to stimulate the BHD and BHC for therapeutic purposes. Heat stimulation by a moxa cone could be varied by changing the physical dimension such as mass, shape, density, and their ingredients. Among various factors, we measured its temperature time courses of two moxa cones of which ingredients were different while varying air flow. Some chemical and thermal analysis of the flosses was also executed for a simple comparison. Influences of the moxibustion on the BHD and BHC were discussed in detail.

2 Method

Moxa cones were made of fine and coarse moxa flosses (DongBang Acupuncture Inc., Korea), respectively. The former was made of dried leaves of a mugwort plant, while the latter was composed of its other parts and the leaves. The mugwort was cultivated on Kwangwha Island, South Korea. A fine moxa cone was 30 mm in height and 23 mm in bottom diameter and coarse ones were 24 and 20 mm, respectively, and their mass was 0.5 g. Small branches were visually observed in the coarse floss. Both are being used for moxibustion treatment in Korea. Moxa cones were formed by using a commercial manual-shaping device and, as a result, some fluctuations in quality were expected. We executed the measurement in a stainless steel bench hood (120×60×90 cm) with a forced ventilation fan located at its top center. Upstream flows of 0–1 m/s in the hood were generated by the fan and measured by a wind meter (Kestrel 1000). Air flows of 0, 0.5, and 0.9 (m/s) were applied for our measurements. Temperature of 21–22°C and humidity of 43–45% in the hood were maintained throughout the measurement. Temperature of the bottom center of a burning moxa cone was measured by an automatic acquisition system composed of an interface and thermocouple input module with 0.08 mm thermocouples. Acquisition was done for 360 s with a frequency of 1 Hz and repeated three times for each moxa and air flow. More details could be found elsewhere [5]. Five replications for each measurement were executed. For each measurement, a fresh moxa cone was prepared based on the specification noted earlier. The moxa cone was placed on a thermocouple installed on a hot plate in such a way that the bottom center of the cone is coincident with the tip of the thermocouple. Temperature of the hot plate was maintained at 34°C, which mimicked the human skin temperature. Another thermocouple was located at 20 mm in the radial direction from the bottom center and 5 mm in height from the surface of the plate. It was to measure temperature evoked by the radiation from a burning cone and convection. A cone was ignited at its apex with a dim fire on a wooden toothpick. Just after the hood door closed, the fan was turned on and data acquisition was started. Residue of the combustion was collected after the completion of combustion and weighed for a simple comparison.

To study chemical and thermal difference between fine and coarse flosses, we analyzed the basic elements in those flosses and thermal behavior once. For element analysis, we used an automatic elemental analyzer (Flash EA1112) and compared the amount of N, C, H, and S of each two samples. For thermal analysis, we employed thermal galvimetric analyzer (TGA Q 5000 IR/SDT Q600) to measure the weight change and heat flow as a function of time and heating temperature. To compare dependencies on air flows, linear regression analysis was done with GraphPad Prism 5.02 (Graph Pad Software, Inc., USA). Values for p of less than 0.05 were considered as statistically significant. Means and standard deviations were calculated from five results from each experiment set.

3 Results

Moxa cones made of the both flosses showed distinctive differences in thermal behaviors. For analysis, the onset time is defined as the time when moxibustion temperature starts higher than 35°C and the peak time is the time elapsed from the onset time to the maximum temperature. The duration is when the temperature was over 35°C during each experiment. The reason that we selected the temperature of 35°C as an indication of the effect of burning cones was based on the consideration of the hot plate temperature at 34°C. According to the results, the onset time of both cones decreased with a similar rate ($p=0.1588$) from about 60 to 50 s with an increase in the air flow. The time much relates to physical properties of moxa cones

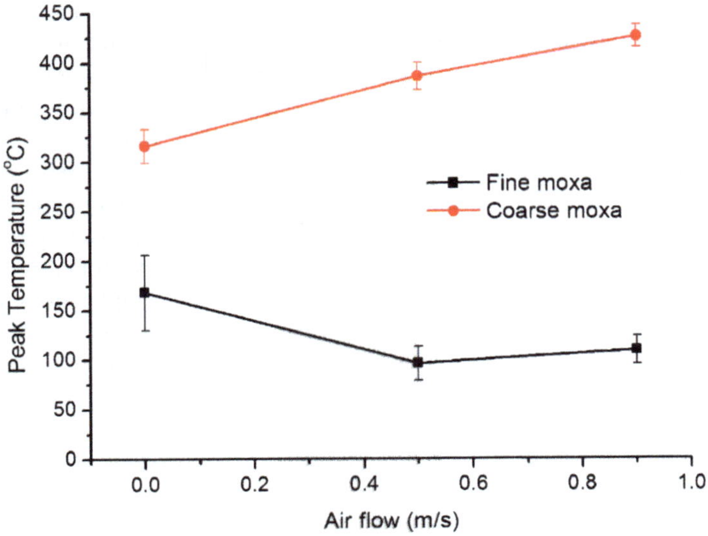

Fig. 43.1 Comparison of the maximum temperature between coarse and fine moxa floss

Table 43.1 Thermal properties of both moxa cones with air flow

	Duration (s)		Peak time (s)		Residue ratio (%)	
Air flow (m/s)	Coarse	Fine	Coarse	Fine	Coarse	Fine
0.0	404±32	231±12	275±10	213±11	18.47	15.28
0.5	325±9	192±17	230±12	180±13	16.01	11.54
0.9	209±10	156±4	174±6	129±14	12.86	10.63
p-Value	<0.0001		0.17		0.643	

p-Values are obtained from the comparison between linear coefficients for each set

and is not important in clinical viewpoint due to low temperature for the onset time. Air flow dependency of the maximum temperature demonstrated a peculiar behavior. The temperature of the fine moxa cones seemed to decrease with increasing air flow, while that of the coarse ones showed an opposite trend as shown in Fig. 43.1. Temperature difference was so significant that the maximum temperature of the coarse moxa cones showed in a range from 320 to 430°C, while that of the fine did from 170 to 100°C. The duration and peak time of both cones decreased with an increase in the air flow ($p=0.1721$), even though they were different in their magnitudes as summarized in Table 43.1. The results clearly show the significant difference between them in terms of maximum temperature, duration, and peak time. Residue ratios (after total mass/before total mass) in Fig. 43.1 showed that the fine floss had less residual compared to the coarse one and roughly indicated the composition difference between two flosses.

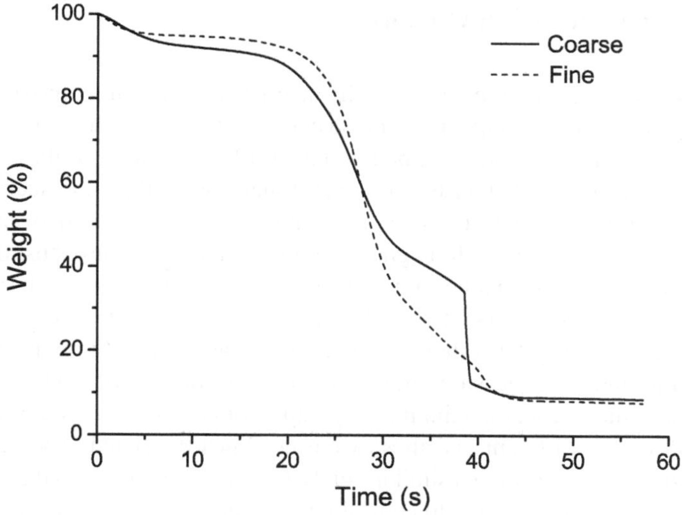

Fig. 43.2 Pryoanalytic comparison between coarse and fine moxa flosses

We also measured the side temperature of the fine cones to check any heating effect caused by radiation and convection. The results implied the existence of two regions demarcated by the air flow of 0.5 m/s. Below 0.5 m/s of the flow, the temperature and duration were increased with the flow. No temperature increase up to 35°C was observed with zero air flow. Beyond that, however, both quantities started decreasing from 92 to 76°C and 190 to 160 s, respectively. There was a signification variation in the temperature time course, indicating some effects of air convection. Average values of the temperature were higher than 50°C for the duration with the flow of 0.5 m/s and faster. Therefore, therapeutic effects could be obtained at the vicinity as well as the bottom of the moxa cone by varying the air flow in a clinical setting.

According to the pyrolysis and element analysis, there was some difference in the contents of C, H, N, and S. The average values of N, C, and H of the coarse moxa floss were 2.49, 43.78, and 6.03, respectively, while those of the fine one were 0.73, 43.95, and 6.47. S was not detected by the test. The content of N was higher in the coarse floss, while others were not much different. As shown in Fig. 43.2, the change rate in the weight was higher for the coarse floss at the initial stage, which means higher composition rate of fluid and gases, mostly water and essential oils. At about 17 min, the slopes of the curves changed and fine floss showed rapider change rate than the coarse one. It implies that there were purer materials in the fine one than other. The total amount of residues for fine floss is lower compared to coarse one, which is a good agreement with the residue ratio in Table 43.1.

4 Discussion and Conclusion

In our study, we measured the thermal characteristics of the direct moxibustion by varying air flow and the compound of moxa cones. In the TCM, moxibustion utilizes combustion heat to stimulate meridian points or a BHC. Nature of the stimulation could be varied by controlling the physical dimension of the cone such as mass, shape, and density. It also depends on the existence of buffer layers between a moxa cone and the human skin such as ginger and air for indirect moxibustion. It was claimed that the thermodynamics of moxibustion needs to be varied with symptoms. Practically, controlling temperature or heat amount of moxibustion as a function of time is not easy in hospital, which might cause inconsistency of moxibustion treatment. In this study, we measured the temperature of moxibustion with various air low. Our data showed that maximum temperature obtained from each moxibustion with the flow was different and those were not linearly related. The duration of moxibustion also demonstrated similar trends. Our study emphasized that air flow is one of the essential parameters to control temperature of moxibustion and, in turn, to make moxibustion more effective. Supplementation/draining are claimed to be important to obtain a maximum efficacy in TCM. For moxibustion treatments, those manipulations are achieved by supplying additional air flow while a moxa cone is burning. Therefore, study on the effects of air flow is imperative and this is the reason why we measured thermal properties of moxa cones with air flow.

The heat stimulus of moxibustion has an important implication for the study of Bonghan theory. According to the theory of the BHD [1], Kyungrak system has definite bioelectrical features. The BHC is located at the Kyunghyul position in the reticular layer of the skin. It is an oval structure with a long diameter of 1–3 mm and a short diameter of 0.5–1 mm, and its long axis stands vertical to the surface of the skin. Bioelectric changes were occurred in the superficial BHC when an electrode was inserted into the BHC and the changes were too peculiar that none was observed in other part of the skin. As mentioned earlier, the electric changes were induced by needle and moxibustion also in different rates, depending on the varieties of the BHCs and their state before induction. For this change, temperature was a crucial factor with aspects that the changes disappeared when the temperature of the environment was dropped to below 27°C and appeared when the temperature is raised again to 39°C immediately for a rabbit. They concluded that there are complicated mutual relations between the activity of the BHC and the intensity of stimulus, and the effect is most apparent as a stimulus of a proper intensity is administrated to the junctions of the given BHC. This conclusion strongly suggests that proper temperature control of moxibustion is required to affect or induce the change and there could be an optimum condition for each thermal stimulus depending on the BHC and its state. Understanding of the thermal properties of traditional moxibustion might give us some clues on the proper stimuli. According to our results and others [5], temperature of moxibustion is difficult to control and is subject to various parameters such as moxa floss, size, humidity, configuration, air flow, and so on. Therefore, reliable and repeatable moxibustion methods should be introduced for the mechanism research as well as clinical applications.

Compared with acupuncture, moxibustion can cover a wider area of the skin and requires less accuracy in position due to its bigger size while there is some difficulty in temperature control. For acupuncture, BHCs should be stimulated by a needle with a diameter of less than 0.3 mm. By considering the size and nonrigidity of a BHC, and its location, it could be hard for the needle to touch a BHC directly. It has been reported that a needle applied to a superficial BHC generates the peculiar movement of the superficial BHC, known as the Kim Se Wook phenomenon [1], and the contractile action of the smooth muscles of the outer layer of the BHC is responsible for that. However, reported cases have been really scarce so far compared with the huge number of acupuncture practice in Korea. This rarity could imply the difficulty in direct stimulation of the BHCs with a needle. Therefore, for study of the superficial primo-vessel, moxibustion stimuli have some compelling advantages over acupuncture method. This argument could be valid for therapeutic interventions in the TCM also.

Acknowledgments This work was supported by the Korea Science and Engineering Foundation (KOSEF) grant funded by the Korea government (MEST) (R11-2005-014).

References

1. Kim BH (1963) On the Kyungrak system. J Acad Med Sci DPR Korea 5:1–41
2. Shin HS, Johng HM, Lee BC et al (2005) Feulgen reaction study of novel threadlike structures (Bonghan ducts) on the surfaces of mammalian organs. J Anat Rec B 284:35–40
3. Kim BH (1962) Research on the entities of Kyungrak. J Cho Med 9(1):5–13
4. Ahn AC, Park M, Shaw JR et al (2010) Electrical impedance of acupuncture meridians: the relevance of subcutaneous collagenous bands. PLoS One 5(7):e11907
5. Yi SH (2009) Thermal properties of direct and indirect moxibustion. J Acupunct Meridian Stud 2(4):273–279

Index

K.-S. Soh et al. (eds.), *The Primo Vascular System: Its Role in Cancer and Regeneration*, DOI 10.1007/978-1-4614-0601-3, © Springer Science+Business Media, LLC 2012

MIX
Papier aus verantwortungsvollen Quellen
Paper from responsible sources
FSC® C105338

If you have any concerns about our products,
you can contact us on
ProductSafety@springernature.com

In case Publisher is established outside the EU,
the EU authorized representative is:
Springer Nature Customer Service Center GmbH
Europaplatz 3, 69115 Heidelberg, Germany

Printed by Libri Plureos GmbH
in Hamburg, Germany